D1009533

Also by Michael Riordan

The Solar Home Book (with Bruce Anderson)

THE HUNTING OF THE QUARK

A TRUE STORY OF MODERN PHYSICS

Michael Riordan

A Touchstone Book
Published by Simon & Schuster, Inc.
NEW YORK • LONDON • TORONTO • SYDNEY • TOKYO

Simon and Schuster/Touchstone Books
Published by Simon & Schuster, Inc.
Simon & Schuster Building
Rockefeller Center
1230 Avenue of the Americas
New York, NY 10020
SIMON AND SCHUSTER/TOUCHSTONE and
colophons are registered trademarks of Simon & Schuster, Inc.

Designed by Irving Perkins Associates
Manufactured in the United States of America

1 2 3 4 5 6 7 8 9 10
1 2 3 4 5 6 7 8 9 10 (pbk.)

Library of Congress Cataloging-in-Publication Data

Riordan, Michael.
The hunting of the quark.

(A Touchstone book)
Bibliography: p.
Includes index.
1. Quarks—History. 2. Nuclear physics—History.
3. Particles (Nuclear physics)—History. I. Title.
QC793.5.Q252R56 1987 539.7'21 87-16530
ISBN 0-671-50466-5
ISBN 0-671-64884-5 (pbk.)

Drawings by Walter Zawojski appear on pages: 25, 26, 71, 93, 116, 159, 196, 231, 243, 326, 344, 346.

Drawings by Russell Tkaczyk appear on pages: 46, 56, 80, 90, 94, 132, 138, 163, 200, 208, 213, 219, 234, 253, 324, 340, 341, 357.

To Linda

CONTENTS

PROLOGUE 9
1. DOWN THE RABBIT HOLE 15
2. THE PARTICLE KINGDOM 42
3. THE TAMING OF THE ZOO 73
4. A QUIRK OF IMAGINATION 100
5. THE BIRTH OF A MONSTER 122
6. A POINT OF DEPARTURE 136
7. A MATTER OF SCALE 156
8. THE ROAD TO OBJECTIVITY 169

MESOLOGUE 189
9. THROUGH THE LOOKING GLASS 194
10. THE DYE THAT BINDS 225
11. THE MOMENT OF CREATION 245
12. THE NOVEMBER REVOLUTION 262
13. A MATTER OF TASTE 293
14. SIX OF ONE . . . 322
15. THE POWER OF GLUE 335
16. THE ROAD TO UNITY? 355

EPILOGUE 364
NOTES 370
INTERVIEWS AND CONVERSATIONS 382
ACKNOWLEDGMENTS 388
INDEX 390

If at every turn we had to construct science anew out of science alone, without the guidance of style and knowledge in their widest sense, how could we hope to catch this complex and infinitely fascinating world with our minds at all?

—Gerald Holton

PROLOGUE

THE 1970s were heady times for high-energy physics. A surprising new layer of matter was discovered during those years, and three of Nature's four fundamental forces began to be viewed as one. When the decade began, most physicists considered protons and neutrons—the stuff of atomic nuclei—to be elementary building blocks of matter. But by its close we knew these subatomic particles to be built from three even tinier objects, dubbed quarks by Caltech theorist Murray Gell-Mann.

Nothing quite like these quarks had ever been encountered before. Not only do they sport electric charges a fraction of that on a proton or an electron, but they also never appear alone in Nature. No scientist has ever witnessed a single solitary quark in a particle detector. When first proposed in 1964, quarks were ridiculed by all but the few eccentrics who went searching for them. After these efforts failed to unearth anything unusual, most physicists thought the matter settled for good.

As often happens in science, the "discovery" of quarks occurred by accident, at the Stanford Linear Accelerator Center in California. Odd things began happening there in a series of experiments on which I was working as an MIT graduate student and research physicist. From 1968 to 1973, we bombarded protons and neutrons with high-energy electrons that, to almost everybody's surprise, ricocheted off at abrupt angles—as if they had struck something hard and tiny inside. Perhaps the quarks?

During the ensuing years our group and others accumulated a mountain of indirect evidence that, yes, we had indeed struck quarks at the heart of the atomic nucleus. But their persistent refusal to come out in the open—to show up in our particle detectors—was troubling. To the skeptics, quarks remained just a mathematical curiosity able to explain mounds of experimental data without possessing any innate "reality" as objects of human experience. Who could believe, after all, in something that absolutely refused to reveal itself?

There the matter might have rested if not for a pair of simultaneous discoveries that occurred in 1974 at Stanford and the Brookhaven National Laboratory near New York. A surprising new particle, called the J or psi depending on one's loyalties, touched off a great flurry of speculation and experimentation. When the dust had settled two years later, it was obvious that we could explain this crazy corpuscle only by invoking a new and different kind of quark, whimsically named the "charmed" quark by Harvard theorist Sheldon Glashow. After these momentous events, quarks finally became "real" in the eyes of particle physicists. The doubters could doubt no more.

Although most of us then perceived it only dimly, if at all, we had been soldiers in a classic scientific revolution. The dominant view of nuclear matter was completely overthrown in the twelve years from 1964 to 1976. Called the "bootstrap model," this theory taught that protons, neutrons, and other visible particles were among the smallest units of existence. Supposedly we had finally reached the innermost layer of the cosmic onion, the lowest rung of the quantum ladder, where every particle was built from all others. The many successes of the quark theory, culminating in the 1974 discoveries and their resolution in terms of another quark, pitched this bootstrap idea into oblivion.

On another front, physicists were beginning to realize Albert Einstein's fondest dream: a unified field theory, in which all the forces of Nature are merely different manifestations of a single, universal Force. In the late 1960s Steven Weinberg at MIT and Abdus Salam at London's Imperial College proposed that the electromagnetic force holding atoms together and the weak nuclear force causing radioactivity were but two different forms of one and the same "electroweak" force. Their radical new idea became established dogma after it received striking experimental proof during the 1970s.

Events were marching ahead at a pace unseen since half a century earlier, during the 1920s, when quantum mechanics had resolved the puzzling structure of the atom. In 1973, several theorists proposed that a new force like the electromagnetic force might explain how quarks are locked together. Quantum chromodynamics, as this theory became widely known, promised a natural reason for the fact that quarks seemed to "exist" only *inside* protons, neutrons, and other subnuclear motes. Here, at last, was a possible solution to the baffling absence of visible quarks.

By the end of the 1970s, a remarkably simple, beautiful picture of the subatomic world had taken shape. There were still worrisome gaps and cracks in the theory, but the major pieces of grand synthesis seemed securely in place. The many subnuclear particles in Nature were the various combinations of a few tiny quarks—themselves close cousins of the familiar electrons coursing through our home appliances. What's more, three of the four supposedly "fundamental" forces now seemed very closely related to one another. And there were theories afoot that might even bring gravity into the fold, too, for an all-encompassing unity. If *that* didn't make Einstein bolt up in his grave, nothing would!

The traditional account of the scientific enterprise depicts it as a fairly orderly process. Based on the available evidence, so it goes, scientists develop a hypothesis to explain existing observations and to predict the results of additional measurements. They then do experiments to test these predictions. If it passes a number of such tests, the hypothesis becomes an accepted theory—or perhaps even a hallowed Law of Nature. The scientists themselves are merely dispassionate arbiters of this rigid process, extracting "objective" truth from the mire of worldly phenomena so tainted by opinion and prejudice.

Though it contains elements of truth, this scenario gives an incomplete picture of the modern scientific process. It may well describe periods of "normal science," to borrow a phrase from Thomas Kuhn, those times when progress is slow and gradual. But it fails to encompass what happens during a great scientific revolution—when an entire field is undergoing rapid conversion from one accepted paradigm to another, radically new world view.

To many of us who lived through this most recent revolution, such a mythical scenario seemed wholly inadequate in describing what happened. The blind alleys taken and their attendant confusions, the completely unexpected results, and the sudden leaps of insight and imagination—all the subjective elements involved in actually *doing* physics—had only token roles in such a script. So, like any scientist mulling over some unruly data that just would not fit his theory, I began to doubt this simple-minded picture of scientific change and to cast about for alternatives.

In thinking over these shortcomings, I ultimately decided to write an account of the quark discovery, using it as the central thread in a history of particle physics. And I wanted this to be a book about *experimental* science, a science grounded in observation and measurement. Most accounts of particle physics have presented quarks as a profound theoretical triumph—as if they had sprung full-formed from the mind of man, like Athena from the brow of Zeus. But I knew better. Quarks were the mutual offspring of *both* theory and experiment; each had played a critical role in bringing these motes to life.

Ever since the birth of quantum mechanics, measurement has played an ambiguous role in physics, and the sharp separation between subject and object has evaporated. In the subatomic realms explored by high-energy probes, we encounter active worlds in violent collision—not some passive, unmolested "objects" with a completely independent existence. This is an environment where fashion and prejudice thrive, as they did in the 1960s, when there was a bewildering variety of contradictory ideas advocated about the ultimate nature of matter.

During the twentieth century, too, the measuring process has grown tremendously in scale. Where experiments previously filled a laboratory bench, they now require enormous accelerators with dimensions in miles. And the human organizations building and operating our colossal detectors have witnessed yet another explosion in size. Under such conditions, experiment is a highly qualified operation. A long chain of reasoning and argumentation underlies any information our equipment has to offer. Its links can be philosophical as well as physical—and can depend a lot upon the individuals involved and their own personal styles.

Given these caveats, it is a small miracle that particle physicists could ever agree on one interpretation of subatomic reality. But that is what essentially happened during the 1970s. A worldwide community of strong-willed and highly opinionated individuals reached consensus on the structure of matter at the deepest levels accessible to human observation. It was a truly fascinating episode in the history of science.

In this book I attempt to portray physics as it really happened, the work of real people who carry unavoidable stylistic baggage. I want to show *how* our mental image of the subatomic world was transformed between 1964 and 1980. The role of fashion and prejudice—the fond hopes and fervent beliefs about how Nature *ought* to be—is an important part of this story. Thus, I also follow some of the wrong turns made along the way, recounting many of the false starts and blind alleys. Discovery is a hit-or-miss, inductive process, not a logical parade of theory and experiment that reaches its conclusions deductively.

As much as anything else, it was the steady advance of experimental technique, the relentless pressure of technical improvements, that led us to "discover" quarks. Before theorists could really think about Nature on such a tiny scale, new tools had to be developed to experience things there. One special "tool" deserves particular attention: the Stanford Linear Accelerator. More than any other, this huge atom smasher was the knife that sliced open the subnuclear world and gave us our first real glimpse of quarks. Two of the four or five most revealing experiments occurred there, and much of the relevant theory was formulated right on the premises.

I have based this account on my own experiences of the late 1960s and early 1970s, a time when high-energy physics was turned upside down by results emanating from Stanford. To describe the important work occurring elsewhere, I have spoken with over a hundred physicists throughout the world; their memories and insights are also included.

The chain of events I recount follows the evolution of my own thinking during this period of tremendous ferment—how I wrestled with and finally came to accept these curious quarks as "real" objects of human experience. Thus the discovery of quarks is emphasized more than the unification of forces, because this is the more familiar viewpoint—the one I can explain from my own perspective.

Such an approach helps me relate the great sense of wonder we all felt during those years. Subnuclear matter was a profound mystery that took many hundreds of the best minds in the world over a decade to solve. Parts of that mystery, in fact, remain unsolved even today.

CHAPTER 1

DOWN THE RABBIT HOLE

All things being consider'd, it seems probable to me, that God in the Beginning form'd Matter in solid, massy, hard, impenetrable, moveable Particles of such Sizes and Figures, and with such other Properties, and in such Proportion to Space, as most conduced to the End for which he form'd them; and that these primitive Particles, being solids, are incomparably harder than any porous Bodies compounded of them; even so very hard as never to wear or break in pieces; no ordinary Power being able to divide what God Himself made one in the first Creation.

—Isaac Newton, OPTICKS

THIRTY miles south of San Francisco, an incredibly long, narrow structure knifes its way across the sprawling Stanford University campus. Motorists speeding along Interstate 280 pass right over this mysterious building, most of them oblivious to what is happening inside. The casual observer might well mistake it for a two-mile-long chain of boxcars rumbling through the scrub oak and manzanita of these parched California hills. On winter mornings, wispy clouds swirling skyward above it reinforce this impression.

This curious artifact is the Stanford Linear Accelerator, a

powerful atom smasher whose principal cargo is high-energy electrons. Built in the midsixties at a cost of $114 million, it is still the world's largest electron microscope. It pushes these subatomic particles to phenomenal energies—high enough, in fact, to slice apart the tiny protons and neutrons found inside atomic nuclei.

The electrons begin their dash at the base of the Santa Cruz Mountains, close to the San Andreas Fault. They are accelerated to high energy inside a long copper tube buried deep underground beneath the visible superstructure, which houses the "klystrons" that feed microwave power to the tube. Like a surfer perched on a surfboard, a bunch (or "pulse") of electrons rides the leading edge of an electromagnetic wave as it surges down the tube at essentially the speed of light. Microseconds later, each electron has gained upward of 20 billion volts, from the energy pumped into the tube—this is about a *million* times

The Stanford Linear Accelerator Center, as seen in 1970. The large concrete building in the center is End Station A, site of the electron scattering experiments.

the energy of the electrons hitting the phosphor screen in a common television set.

Imagine, then, the total power of a *trillion* such electrons—a typical pulse roaring out of the machine and into the "beam switchyard." There computer-controlled electromagnets deftly and gingerly divert each pulse along imaginary magnetic tracks to its appointed destination. Electrons headed for End Station A, the main experimental area during the accelerator's heyday, go screeching through a turn of 24 degrees, losing a little energy as they veer left.

These pulses emerge into a vast concrete cavern some eight stories high, where they slam into targets of hydrogen, copper, iron, or other material deliberately placed in their path. Most of the electrons just tear straight through undisturbed, but a few of them *are* deflected by the target nuclei. Two magnetic "spectrometers," each resembling a tremendous cyclops, glower at the target from different vantages, ever watchful for these scattered electrons. Built with hundreds of tons apiece of concrete, lead, and iron and balanced on sets of circular steel rails, both spectrometers can pivot about the target and view it from many different angles. Each has an array of magnets to deflect particles of unwanted energy and to focus the rest on rows of detectors huddling in a concrete "cave" behind the last magnet.

These detectors signal the passage of charged particles and distinguish electrons from the other subatomic garbage rattling up into the cave. A small flash of blue light in one of many plastic rods occurs whenever a particle passes through; then a sensitive photocell glued to the end of this rod responds to this flash by firing off an electronic pulse. Other detectors generate more complex signals that reveal whether or not that particle was indeed an electron.

These signals race out of the cave, back down the spectrometer carriage, and up the forward wall of End Station A through a fat bundle of green cables. In the "counting house" high above the experimental floor, the raw signals are processed by banks of electronic modules, and the results recorded by a minicomputer. In the free time between pulses, it also monitors the apparatus and gives physicists a preliminary look at the experimental results.

This complex system has striking parallels with human vision. Certainly all the parts are there. Electrons instead of light bounce off the subatomic objects being studied. Lead apertures on the two spectrometers form the irises—and the magnets the lenses—of two enormous eyes viewing these objects. Particle detectors in each cave resemble the rods and cones of our retinas. Electronic signals course up the two bundles of cables exactly as they do in our optic nerves. And the computer supplies the brains of the operation, converting raw electronic pulses into a more convenient geometrical image of the world inside protons and neutrons.

Using the linear accelerator and these two spectrometers like an extremely powerful microscope, physicists caught the first glimpse of even tinier motes swarming around inside these nuclear building blocks. Beginning in 1967, this work led to a striking new picture of the subnuclear world that eventually revolutionized the entire field of particle physics.

The new image drew heavily upon a long "atomic" tradition, dating from the Greeks, that has dominated the way we think of matter in this century. As more and more powerful means to subdivide matter have been developed, scientists have encountered successive layers of supposedly "elementary" building blocks, only to find them divisible in turn when sharper knives eventually became available. Molecules are known to be built from atoms, atoms from electrons and nuclei, nuclei from protons and neutrons—which are themselves made of quarks. Each step of this seemingly relentless descent into the heart of matter has relied heavily on the insights that came before.

The idea of elementary particles is as old as Western civilization itself. Our need to find a simple basis for all the many complexities of Nature has often been very strong. Time and again, elementary particles have been proposed to explain Nature's seemingly bewildering diversity.

Troubled by the apparent "chaos" of the everyday world, ancient Greek philosophers sought an underlying order, or "cosmos," in Nature. Leucippus and Democritus viewed the world as composed of tiny particles, or "atoms"—from the Greek word *atomos* for "uncuttable" or "indivisible." Accordingly, all the diverse objects of everyday life were just the many different combinations of a few basic atoms, which had various shapes, sizes, and weights. Impermanence in Nature was interpreted as the result of changing combinations of atoms, but the atoms themselves remained permanent and unchanging. For Democritus, atoms were everything—the primary reality behind the world of the five senses. "Nothing exists," he wrote, "but atoms and the void."

Well before Leucippus and Democritus, however, Pythagoras and his acolytes had taught that primary reality lay not in the everyday world of the senses but in the abstract realm of pure thought. In mathematics, they claimed, one glimpsed this perfect reality—of which our sensory world is but an imperfect reflection. Whole numbers were the chief deities in their pantheon. The Pythagoreans were fascinated by the five regular solids—symmetrical objects like the cube whose faces are all the same regular polygon. Four of these solids they associated with the four elements then imagined to constitute the world: earth, air, water, and fire. The fifth was identified with an unknown heavenly substance—the ether.

This struggle between the material and the ideal has been repeated often in history, but the idealists won the first battle handily. Plato's parable of the cave, in which everyday objects are but shadows of the true reality, immortalized for all time the Pythagorean notions of reality. Only fragmentary writings of Democritus have survived, while Plato's *Dialogues* and *Republic* are the staple diet of freshman courses in Western Civilization. Since Plato, we have held a largely dualistic idea of reality—putting mind before matter, form over substance, and heaven above earth.

Aristotle had another approach. Although a disciple of Plato, he was far more comfortable with the complexity and diversity of natural phenomena. For him, matter was an unformed "potentia" that became real only when it assumed specific forms. And he understood substance not as a static entity but as *process*—as becoming, not being. Aristotle rejected attempts to reduce Nature to basic elements or atoms. A firm believer in observation, he was content to categorize and classify Nature in her many manifestations. Stable and imperceptible, atoms had no place in Aristotle's universe.

For almost two thousand years, the philosophies of Plato and

Aristotle held sway in Europe. Christianity held both in high regard—especially Plato's. His doctrine of a higher reality divorced from sensory experience found special favor in a growing world religion that taught of an immortal soul yearning for heaven. Atomic theories of Nature surfaced occasionally, however; in *De Rerum Natura,* the Roman poet Lucretius argued that

> *. . . there are things with bodies*
> *Solid and everlasting; these we call*
> *Seeds of things, firstlings, atoms, and in them lies*
> *The sum of all created things.*

But otherworldly concerns had already begun to dominate Mediterranean life. By the time Constantine made Christianity the official religion of the Empire in the fourth century A.D., the victory was complete. Divine causation now provided the only dynamic of a universe modeled on Aristotelian categories. Good Christians abandoned speculative philosophy to gird themselves for the Apocalypse.

During the Renaissance, atomism flourished again amidst the general revival of Greek and Roman ideas. In one form or another, Galileo, Francis Bacon, René Descartes, and Isaac Newton all considered atoms the basic building blocks of matter. Indivisible particles and the void found important roles in a clockwork universe perceived as a Great Machine. Newton thought atoms were hard spheres that interacted with each other through forces like gravity and electricity. "Have not the small Particles of Bodies certain Powers, Virtues or Forces," he asked, "by which they act at a distance . . . upon one another for producing a great part of the Phaenomena of Nature?"

The days were past, however, when one could rely on the mere word of hallowed authorities. During the Enlightenment a growing empirical bias demanded experimental proof of such "metaphysical" speculations. But the measuring techniques of seventeenth- and eighteenth-century physicists were wholly inadequate, and the search for fundamental entities was left instead to chemists like Robert Boyle, Antoine Lavoisier, and John Dalton. In his book *The Sceptical Chymist,* written in 1661,

Boyle introduced the modern idea of an "element" as a substance, like gold, that could not be further reduced into simpler substances.

Chemists soon identified a growing list of basic elements—including hydrogen, nitrogen, and oxygen. When Lavoisier and others developed quantitative chemisty late in the 1700s, they soon recognized that these elements combined in specific ratios of whole numbers to form compounds. For example, nine grams of water can always be decomposed into one gram of hydrogen and eight grams of oxygen.

Based on these discoveries, John Dalton announced his Modern Atomic Theory in 1808. He proposed that each element is composed of many identical atoms each with the same fixed mass, called its "atomic mass." Two or more atoms combine in very specific ways to form a "molecule" of a particular compound. Thus, the individual masses of the various elements making up a compound always occur in the same fixed proportions, as observed. For example, two atoms of hydrogen (each with atomic mass = 1 unit, for a total of 2) combine with one atom of oxygen (with atomic mass = 16) to make a single molecule of the compound water (H_2O). Hence the observed mass ratio of 2:16, or 1:8.

Dalton's theory did not *prove* the existence of atoms. But by using them to explain the regular patterns observed in chemical reactions, he made atomism a far more plausible idea. "The existence of these ultimate particles can scarce be doubted," he wrote in 1810, "though they are probably much too small ever to be exhibited."

The ultimate test would be what *else* the theory could explain, and here it performed admirably. The kinetic theory of gases—which was based on the atomic theory and took a gas to be composed of atoms and molecules in random, chaotic motion—indeed explained a number of well-known physical laws about the temperature, pressure, and volume of gases. Pressure, for example, was shown by the British physicist James Joule to be caused by speeding gas molecules colliding with the walls of their container. The kinetic theory eventually allowed chemists to estimate the size of atoms. As expected, they were very tiny indeed—about ten billionths of a centimeter (10^{-8} cm) in diameter. But nobody had yet *seen* an atom in an experimental apparatus. All the detailed arguments, based as they were on abstract assumptions and mathematical statistics, were unconvincing to certain influential scientists of the late nineteenth century.

Fortunately, there was a way to "see" the effects of single atoms and molecules directly, after all. In 1827, the English botanist Robert Brown, while examining microscopic pollen grains suspended in water, had noticed a random, trembling motion of these tiny particles—an effect that came to be known as Brownian motion. With the atomic theory gaining favor in the late 1800s, some scientists suggested that this motion occurred because individual water molecules slammed into the tiny pollen grains. Then in 1905, in one of the three papers that brought him immediate worldwide recognition, Albert Einstein used the atomic theory to make detailed predictions of Brownian motion. Almost a century after Dalton first proposed it, the modern atomic theory had finally convinced its last skeptics. Seeing *was* believing.

In their moment of triumph, however, atoms were in deep trouble. Their purported role as the structureless, irreducible units of existence was coming under increasing attack. By the turn of the century, there were far too many different atoms for these motes to be considered the true elementary building blocks of Nature. If atoms were the simple basis of existence, why were there almost a hundred kinds? And why were there completely new particles, like the electron, that defied explanation within the atomic theory?

Nineteenth-century scientists had struggled to find order in this growing chaos. By 1869, the Russian chemist Dmitri Mendeleev had organized all the known elements—more than sixty at the time—into a table of rows and columns now called the Periodic Table. The elements in any single column possess similar chemical properties. Observing a few gaps in the table, Mendeleev predicted the existence and properties of three new elements (now called gallium, germanium, and scandium) that were then completely unknown. After their discovery, his table became firmly established in scientific circles. But the table by itself *explained* little. It was just a convenient way to organize the growing variety of known elements.

In 1897, the British physicist J. J. Thomson discovered the electron. More accurately, he showed that the well-known radiation emanating from a cathode-ray tube (the forerunner of modern TV tubes and video display monitors) was a stream of charged *particles*—not waves, as many scientists had argued. He proved this hypothesis by showing that both electric and magnetic fields deflected these cathode rays from their normal straight-line paths. Ergo the rays had to have an electric charge, a property most uncharacteristic of normal electromagnetic waves. What's more, Thomson discovered that the mass of these "electrons" was about one two-thousandth (later measurements put it at 1/1836) of the mass of a hydrogen atom. Here was a curious particle with an electric charge and a mass that was *minuscule* by comparison with the smallest atom.

Thomson's discovery of the electron came at a time of great ferment in physics. In a remarkable ten-year period stretching from 1895 to 1905, Konrad Roentgen identified X rays, Henri Becquerel discovered natural radioactivity, Max Planck proposed his revolutionary quantum theory, and Albert Einstein explained the photoelectric effect and published his Special Theory of Relativity. Nature was proving to be *granular* at a level deeper still than the atomic. Even *light* seemed to behave like a particle! The familiar, continuous world of classical, Newtonian physics was cracking up into a weird, fragmented world governed by relativity and the quantum theory.

In art, meanwhile, the abstract cubist paintings of Georges Braque and Pablo Picasso were evolving from antecedents in French Impressionism. Edvard Munch, Emil Nolde, and other German Expressionists were beginning to stretch and contort the human figure in wholly new directions. The familiar three-dimensional perspective that had dominated Western art since the Renaissance was disintegrating practically overnight.

In 1898, Thomson proposed an atomic structure that incorporated the newly discovered electrons. He considered the atom as a diffuse ball of positively charged matter about ten billionths of a centimeter across, with the far tinier electrons embedded within, "like raisins in a pudding." The total positive charge, he imagined, equaled the sum of the negative charge of all the electrons—leaving the atom electrically neutral. Because unlike charges attract, the electrons would cling tenaciously to the atom. But strong forces could readily pry some of them loose, leaving behind a positively charged particle called an ion. An atom could also attract extra electrons into its grasp—making it a negatively charged ion. Thomson's model explained a lot of confusing phenomena in the chemistry and physics of his day. It went largely unchallenged for a decade.

The challenge came, finally, from Ernest Rutherford and his colleagues. A robust New Zealander who had studied under Thomson at Cambridge University, Rutherford had been examining natural radioactivity with the chemist Frederick Soddy since 1898. They established that part of the radiation coming from the element radium was a stream of heavy, positively charged particles, which they named alpha particles. Further experiments showed that these particles were actually helium atoms shorn of their two electrons—doubly charged helium ions.

In 1907, Rutherford began a long series of experiments at the University of Manchester in which he used these alpha particles to study atomic structure. Using radium as his source, he fired a narrow beam of alphas at a thin gold foil. If gold atoms were anything like Thomson's diffuse balls of matter, he reasoned, then these tiny motes should speed right on through with little or no deflection. By analogy, imagine you fired a rifle bullet at a bag of snowballs; you would hardly be surprised if the bullet tore straight through without any deflection whatsoever. As expected, the vast majority of alpha particles penetrated the foil with little change in direction.

But a few particles *were* deflected, or "scattered," at angles greater than 90 degrees. Rutherford's young assistant, Ernest Marsden, first noticed these odd events during careful observations of a zinc sulfide phosphorescent screen placed on the *same* side of the foil as the source of alphas. When an alpha struck the screen, it produced a small flash of light that he noticed and recorded. About one in every ten thousand alpha particles were actually scattered backward. Puzzled, Marsden showed the results to Rutherford.

"It was quite the most incredible event that has ever happened to me in my life," Rutherford later recalled. "It was almost as if you fired a fifteen-inch shell at a piece of tissue paper and it came back and hit you!"

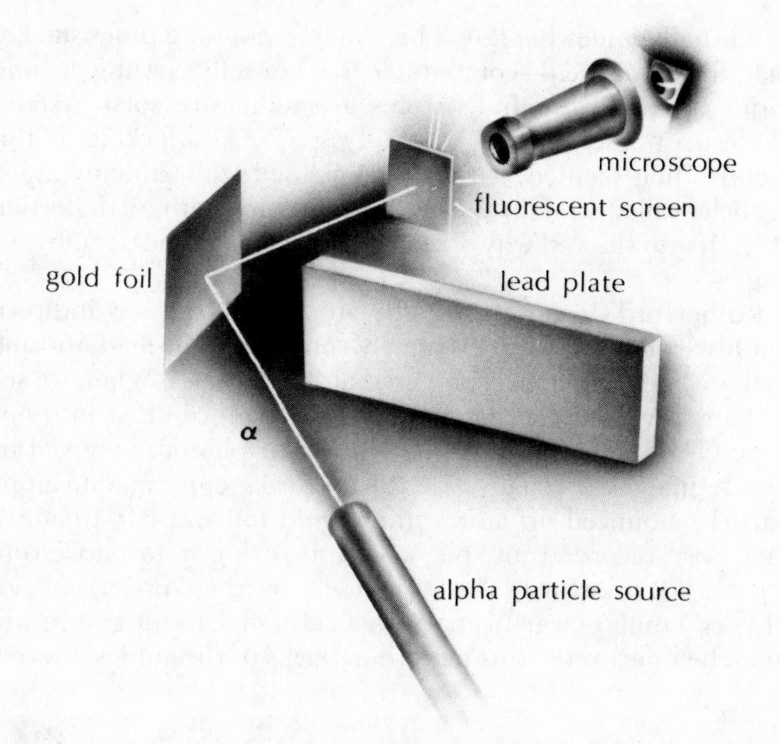

The apparatus Ernest Marsden first used to discover that alpha particles were deflected backward by gold foil.

To continue the snowball analogy, it was as if a bullet had pierced the bag of snowballs and come blazing back toward the rifle. The only conceivable explanation would be that the bullet had struck something small and hard—like a steel ball bearing—embedded inside one of the snowballs. Most of the time, bullets would tear straight through the bag, but once in a while one would strike metal and ricochet backward.

In 1911, Rutherford presented a new model of atomic structure to explain these curious results. Almost all of the atom's mass, he declared in a famous paper published in the *Philosophical Magazine*, "is concentrated into a minute center or nucleus," an unbelievably tiny object located at the center of the atom. The electrons in Rutherford's atom orbited this nucleus like planets about the sun.

From more detailed measurements made by Marsden and Hans Geiger, Rutherford deduced that this nucleus was no more than one *trillionth* of a centimeter (10^{-12} cm) in diameter, and usually somewhat less. This was ten thousand times smaller than the atom itself—comparable to a housefly buzzing around inside a huge cathedral. Atoms are miniature solar systems made up predominantly of empty space. As shocking as this picture then seemed, it readily explained why so many alpha particles sailed straight through the gold foil with no deflection at all. It also showed why occasional alphas rebounded at drastic angles.

Rutherford's evidence for the atomic nucleus was indirect, but his arguments were extremely convincing. Instead of using light to "see" the nucleus, he used alpha particles. When we see a chair, for example, particles of light bounce off it into our eyes. Our brains interpret the pattern of photons received by our retinas as a "chair." In Rutherford's experiment, alpha particles bounced off atoms in the gold foil and left a pattern that was recorded by his assistants using phosphorescent screens—Rutherford's "eyes." Aided by the known laws of physics, some powerful mathematical tools, and a good intuition, he interpreted this pattern in terms of the atom's internal

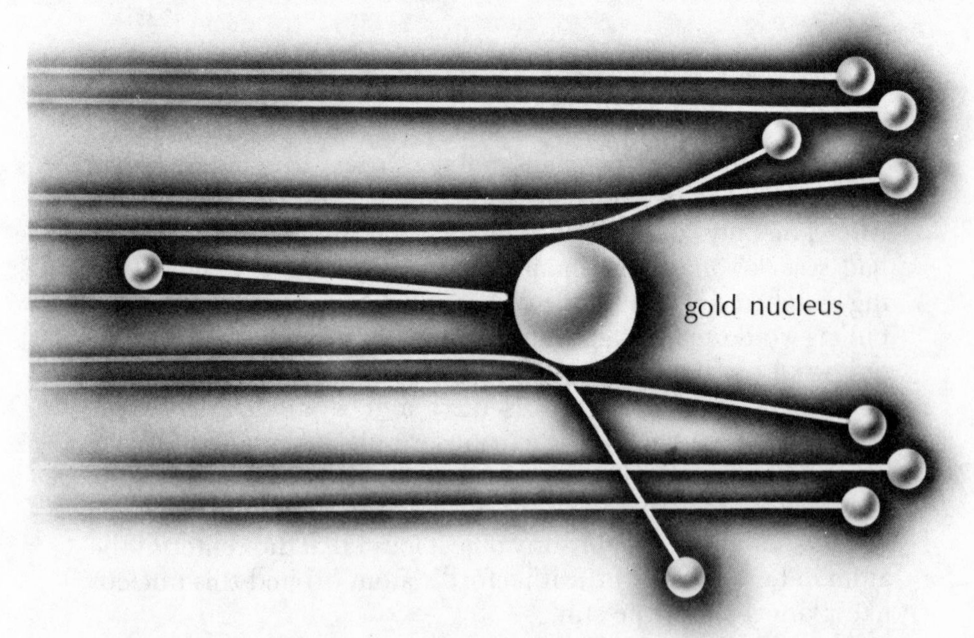

Scattering of alpha particles by an atomic nucleus. The vast majority sail right on by, but those few passing very close to it ricochet at large angles.

structure—just as our brains ascribe a certain structure to the chair. This way, Rutherford was able to deduce the atomic nucleus and to estimate its size.

At a dinner party Rutherford was asked whether he believed his alpha particles really existed, after all. "Not exist? Not exist?" he recoiled indignantly. "Why, I can see the little beggars right there in front of me as plainly as I can see that spoon!" For him, seeing was believing. His was just a new and different way of "seeing"—but a way that would dominate nuclear and particle physics for decades to come.

The Rutherford atom, however, suffered one critical flaw. According to Maxwell's Equations (the universally accepted canons of classical electricity and magnetism), an electron whirling about a nucleus should emit radiation copiously. Thus, it should lose energy and eventually spiral into the nucleus. Glowing intensely, such an atom would survive only a billionth of a second—hardly the stable, indestructible particle needed as the basis of all existence!

A way out of this dilemma was discovered by the Danish physicist Niels Bohr. After earning his Ph.D. from the University of Copenhagen in 1911, Bohr came to England for postdoctoral work at Cambridge and then Manchester, where Rutherford took a liking to him. The speculative young Dane quickly impressed everybody in this Midlands hotbed of British empiricism with his intuitive, imaginative approach to atomic physics.

In a *Philosophical Magazine* article published in April 1913, Bohr proposed a theoretical model for the atom in which the electrons could move only in a specific set of special orbits. In the exquisite words of Banesh Hoffmann, "No longer could an electron roam fancy-free wherever it wished but, more like a trolley car than a bus, it must keep to the tracks laid down by Bohr." He assigned an energy level to each track—much like ascending rungs on a ladder. An electron could jump to a lower (or higher) track by emitting (or absorbing) a single unit of light energy *exactly* equal to the difference between the two levels. But there *was* a lowest level. Once there, the electron could drop no further—sparing the atom its inglorious demise.

More than a decade earlier, in 1900, a German professor of

theoretical physics had paved the way for Bohr. Almost apologetically, Max Planck proposed that matter emits and absorbs light energy in neat little packets or bundles—*not* continuously, as everyone else then believed. The energy E of a particular bundle, which he called a quantum, could be obtained from its vibrational frequency f (corresponding to the *color* of the light) according to the formula $E = hf$, where h is the now famous "Planck's constant" so central to modern physics.

At the time, this formula and the theory behind it were unspeakable heresy to classical physicists enamored of a smooth, continuous world. Try to imagine an ocean wave crashing against a long seashore and depositing *all* its water inside a single cove; this is essentially the quantum idea, viewed on a macroscopic level. Even Planck was somewhat horrified by the Pandora's Box he had opened; he spent the next five years trying to heal the wound he had inflicted on classical physics. But to no avail. His quantum hypothesis was marvelously successful in averting the "ultraviolet catastrophe" then plaguing theoretical calculations of radiation from matter. As Planck later recounted, "It was clear to me that classical physics could offer no solution to this problem and would have meant that all energy would eventually transfer from matter into radiation."

In 1905, Albert Einstein took the one crucial step that Planck was too timid to take. In explaining the photoelectric effect, whereby light knocks electrons out of certain metals, he proposed that a bundle of light energy *continues* to behave as a bundle, even after it escapes from matter. This was no mere heresy; it was out-and-out treason! Here was an obscure Swiss patent clerk claiming that light was a *particle*, after two centuries of experiments had proved it was a wave. But in a very simple and straightforward manner, Einstein showed how these bundles of light, or "photons," could explain all the experimental details of the photoelectric effect. It was mainly this explanation, and not his theory of relativity, that earned him the Nobel prize in 1921.

Armed with the revelations of Planck and Einstein, Niels Bohr sallied forth into the Rutherford atom. "It seems necessary to introduce," he proclaimed in his 1913 paper, "a quantity foreign to classical electrodynamics, Planck's constant, or as it is often called, the elementary quantum of action." In Bohr's theory, the electrons whirling about a nucleus could only occupy specific orbits or "tracks," each denoted by a corresponding whole number n (equals 1, 2, 3, . . .) called its "principal quantum number."

Bohr also supplied a specific mechanism whereby his atoms took in and gave off photons as the electrons leapt from one track to another. His model could reproduce almost exactly the spectrum of light frequencies—the distribution of colors—emitted by the simplest atom, hydrogen. Actually, this was no great feat. In building his model, Bohr had borrowed heavily from the well-known spectrum of hydrogen. Small wonder, then, that it came out right. But the real test of the Bohr model would come with the light spectra emitted by *other* atoms. Could it reproduce these, too?

Try as they might, however, atomic scientists could not reproduce the spectrum of helium, the simplest atom after hydrogen. They were not even close. By the early 1920s, the Bohr theory was on a steady slide, unable to explain a growing list of atomic phenomena. Still, Bohr's theory had been remarkably successful for hydrogen, rescuing Rutherford's planetary model of the atom. Bohr was awarded the 1922 Nobel prize for this outstanding contribution to physics, a contribution immortalized by its mention in almost every college textbook on atomic physics and quantum mechanics.

In retrospect, the Bohr-Rutherford model of the hydrogen atom, with its curious admixture of classical and quantum ideas, was about the furthest one could push a mechanical model into the strange new "subatomic" world inside the atom. The theory had served scientists well for a decade. And it had helped bring many a skeptical physicist into the quantum fold. What they now needed was a *deeper* theory, one that reconciled the seemingly contradictory wave and particle properties of light.

Wave or particle? The debate about the nature of light had surged back and forth for centuries. In his *Opticks,* Isaac Newton had argued vociferously on behalf of its particle nature. After his death, however, wave partisans gradually gained the upper hand. Such wavelike phenomena as refraction (the bending of a light ray by a prism) and diffraction (the spread of light rays around corners) were difficult to imagine if light were a localized, corpuscular entity. Interference patterns proved impossible. These alternating light and dark bands occur when light from two point sources is superimposed; they can be explained only if light is an extended, oscillatory phenomenon like a wave.

By the mid-1800s, there was little remaining doubt: light was indeed a wave, after all. In 1864, Maxwell's Equations extended the wave interpretation to the entire electromagnetic spectrum. Infrared and ultraviolet radiation, as well as radio waves and X rays, all proved to be wave phenomena, too.

There the matter stood until Einstein surprised the physics community with his particle interpretation of the photoelectric effect. Subsequent experiments verified his predictions to an uncanny degree of accuracy. The final blow came in 1923. The American physicist Arthur H. Compton fired a beam of X rays (a high-frequency cousin of ultraviolet light) into a cloud chamber and watched them carom off atomic electrons in the gas. Like the cue ball in a game of pool, an X ray lost energy during a collision and emerged weakened. Like the object ball headed for the pocket, the rebounding electron always came away with the missing energy. The only way that Compton could interpret this curious observation was to conclude that light was Einstein's photon, a localized entity that smashed into the electron and gave it some energy before ricocheting away. Waves did not act like this.

So here was a truly vexing quandary. Depending upon the experimental conditions, light (or its invisible cousins) could behave as *either* a wave or a particle. When fired at a crystal surface, the same X rays that had behaved like particles in Compton's experiments developed an interference pattern distinctly characteristic of waves. It was almost as if light had a mind of its own! Such a "dual" nature had no precedent in physics. This was something altogether new.

Enter Louis de Broglie. Late in 1924, this French nobleman published his doctoral thesis, in which he proposed that *matter* might sometimes act like a wave! His argument was actually quite simple. Almost two decades earlier, Einstein had established—in his famous equation $E = mc^2$—that energy and matter were just different forms of each other. As light energy always had an associated vibrational frequency, de Broglie reasoned, "matter energy" should also vibrate or pulsate. Then a particle of matter drifting through space, vibrating as it went, should behave somewhat like a "matter wave." Thus, its wavelength λ (the Greek letter *lambda*), he predicted, would be Planck's constant divided by the particle's momentum p, or $\lambda = h/p$. Planck's constant seemed to be turning up everywhere one looked.

Graduate students in theoretical physics often propose pretty ideas like this, but few with any firm connection to reality. De Broglie's idea was different. In 1925, quite by chance, Clinton Davisson and Lester Germer at Bell Telephone Labs in New York were firing electrons at a crystal of the metal nickel. They discovered the very same interference patterns one got by scattering X rays from crystals. These electrons were behaving like waves!

The sharp distinction between wave and particle, so central to classical physics, was now a thing of the past. Everything in existence had both a wave nature *and* a particle nature. Depending on how you did your experiment, one or the other might manifest itself. And, after Einstein, matter and energy were just different forms of each other. Where could one go to find solid ground amid these shifting quicksands?

Physicists love a paradox like this. A paradox is a sure sign that Nature is trying to tell us something completely new and different about herself. Revolutionary insights are at hand. A paradox usually occurs because the way we *think* about "reality" is seriously flawed in a fundamental way. Its resolution promises a radical new vision of the universe and our place within it.

In 1905, Albert Einstein resolved a paradox that had been vexing scientists since the 1887 experiment of Michelson and Morley, who showed that light travels at the same speed, relative to an observer, no matter how that observer is moving. Common sense tells you that light should seem to travel faster if you are moving toward an approaching beam—and slower if you are moving away. But this was simply not the case in actual practice. Einstein's solution of this paradox, his Special Theory of Relativity, banished absolute length, time, and velocity from the universe.

After de Broglie's crucial insight, the solution of the wave-particle paradox came swiftly. Physicists followed two separate paths, one pioneered by the young German Werner Heisenberg, the other by the Austrian Erwin Schrödinger. In the summer of 1925, Heisenberg—then a postdoc at Göttingen University—developed a new mathematical formalism called "matrix mechanics." In it, the equations of Newtonian mechanics were replaced by similar equations involving "matrices," two-dimensional arrays of numbers resembling mathematical tables. Later that year, Einstein's praise brought de Broglie's work on matter waves to the attention of Schrödinger, then a professor of theoretical physics at the University of Zurich. In a few short months, he developed the now-famous Schrödinger Wave Equation to describe the motion of these matter waves, which he represented by the Greek letter ψ (*psi*).

With only a few other assumptions, Schrödinger was able to *derive* the frequencies of light emitted by hydrogen atoms, a feat that Bohr had never accomplished. He published his results in January of 1926. That very month, the Austrian Wolfgang Pauli and Paul Dirac at Cambridge independently published their own solutions to the spectrum of the hydrogen atom based on Heisenberg's matrix methods. The revolution had begun in earnest.

Schrödinger's approach treated matter as waves, while Heisenberg's treated it as particles. The Austrian now took the offensive and showed how his wave equation could, after detailed calculations, reproduce the German's curious matrices. The German used his matrix approach to calculate the spectrum of the hitherto inscrutable helium atom. But Dirac outflanked them both and developed a new, more general theory that included the two previous approaches as special cases. Mathematically speaking, wave and particle behavior fell out of the very same theory.

Max Born, who a year earlier had brought young Heisenberg to Göttingen, now joined the fray. In the summer of 1926, he interpreted Schrödinger's curious wave function ψ: the square of its amplitude (or wave height) gave you the *probability* of locating a particle at some chosen place and time. In essence, ψ was a "probability function" that specified your chances of ever finding the particle in your experimental apparatus. In advance, you could never know the exact results for sure.

Here is where randomness entered physics, where quantum mechanics abandoned the deterministic, clockwork universe of classical Newtonian physics. In the subatomic world, Nature resembles a floating crap game. At the time many physicists had terrible problems with Born's probability interpretation of the wave function ψ. The loudest among them, Einstein protested that "God does not play dice!" Even Schrödinger had second thoughts about the mischief he had wrought. "If I had known all this *Herumspringerei,* all this jumping about, to which my equation was going to give rise," he confided, "I would never have had anything to do with it in the first place."

Philosophers at heart, Bohr and Heisenberg anguished together over the apparent contradictions and outright absurdities they confronted in the interpretation of quantum mechanics. As Heisenberg later recounted,

> I remember discussions with Bohr, which went through many hours till very late at night and ended almost in despair; and when at the end of the discussion I went alone for a walk in the neighboring park, I repeated to myself again and again: Can Nature be as absurd as it seemed to us in these atomic experiments?

The two giants of quantum theory wrestled with these problems into early 1927.

In February, Heisenberg had the crucial physical insight that made all the abstract mathematics somewhat comprehensible. The subatomic world inside atoms at last began to make sense. The apparent contradictions arose, he realized, when one blindly tried to extend classical concepts derived from everyday experience—like position, velocity, energy, and time—into this tiny realm. In ascertaining the position of an electron, for example, you have to make a *measurement:* you have to hit it with a photon or another electron and detect the rebounding mote in your apparatus. Of course such a collision gives the electron you are studying an indeterminate kick, so that you cannot simultaneously measure its velocity with infinite accuracy. Heisenberg's words say it best:

> One could speak of the position and the velocity of an electron as in Newtonian mechanics, and one could observe and measure

> these quantities. But one could not fix both quantities simultaneously with an arbitrarily high accuracy. Actually the product of the two inaccuracies turned out to be not less than Planck's constant divided by the mass of the particle.

This is the famed Heisenberg Uncertainty Principle. Here was the true source of the indeterminacy implied by Born's probability interpretation of the wave function. There are fundamental *limits* on the accuracy to which we can simultaneously measure certain properties of subatomic particles. Heisenberg's strange principle flew in the face of classical physics, which had always assumed one could subdivide space and time indefinitely, with infinite accuracy. This assumption was the fundamental flaw in the way classical physicists *thought* about reality, the true source of the apparent wave-particle paradox.

After Heisenberg's decisive victory, Bohr came in to mop up. He introduced the idea of "complementarity," whereby no single image of reality—wave or particle—can suffice to explain the subatomic world in its entirety. The wave picture and the particle picture, he argued, are two exclusive, *complementary* aspects of the same fundamental reality, which somehow lies beyond the grasp of any single viewpoint. A specific measurement can show us only one facet of the gem. And the act of measurement is the crucial step. In setting up an experiment, we determine in advance which facet we will observe. The act of measurement drastically alters what we measure, so that we cannot return and accurately measure other complementary properties.

Bohr's Complementarity Principle introduced a subjective element into the interpretation of the subatomic world. To some extent, we predetermine the results of our own measurements. This blurring of the heretofore sharp distinction between the subjective and objective realms troubled many a classical physicist—Einstein the most prominent among them—wedded to the notion of an objective, knowable world. Warm and close friends, Bohr and Einstein nevertheless spent the rest of their days arguing about the meaning and validity of quantum mechanics.

The final battle of the quantum revolution occurred during the autumn of 1927 at the Solvay Conference in Brussels. There Bohr presented a paper summarizing the physical

Albert Einstein and Niels Bohr in the late 1920s.

interpretation of quantum mechanics developed in Copenhagen by himself, Werner Heisenberg, and Wolfgang Pauli. Again and again, Einstein protested the indeterminism essential to their approach. Again and again, Bohr and Heisenberg refuted his arguments. After much heated debate, most of the scientists there accepted this interpretation of quantum mechanics. Called the "Copenhagen interpretation," it has survived to this day and forms the basis of how atomic, nuclear, and particle physicists think about their tiny worlds. In essence, the subatomic world is an indeterminate world whose properties depend in part on how we choose to measure them.

After the Solvay Conference, the quantum revolution was essentially over, except for a few small skirmishes. The road to Stockholm was soon crowded with quantum physicists hurrying to collect their Nobel prizes: Compton in 1927, de Broglie in 1929, Heisenberg in 1932, Dirac and Schrödinger in 1933, and Davisson in 1937. Somewhat later, after Hitler had been subdued and World War II had subsided, Born and Pauli got theirs, too.

At first glance, the subatomic world seems a schizoid realm, a modern Wonderland. The bizarre, the fantastic, and the downright absurd are the rule here, not the exception. Particles act like waves—and waves like particles. Space and time, energy and momentum have lost their traditional meaning. The inhabitants are quirky at best; if you try to examine one too closely, it leaps away from your scrutiny. Objectivity evaporates into a mystical sigh.

Fortunately, we have a reliable guide to this nightmarish world. The Copenhagen interpretation of quantum mechanics leads us down a tentative path. Its twin principles of uncertainty and complementarity tell us what we can and cannot know. Heisenberg's Uncertainty Principle warns us to watch our language in this Lewis Carroll landscape. Everyday words and concepts *can* be useful here, as long as we are careful not to forget their innate limitations. And Bohr's Complementarity Principle counsels us to be circumspect in dealing with the natives. No one viewpoint contains the whole truth about them. Together, our two guides make the subatomic world a comprehensible place.

Consider, for example, the idea of an electron orbit. You can probably imagine, as did Bohr at first, an electron following a stately circular path about a nucleus, smooth and continuous from one point to the next. All of us have seen the atom pictured like this. The seal of the old United States Atomic Energy Commission had such a picture of a lithium atom upon it—with three electrons revolving serenely about a nucleus.

"But you are misusing everyday concepts in the subatomic

world," Heisenberg objects. "If you wish to speak of an electron orbit, then show me measurements of its position and velocity along its entire path."

Okay, you accept his challenge. You decide to shoot in another particle, say a photon, and let it bounce off the orbiting electron. By capturing the rebounding photon, say in a photocell, you can then determine the electron's position at the moment of impact. All seems well and good. But in order to study the fine details of the electron's motion, you need to use photons with a wavelength much *smaller* than the size of the atom. Small wavelength means high frequency, and high frequency means high energy—according to Planck's formula. Visible light just will not do. It turns out that you have to use powerful X rays if you want any chance of "seeing" the fine details of the electron orbit.

But wait. This is just Compton's experiment: bouncing X rays off atomic electrons like billiard balls. And remember what that does? The X ray clobbers the hapless electron and knocks it completely out of orbit. It's like a blind man trying to determine the exact position of a plate glass window by throwing a huge boulder at it. He may discover where the window *was,* but he completely destroys the very thing he was trying to measure.

Heisenberg is doubled over with laughter. "The electron 'orbit' is merely a figment of your classical imagination," he chides, "a throwback to concepts borrowed from everyday experience."

Taking pity, kindly old Bohr now comes to your rescue. "Stop thinking of electrons here as particles," he counsels, "and consider them instead as de Broglie's matter waves." So you haul out Schrödinger's Wave Equation, insert the proper conditions, and crank out the appropriate wave function ψ. The square of ψ at any point gives you the odds of finding an electron there—say one in ten. The electron now appears to you as a "smear" or distribution of electronic matter centered about the nucleus. Where the distribution is dense, you have a better chance of finding the electron. With a few more calculations you also discover that this wave function reproduces the energy levels for the hydrogen atom.

In barest essence, this is the way physicists learn to think about the subatomic world. Consciously or not, the Uncertainty and Complementarity Principles guide our thought processes. We take extreme care when using everyday language and classical ideas in this strange domain. They *have* their place here, but you must be very careful in using them—lest you slip into absurdity and contradiction. Physicists may speak of a "particle," for example, but we really mean a very fuzzy object we had better not pigeonhole too narrowly. And we are also careful to remember its wave nature when performing our calculations. Dancing back and forth between wave and particle pictures of the subatomic world, we begin to form a composite mental image that corresponds fairly well to the reality measured by our instruments. When peering into the subatomic world, we have to use stereoscopic vision.

But then you might well ask, "What is this strange 'wave function' ψ, anyway?" A good question. In fact, it's a truly weird beast with no parallel in the everyday world of our senses. Mathematically speaking, ψ is a probability function, a whole field of complex numbers whose square at any point gives you the chances of finding a particle there. We might liken such a matter wave to a crime wave (rather than, say, a water wave). During a crime wave, your chances of getting mugged increase, but you are never *sure* of getting mugged until it actually occurs.

In essence, ψ is an abstract mathematical entity with no concrete reality of its own. It represents the *potential* of something to be found in a specific place at a given time. "The probability function does not in itself represent the course of events in the course of time," warned Heisenberg. "It represents a tendency for events and a knowledge of events." Tendency becomes reality when we make a measurement. The act of observation intrudes upon the placid subatomic world and compels the thing in question to take a firm stand. Heisenberg again: "The transition from the 'possible' to the actual takes place during the act of measurement."

In a sense, the wave function incorporates both subjective and objective features. As Heisenberg stressed, the wave function embodies Aristotle's concept of matter as an unformed "potentia" that becomes fully real only when it assumes a specific form. The act of measurement—the conscious act of a willing subject—forces matter to choose among its many possibilities.

In the subatomic world, therefore, "reality" is neither subjective nor objective, but *interactive.* In the macroscopic world of the senses, by contrast, scientists can pretend that we do not disturb an object when measuring it. What little disturbance there is can be easily ignored because it is so small. But in the subatomic world, we have to use some probe that interferes—often very violently—with the "object" of our study. Instead of measuring the properties of the "thing-in-itself," we measure how the thing interacts with our probe (and, ultimately, with *us*). What we learn about the target can be quite different, depending on our choice of probe and experimental arrangement.

Subatomic reality is a lot like that of a rainbow, whose position is defined only relative to an observer. This is not an objective property of the rainbow-in-itself but involves such subjective elements as the observer's own position. Like the rainbow, a subatomic particle becomes fully "real" only through the process of measurement.

Since the time of Francis Bacon three centuries earlier, scientists had firmly believed in the inherent orderliness of Nature. The molecules might dance, but they must do so according to natural laws that man could discover by careful experiments and the use of mathematical logic. The universe was not some cruel cosmic joke played on a hapless humanity. As Alfred North Whitehead put it in 1925, "There can be no living science unless there is a widespread conviction in the existence of an *Order of Things* and, in particular, of an *Order of Nature.*" This belief in an orderly universe was an article of faith shared by all classical physicists.

Quantum mechanics threatened to shake this faith to its very foundations. That is why Einstein protested the Copenhagen interpretation so vehemently. Instead of order being an inherent feature of Nature, it now seemed to arise only when scientists used their tools to measure what is fundamentally arbitrary and chaotic. How else to interpret the puzzling randomness of the subatomic world? Whatever order one could find seemed to be imparted by observers only during the act of measurement.

An entire school of thought has grown up around this notion. A noted theoretical astrophysicist, John Archibald Wheeler, one of Bohr's close colleagues during the 1930s, likes to speak of an "observer-created universe" in which things "exist" only after being seen or measured. This is the old, familiar argument about a tree falling in a deserted forest: if nobody heard it happen, could the tree really have made a noise?

Eastern mystics have seized upon these same notions to stress the parallels between particle physics and Oriental philosophies. In particle physics, they claim, we learn truths only about our own mental processes and not about some nebulous "objective" world—truths supposedly known for centuries by gurus, swamis, and other self-appointed mystics.

Solipsistic arguments like these try to restore humanity to the center of the universe by denying the independent existence of everything else. And they ignore the obvious fact that there *are* objective features of the subatomic world that exist independently of any particular observer.

Take the electron again. Every single electron that scientists have observed to date has always revealed the exact same value for its electrical charge. No matter who made the measurement or how they made it, the result has always been essentially the same.

This granularity of the electron's charge was first observed in 1910 by the American physicist Robert A. Millikan. In an elegant experiment, he suspended fine droplets of oil in an electric field. The droplets acted as if they all possessed a charge that was an integral multiple of a fundamental unit charge (now denoted by $e = 1.6 \times 10^{-19}$ coulombs). Millikan concluded that each droplet carried a small excess or deficit of electrons—all with exactly the same unit charge.

Likewise, every electron has always revealed the same mass and spin. Millikan's measurements enabled others to calculate the mass (denoted by $m = 9.1 \times 10^{-28}$ gram) from J. J. Thomson's earlier work. Electrons have never been found with any different value. And the intrinsic angular momentum—the "spin" of an electron about some axis—has always proved to be Planck's constant h divided by 4π.

Billions and billions of electrons have been sighted since

Millikan's day, but no measurement of their properties has ever turned up any different values for their charge, mass, and spin. All electrons are *identical.* Our modern electronic civilization depends in large part on this experimental fact. The sharp color picture on our television sets, for example, would be impossible if electrons differed from one to another.

For physicists, such repeatability is convincing proof that we are dealing with the intrinsic properties of matter "out there" and not just some figment of our collective imagination. We call these intrinsic properties "quantum numbers"; they establish a particle's very identity. Besides charge, mass, and spin there are many others—parity, strangeness, charm, color, beauty, and even truth—that you will encounter later in this book. Like the colors and symmetries of a rainbow, these quantum numbers—and not space or time coordinates—are the properties that observers can discuss meaningfully. "Where?" and "When?" are relatively unimportant questions.

What is truly remarkable is the *durability* of these quantum numbers and other properties of subatomic particles. As in classical physics, the total energy and total momentum of the particles leaving a collision are the same as they are for the particles entering it. We say energy and momentum are "conserved" in subatomic collisions: they do not change. Most of the other quantum numbers are conserved, too. The total charge, for example, stays the same after a collision—as does the spin. Conservation laws like these help to convince us that the subatomic world is not an entirely capricious realm.

In studying the subatomic world, we are all blind men playing billiards. With high-energy particles as messengers, we grope for knowledge of its uncertain terrain. Space and time have lost their traditional meaning here. Fortunately, however, there *are* a few certainties in this paradoxical realm: the quantum numbers of our particles and the conservation laws they obey. These are our guideposts. Quantum mechanics lights our path—indicating where we can or cannot walk. It is not a perfect guide, but like the blind man's cane it gets us past pitfalls and obstructions to our eventual destination.

CHAPTER 2

THE PARTICLE KINGDOM

There are therefore Agents in Nature able to make the Particles of Bodies stick together by very strong Attractions. And it is the Business of experimental Philosophy to find them out.

—Isaac Newton, OPTICKS

IN 1919 Ernest Rutherford returned to Cambridge to succeed his old professor, J. J. Thomson, as Cavendish Professor of Physics, Britain's most prestigious and powerful chair. He gathered about him an excellent group of experimenters and made the Cavendish Labs the undisputed world leader in the new field of nuclear physics, a field he had pioneered with his 1911 discovery of the atomic nucleus. Under his fatherly guidance the Cavendish produced Nobel laureates James Chadwick, John Cockcroft, Peter Kapitza, Cecil Powell, and Ernest Walton. Later, they fondly recalled Rutherford's booming voice, which could easily upset their fragile equipment.

Undaunted by the quandaries of his day over quantum theory and atomic structure, Rutherford forged ahead with his trail-blazing alpha particle experiments. The son of a small farmer and utility man, he was uncomfortable with the complex arguments of learned theoretical physicists and far more at home poking around in the laboratory. About the only theorist

J. A. Ratcliffe and Ernest Rutherford at the Cavendish Labs during the 1930s.

he paid much heed was his old postdoc and good friend, Niels Bohr.

In his first year back at Cambridge, Rutherford and James Chadwick discovered that the atomic nucleus had its own internal structure. They were firing alpha particles into a tube of nitrogen gas. To their surprise, they found hydrogen atoms remaining in the tube afterward. The two scientists could only conclude that the nucleus of nitrogen was *itself* composite—and that the alpha particles were occasionally breaking off pieces identical to the nucleus of a hydrogen atom. Rutherford dubbed the hydrogen nucleus the "proton," from the Greek word *protos* for "first," and predicted that all the heavier atomic nuclei were built up from protons, the simplest available. His proposal was a good guess, based on the evidence available to him, but it was not quite right.

The main problem with this idea was that the heavier nuclei have too little charge. Nitrogen, for example, is about fourteen times as heavy as hydrogen, but its nucleus has only *seven* times the positive electrical charge. Fourteen protons should have given fourteen times the proton charge, too. One way around this difficulty, favored by many physicists during the 1920s, was to assume that the nitrogen nucleus was composed of fourteen protons and seven electrons. The electrons would contribute negligible mass, but their negative charge would lower the net positive charge of a nitrogen nucleus to seven times the proton charge, as observed. And by putting electrons in the nucleus like this, they could explain why high-energy electrons (then called "beta particles") emerged from radioactive "beta decays" of heavy nuclei like uranium.

But Rutherford would have none of this. With characteristic insight, he proposed a different solution: there should be an uncharged, *neutral* particle about equal in mass to the proton. Hints of such a particle were found in the late 1920s, but it was not until 1932 that Chadwick had firm evidence in hand. Christened the "neutron," it has a mass 1839 times that of an electron, slightly heavier than the proton at 1836.

The discovery of the neutron helped resolve the long-standing puzzle about the mass and charge of nuclei heavier than hydrogen. The nitrogen nucleus, for example, is composed of seven protons (which give it a positive charge of seven units) and seven neutrons—for a total mass about fourteen times the proton mass. Alpha particles, or helium nuclei, contain two protons and two neutrons, for a total mass four times that of hydrogen. The neutron helped explain other puzzling facts about the many different atomic nuclei known, which were finally understood as the various possible combinations of protons and neutrons—collectively known as "nucleons," the denizens of the atomic nucleus. Beta decay resulted from the slow disintegration of a neutron into a proton and an electron, with the electron subsequently emerging from the nucleus.

No sooner had one riddle been solved, however, than another reared its head. To pack and hold these massive nucleons in such a tiny volume as that occupied by an atomic nucleus required some kind of terribly strong force. It could not be the electric force, because neutrons were neutral, uncharged particles that would not even feel such a force. And besides, the electric force was far too weak—about a hundred times too weak—to do the job. To compound the mystery, this strong force acted only over extremely short distances about the size of a tiny nucleus. Its pull did not extend to infinity like the familiar forces of gravity and electromagnetism.

Japanese theorist Hideki Yukawa suggested a possible solution in February 1935. This slight, bookish man was teaching at Osaka University when he published an epochal paper. Based on its extremely short range, he predicted that the strong force was transmitted through space by a hypothetical new particle weighing about 200 times the electron, or one ninth the proton. In formulating his theory, Yukawa used Heisenberg's Uncertainty Principle and argued by analogy with the electromagnetic force.

After the marriage of quantum mechanics and relativity in the late 1920s, one could visualize the force between two subatomic particles as being "carried" by another particle that passes between them. This principle was the starting point of Yukawa's theory. For example, the electromagnetic force acting between an electron e and a proton p can be pictured as the transfer of a photon, or particle of pure electromagnetic energy, between them—like a ball passed between two basketball players on a fast break. One particle coughs up a photon, denoted by the γ (the Greek letter *gamma*), which carries energy and momentum that it hands over to the second particle shortly thereafter. Both the electron and proton then react accordingly; if the photon carries enough energy, the proton can even disintegrate.

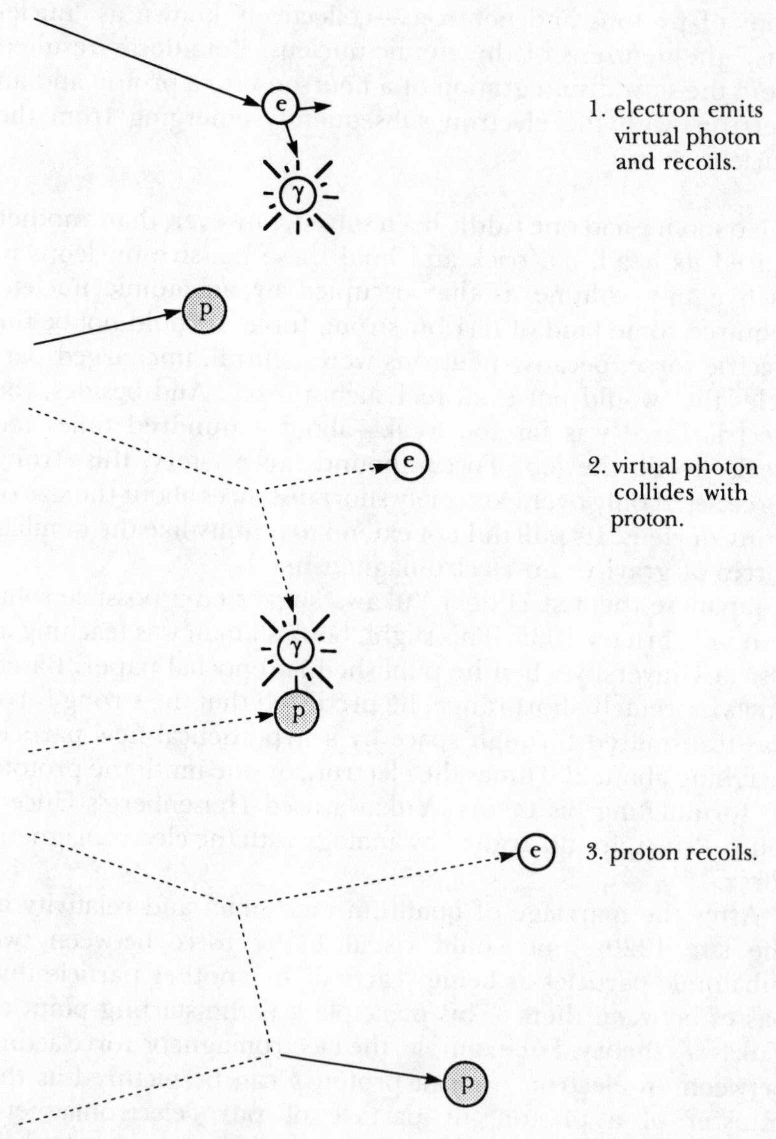

The electromagnetic force transmitted by a "virtual" photon being passed between an electron and a proton.

Physicists call a photon passed like this a "virtual" photon. Unlike real photons that we can perceive with our eyes or other detectors, virtual photons cannot be "seen" directly. They exist only to carry forces between two charged particles.

Heisenberg's Uncertainty Principle governs how long a virtual photon can exist and thus how far it can travel in such an exchange. Like momentum and position, energy and time cannot be measured simultaneously with infinite accuracy. According to the Uncertainty Principle, the product of the two inaccuracies in measuring a particle's energy and its lifetime must always be greater than Planck's constant h. What this means in practice is that very high-energy virtual photons can "exist" only for very short times and therefore can be transmitted only over very short distances. Our two basketball players can pass a basketball the length of the court, but if they toss a medicine ball instead, it will travel only a few feet.

But if the two players throw lighter and lighter objects, like baseballs and tennis balls, they will be able to "interact" over greater and greater distances—perhaps as much as the length of a football field. In the same way, virtual photons can carry the electromagnetic force over tremendous distances as long as their energy is sufficiently small. And because photons have no intrinsic mass, or "rest mass," we can employ arbitrarily weak virtual photons to carry such long-range interactions. In doing so, we are using the equivalence of mass and energy implied by Einstein's famous equation, $E = mc^2$. As photons have zero rest mass m, there is no lower limit to their total energy—and hence no upper limit to the distance over which they can act.

Yukawa's revolutionary proposal, that the strong force is carried by a heavy new particle, explained why it acts only over short distances of about 10^{-13} cm—the size of a proton or neutron. Because this particle had a large rest mass, there *was* a maximum distance over which it could carry the strong force. Unlike the photon, its total energy is always large, so it can act only over very short times and distances. It is much like our two basketball players passing a cannonball back and forth: they almost have to *hand* it to one another.

Actually, Yukawa worked the other way around. From the observed short range of the strong force, he postulated a very massive exchange particle—about 200 times the electron mass or about 100 million electron volts (100 MeV) in equivalent energy units. (One electron volt is the energy an electron gains in passing through a 1-volt potential drop, about what happens in a common flashlight battery.) For comparison, the proton "weighs in" at a rest mass of 938 MeV and the electron at 0.51 MeV, when expressed in energy units.

Experiments had shown that the strong force was the same between two protons as it was between two neutrons and between a proton and a neutron. To conserve electric charge in all these exchanges, Yukawa's particle thus had to come in three varieties: positive, negative, and neutral. Normally, these hypothetical fragments would exist only inside a nucleus, but a sufficiently violent blow should have occasionally knocked one of them loose. After scanning the available experimental literature, however, Yukawa could find no examples of any such particle. "As such a quantum with large mass has never been found by experiment," he concluded, "the above theory seems to be on a wrong line."

Unknown to Yukawa, a group of experimenters at Caltech had just begun to find evidence for such a particle. Under the general direction of Robert Millikan, who had won the 1923 Nobel prize for his measurement of the electron charge, this group was using cloud chambers and other devices to study cosmic rays. They carried these contraptions down mineshafts, flew them in light airplanes and balloons, and even lugged them to the top of Colorado's Pikes Peak and shipped them to the Panama Canal in their exhaustive studies of this extraterrestrial radiation.

The cloud chamber, which had been developed at the Cavendish Labs by C. T. R. Wilson back in 1895, is just what its name suggests. A volatile liquid such as alcohol evaporates and forms a thick vapor inside a closed chamber. When the gas is suddenly cooled or the pressure on the chamber is reduced, say by pulling a piston outward, the gas inside contains more vapor than it can really bear. We say it is "supersaturated" with the vapor; thus clouds begin to form, condensing first upon dust particles and charged ions in the gas. Now a charged particle passing through matter leaves a trail of ions in its wake, so a line of droplets quickly forms to reveal where any particle has just passed. By shining a strong light into the chamber, one can see and even photograph these "tracks." Like the vapor trail behind a jet plane high in the sky, such a track indicates that something has just gone by—in this case, a charged particle.

Carl Anderson, a postdoc under Millikan, had been working alone, taking photographs of cosmic ray tracks in the Caltech cloud chamber, when late on the night of August 2, 1932, he found a track the likes of which he had never seen before. At first glance, this particle behaved like an electron, but its track curled the *wrong* way in the magnetic field applied to the chamber. The only possible interpretation of this odd track was that it revealed a particle identical to the electron in every respect except its charge—which had to be *positive* instead of negative. The next day Anderson showed the curious photo to Millikan, who agreed with his assistant's interpretation. They dubbed it the "positron," for *positive electron,* and Anderson hurriedly published his results. The positron soon turned up in cloud chambers throughout the world; many even appeared in old photos taken well before Anderson's discovery.

Anderson did not know it at the time, but Paul Dirac had formulated a theory in 1928 that required the existence of a positron. Dirac developed a relativistic version of the Schrödinger Wave Equation, now known as the Dirac Equation, that possessed *two* separate solutions corresponding to particles of positive and negative energy. The positive-energy solution could be identified with the familiar electron, but what was this confounding negative-energy solution? How could a particle have an energy less than zero?

At first, theorists didn't believe Dirac's equation and dismissed this odd solution as a strange mathematical quirk, not a real particle. Or, like Dirac, they tried to associate the negative-energy solution with the oppositely charged proton. But then, why were electron and proton masses so vastly different? Characteristically blunt and outspoken, Wolfgang Pauli even opined that "an attempt to save the theory in its present form appears hopeless."

Anderson's 1932 discovery vindicated Dirac. The new positron was quickly identified with the negative-energy solution of the Dirac Equation. Because of relativity, every subatomic particle has such a twin sister called its "antiparticle." The positron is just the antiparticle of the electron.

Anderson, however, was not content to rest on his laurels. Teaming up with graduate student Seth Neddermeyer, he used the cloud chamber to study some penetrating cosmic rays they at first called "X-particles." They learned that these particles came with both negative and positive charge, like the electron and positron, and left similar tracks in the cloud chamber. But the X-particles passed readily through lead and platinum slabs placed inside the cloud chamber. Electrons and positrons, by contrast, stopped abruptly—or at least triggered large showers of other particles.

Anderson and Neddermeyer were unusually cautious about reporting these results. Apart from a few comments made at

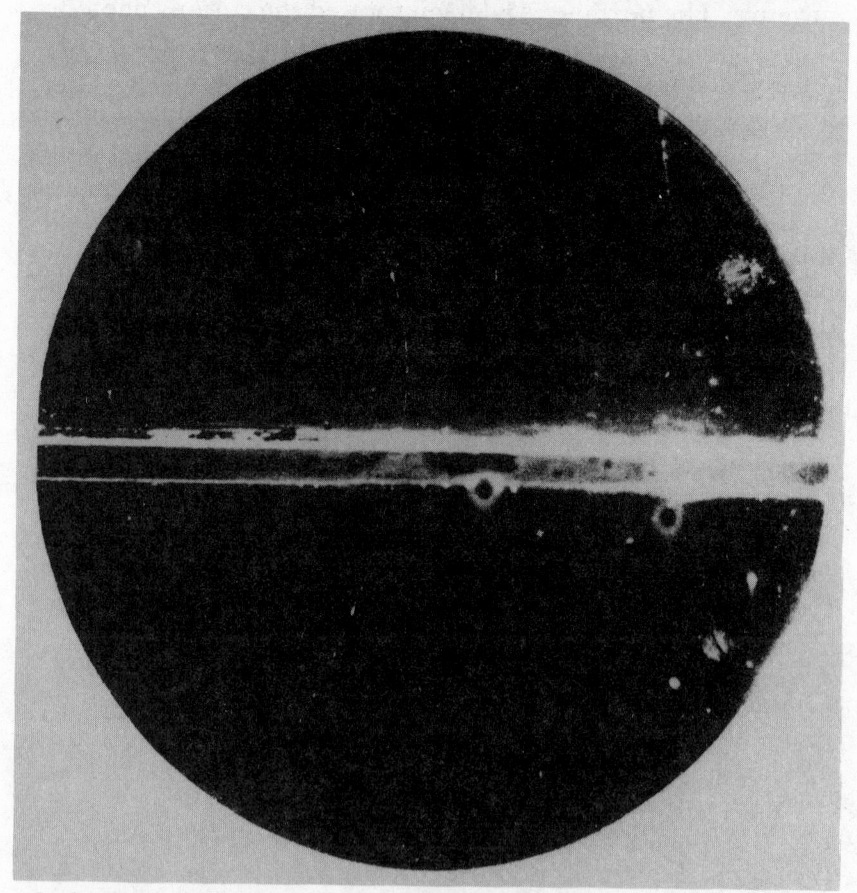

Carl Anderson's first photograph of a positron. This particle, which was rising *in his cloud chamber, curved in the opposite direction as an electron—thus revealing its positive charge.*

the London Conference on Nuclear Physics in 1934, they published nothing for two years. Instead, they performed further tests and studies to confirm their findings. In October 1936, Anderson learned that he would receive the Nobel prize for his positron discovery. Thus emboldened, he reported the new X-particles in a small colloquium at Caltech the next month. He also mentioned them in Stockholm in December: "These highly penetrating particles, although not free positive and negative electrons, appear to consist of both positive and negative particles of unit electric charge, and will provide interesting material for future study."

Sightings of the new particles and measurements of their properties began to come in slowly from Europe, Japan, and America. They seemed to have many of the same properties as electrons, except for a mass intermediate between the electron and proton masses. Although there was a wide variety of values reported, the best measurements of the mass seemed to cluster about 200 electron masses—just the value Yukawa had predicted.

Physicists were elated at this apparent confirmation of Yukawa's bold theory. In a letter to Millikan in 1938, Niels Bohr wrote:

> The story of the discovery of these particles is certainly a most wonderful one. . . . At the moment I do not know whether one shall admire most the ingenuity and foresight of Yukawa or the tenaciousness with which the group in your Institute kept on in tracing the indications of the new effects.

The Caltech group promoted the name "mesotron" for the particle, from the Greek *mesos* for "intermediate" or "middle." This was the name generally used, although the particle was also called the "meson" or "heavy electron" in Europe.

But the mesotron also had a few disquieting properties, most notably its ability to penetrate vast quantities of solid material. As the carrier of the strong force, Yukawa's particle should have interacted very *strongly* with matter, its flight halting abruptly the very moment it passed near an atomic nucleus. In the late 1930s, there was growing doubt that the mesotron was Yukawa's hypothetical force-carrier, after all.

But with the beginning of World War II, further experiments

on cosmic rays slowed while physicists were herded off to secluded laboratories to begin war research. Communication among the international community of physicists was drastically curtailed by the war. "The mesotron theory is today at an impasse," wrote a dejected Yukawa in Kyoto. It was November 1941, a month before Pearl Harbor.

After the war, nuclear physicists returned to their laboratories while the world got its first light sprinkling of an entirely new kind of rain: radioactive fallout. Due largely to their efforts, the United States (and soon Britain, France, and the Soviet Union) now had the means to atomize the planet.

For many physicists, the mesotron puzzle was a key problem to solve. During the war, three Italian scientists had performed an important experiment. Working secretly in a Roman cellar lest they be discovered by the German army, Marcello Conversi, Ettore Pancini, and Oreste Piccione proved beyond doubt that mesotrons interacted hardly at all with atomic nuclei. When they finally released their results in 1947, physicists realized that the mesotron could not possibly be Yukawa's hypothetical carrier of the strong force.

A solution to the mystery came from another quarter: the use of photographic emulsions to detect particle tracks. The technique had actually been developed early in the century, but only after World War II did a British chemical company begin to produce extremely sensitive emulsions able to reveal high-energy cosmic rays. A charged particle passing through the emulsion leaves a trail of ions that induce black grains to form after development. From the number and density of the grains, one can deduce some of the particle's characteristics—like mass and energy.

Using these photographic emulsions, a group at Bristol led by Cecil F. Powell, who had been one of Rutherford's graduate students, began to find convincing evidence for cosmic-ray particles that *did* interact strongly with atomic nuclei. Careful analysis of the tracks of these new particles showed that they were slightly heavier than mesotrons—about 270 electron masses instead of 200. And the heavier particle usually decayed into the mesotron plus some unseen neutral particle.

Using the nomenclature more common to Europe, Powell

christened the new, heavier particles "pi-mesons" (or π-mesons, from the Greek letter *pi*) and the old, lighter particles "mu-mesons" (or μ-mesons, from the Greek letter *mu*). Today, we have shortened these names to "pions" and "muons," while the term "meson" refers to a whole class of particles including the pion—but not the muon—that carry the strong force. Mesons hold the atomic nucleus together. Think of it as a bundle of restless nucleons with pions flitting betwixt and between, keeping them from flying apart.

Here was the final solution to the conundrum. With a rest mass 273 times the electron mass, or 140 MeV, the pion is just the particle Yukawa predicted, albeit slightly heavier. Because it decays very quickly—after about 30 billionths of a second (or 30 nanoseconds)—it could not be detected easily in cloud chambers; with a lifetime about 200 times longer, the muon was far more evident until photographic emulsion techniques were improved. It had taken a dozen years to find the elusive carrier of the strong force.

Yukawa received the 1949 Nobel prize for his brilliant prediction, followed by Powell in 1950 for making the eventual discovery. Yukawa's Nobel, awarded four years after the atomic bomb was dropped on Hiroshima, was the first ever for a Japanese citizen. Ironically, he had anticipated the particle responsible for all the enormous energy wrapped up in atomic nuclei.

So far we have been discussing particles as if they are hard, well-defined objects bashing into one another. Problems arise if we look instead at the "wave picture" of the subatomic world and think of them as bundles of waves spread out in space and time. For example, an electron coasting along its trajectory interacts with its own electromagnetic field wherever its wave function ψ overlaps with this field. In the "particle picture," this effect can be visualized as an electron coughing up a virtual photon and swallowing it again shortly thereafter. If you try to calculate the resulting extra "self-energy" of the electron using Dirac's theory, you get an infinite answer—hardly compatible with the small electron mass seen in experiments. Theorists of the 1930s were duly puzzled by this infinity, but not unduly alarmed: they just assumed it to be an absurdity, ignored it in calculations, and went on about their business.

But in June 1947, two Columbia University scientists presented surprising measurements at the Theoretical Physics Conference at Shelter Island, New York. Using vastly improved microwave techniques derived from wartime radar development, Willis Lamb and Robert Retherford showed that two different states of the hydrogen atom, which supposedly had exactly the same energy in Dirac's theory, in fact possessed slightly *different* energies. It was truly the hit of the whole gathering, the subject of many heated discussions.

On the train returning to Cornell, the German immigrant Hans Bethe, a prime mover of the wartime Manhattan Project, figured out a partial explanation of this "Lamb shift," as the energy difference became known. The infinity in the calculation of an electron's self-energy should be regarded as already

Willis Lamb (standing at left), Richard Feynman (pen in hand), Julian Schwinger (kneeling), and others discussing physics at the Shelter Island Conference, 1947.

buried in its observed mass. So the shift itself was due to other, finite contributions to the self-energy of the hydrogen atom's orbital electron. These one *can* calculate. Ignoring the effects of relativity, Bethe hurriedly made a rough, back-of-the-envelope calculation of the Lamb shift and sent a short paper to the *Physical Review*.

Two brilliant young theorists at the Shelter Island Conference, Julian Schwinger of Harvard and Richard Feynman of Cornell, picked up where Bethe left off. Both in their late twenties and already physics professors at major universities, the two whiz kids independently reformulated quantum mechanics in ways that incorporated both electromagnetic fields and relativity without the nagging infinities of the Dirac theory. Using their revolutionary methods, one can now obtain the Lamb shift to an accuracy of one part in ten million. This is an extraordinary level of accuracy, unequaled in the entire field of physics.

Unknown to the American physics community, the Japanese theorist Sin'itiro Tomonoga had been working on similar problems during and shortly after World War II. His work became familiar to the Americans only during 1948, when he published a short letter in *Physical Review* summarizing his findings. Later that year, the three approaches were shown to be three equivalent versions of a theory of "quantum electrodynamics," or "QED" for short—to this day the most accurate theory in all physics. No significant discrepancies have yet been verified between any precision measurement and its predictions for the interactions of photons, electrons, and positrons. (Years later, Feynman, Schwinger, and Tomonoga shared the 1965 Nobel prize for their landmark theory, one of the firmest foundations of particle physics.)

Feynman's formulation of quantum electrodynamics, the most accessible of the three approaches, has had a particularly profound impact on particle physics. Like all good theory, it did much *more* than just resolve the problems at hand. His simple, intuitive picture of high-energy collisions has deeply influenced the way we visualize the subatomic world. With only a little exaggeration, his methods could be said to permeate the subconscious of particle physicists throughout the world—an essential part of our universal mental baggage.

Born in New York City and raised in Far Rockaway, Feynman brings a brash enthusiasm and impishness to whatever he pursues. His Brooklyn accent and frequent use of the vernacular belie a man of prodigious academic achievements; he is equally at home delivering a physics lecture or hanging out in the local bistros after hours, drumming along with the musicians. "Feynman is a splendid lecturer," noted C. P. Snow, "but in a distinctly different tone, rather as though Groucho Marx was suddenly standing in for a great scientist."

While a graduate student at Princeton during the early 1940s, Feynman began developing a completely new approach to quantum mechanics. Crudely put, his method was a quantum mechanical version of the classical idea that a particle takes the "path of least resistance" in going from point A to point B. In 1949, he finally spelled out his method in two detailed papers, "The Theory of Positrons" and "Space-Time Approach to Quantum Electrodynamics," introducing simple diagrams that are simultaneously a graphical representation of subatomic particle collisions and a convenient shorthand for the arduous calculations involved in predicting the outcome.

The diagram here, for example, depicts the interaction of two electrons by the transfer of a single virtual photon. Each line corresponds both to a particle and to a specific term in the

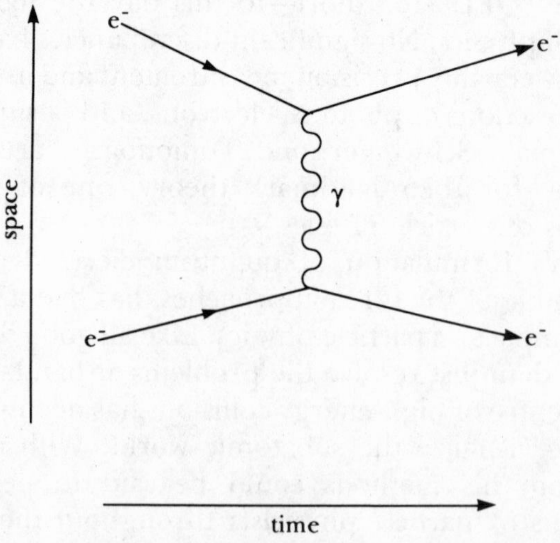

Feynman diagram depicting the simplest collision of two electrons.

complex mathematical expression for the probability of this collision. The straight lines represent charged particles and the wiggly line a virtual photon. Similar diagrams can easily be drawn for other cases where the electrons swap two virtual photons or more; the mathematical expressions for the probability of these events follow immediately.

"Feynman diagrams" (as these graphs came to be known) have proved a great boon to theorists and experimenters alike. The former could calculate any sort of complex particle collision at the drop of a hat—almost unthinkingly. The latter had a straightforward "picture" of the process in question, one that served as a convenient summary of a lot of otherwise confusing mathematics. "Like today's silicon chip," observed Julian Schwinger in 1980, "Feynman diagrams brought calculation to the masses." In one simple package, these diagrams allowed physicists to *visualize* particle collisions and *talk* about them in a common language. They have become the most powerful metaphor of our discipline.

The two great revolutions of twentieth-century physics—quantum mechanics and relativity—have profoundly altered our mental images of subatomic particles. The classical image of an electron, for example, was that of a tiny, inert ball surrounded by electric and magnetic field lines emanating from it like rigid spokes. But after Heisenberg it became heresy to think of an electron located exactly at some point in space. In the wave picture of quantum mechanics, the electron is a smear—a probability function ψ that predicts our chances of finding it nearby. Relativity allows the force fields surrounding it to materialize briefly as particles, too. Virtual photons, electron-positron pairs, and even muon pairs momentarily emerge and disappear, like honeybees swarming angrily about a hive. Instead of the classical image of an electron as dead matter surrounded by rigid field lines, it is a beehive of activity literally seething with potentiality.

The virtual particles swarming about an electron *remain* mere potentialities, however, unless extra energy comes along to boot them into reality. They "exist" because the uncertainty principle allows them to, even though their emergence seems to violate briefly the cherished law of energy conservation. A virtual photon of energy E can emerge for a time period t as long as the product $E \times t$ is less than Planck's constant h. In other words, high-energy virtual photons can play hookey for only a fleeting moment before once again disappearing; low-energy photons have a much longer lease on life. Ultimately, however, all truant photons have to return to the mother electron—unless some other particle happens along to make up the deficit.

A good analogy to help you understand virtual particles is the employee who embezzles company funds to bet on the horses. In effect, he is creating "virtual money" and violating the conservation of dollars. If he takes only a small sum, he can probably use it for a long time before the company auditors find anything amiss. But if he makes a large dip into the till, he'd better put it back quickly lest his embezzlement be discovered and his job abruptly terminated. In either case, however, this virtual money can have a very real impact on the embezzler's total wealth; depending on his luck at the track, it can add or subtract. In a similar way, the virtual particles emitted and reabsorbed by an electron make a real, measurable contribution to its total energy, or mass.

So the physical, observable electron of our experiments is actually a very complicated object, a seething microcosm of the subatomic world itself. Even some of its mass derives from the cloud of virtual particles surrounding it, just as part of a beehive's mass comes from the honeybees themselves. Fortunately, particle physicists have a successful quantum field theory, quantum electrodynamics, to describe the electron's every nuance—including these virtual contributions. As far as we can tell, the "bare" electron at the core of this cloud still behaves as if it were concentrated at a mathematical *point*. Today, its "size" is known to be less than 10^{-16} centimeter, or a thousandth the diameter of a proton.

By the early fifties, physicists were triumphant; quantum electrodynamics could explain the electromagnetic interactions of electrons, positrons, and photons to a breathtaking degree of accuracy. They had fond hopes that a similar sorcery would explain the strong force governing the interactions of neutrons and protons. Perhaps these, too, had their own bare, point particles hiding amid turbulent swarms of virtual pions. The search was on for a quantum field theory of the strong force.

To the dismay of midcentury theorists, however, Nature refused to knuckle under. Experiments again and again revealed the strong force to be far more complex than had previously been anticipated. Whole families of new particles closely related to protons and pions emerged during this period. Each new particle made a big contribution to the virtual cloud swarming about the nucleon. The mathematical equations describing these clouds soon became hopelessly involved, defying the best efforts of theorists to solve them.

The first new particle to surface was the K-meson, or kaon, a cousin of the pion. It had been spotted in 1947 by George Rochester and Clifford Butler at the University of Manchester. While studying cosmic rays in a cloud chamber, they found a pair of tracks emanating from the exact same point—making a shape like the letter V. They interpreted the visible tracks to be the remnants of a neutral, and thus invisible, object they dubbed the "V-particle." Its mass, which could be calculated from the properties of its decay products, was about a thousand times the electron mass, or about 500 MeV in equivalent energy units.

While studying photographic emulsions in 1949, Cecil Powell's group at Bristol found a charged particle of similar mass that decayed into three pions, not two. This they named the τ-meson (from the Greek letter *tau*). Not until 1957 did scientists finally recognize that the τ and these V-particles were merely different states of the same object, now called the kaon, which comes in both positive and negative versions at 494 MeV, as well as two neutral versions at 498 MeV.

In the early 1950s, cosmic-ray physicists found another, even heavier V-particle that seemed a close cousin of the proton. Christened the Λ (the Greek letter *lambda*), this neutral mote had a mass of 1115 MeV, about 20 percent heavier than a proton. It decayed into a proton and a negative pion, which again left two visible tracks in cloud chambers—emanating from the same point.

What *were* all these strange new particles? Theorists hadn't the slightest idea. There seemed no pressing need for any of these odd corpuscles—the V-particles and the muons—in a universe that supposedly was simple and orderly. ("Who ordered that?" asked Columbia's Isidor Rabi when the muon was identified in the late forties.) To make matters worse yet, the V-particles survived a lot longer than anybody could readily explain. Lacking any good explanation, theorists simply dubbed them "strange" particles, a name that has survived to this day as the label for a completely new property of matter.

Cosmic-ray physicists and their cloud chambers had done yeoman service during the thirties and forties, but the mantle was about to pass to physicists with access to the powerful new particle accelerators. These machines, capable of delivering uniform, controlled beams of high-energy electrons, protons, and even pions, were ideal tools for producing new particles and studying their detailed properties by smashing one into another. The field of high-energy physics really began in the late 1940s, when accelerators able to produce pions were first built in the United States. Still suffering from the ravages of World War II, which had deprived the Continent of some of its best minds, Europe now took a back seat to America.

The pioneering spirit in American accelerator building was to be found at Berkeley, in the person of Ernest O. Lawrence. Born and raised in North Dakota, this tall, broad-shouldered, bespectacled graduate of Yale had conceived his novel idea of the "cyclotron" by 1930. A remarkable device, the cyclotron accelerates charged nuclei, usually protons or alpha particles, in spiral paths between two circular poles of a large electromagnet. Twice each revolution, these particles receive a small electric kick that boosts their energy and thus broadens the spiral slightly. Because earlier accelerators allowed charged particles only a single pass, they required extremely strong voltages or huge physical dimensions to achieve high energies. Lawrence's new cyclotron achieved the same results without either disadvantage by repeatedly imparting many small kicks to a single bunch of circulating particles.

Throughout the thirties a growing team of scientists at Berkeley's Radiation Laboratory built larger and larger versions of the cyclotron. These were the days before large government grants, however, so the indefatigable Lawrence spent much of his time on the lecture circuit, touting the virtues of his cyclotron to wealthy patrons, medical foundations, and any influential person who might listen. His first prototype could fit in the palm of his hand, but by 1939 the accelerator group had produced a 60-inch-diameter device that boosted protons to energies above 10 MeV. When war came, Lawrence was planning and building a 184-inch cyclotron that he hoped would deliver protons at the unheard of energy of 100 MeV.

Lawrence's single-minded devotion to building larger and better machines, almost to the exclusion of using them for any physics, was probably an important ingredient in his success. The profession of "accelerator physicist," one of the three major specialties in high-energy physics today, was really born during the thirties at the Radiation Laboratory. Many of our leading accelerator physicists—Luis Alvarez, M. Stanley Livingston, Edwin McMillan, and Robert R. Wilson the most prominent among them—learned the trade at Lawrence's side in those germinal years. In his honor, the place is now called the Lawrence Radiation Laboratory, still an important center of high-energy physics.

After the war, Berkeley became the focus of the growing profession. Edwin McMillan returned from working on the bomb in Los Alamos with his "phase stability" principle, which allowed Lawrence's dream of a 100 MeV proton cyclotron to become reality. This principle enabled them to bypass certain barriers previously thought to limit cyclotron energies. Renamed the "synchro-cyclotron," the device soon pushed protons to *hundreds* of MeV. In 1949, Berkeley scientists isolated the neutral pion, the first new particle to be found at an accelerator instead of in cosmic-ray experiments. The mantle was indeed passing.

Not to be outdone, a consortium of eastern universities was busily planning a much larger machine, capable of delivering protons at over a *billion* electron volts, or 1000 MeV. Built under the direction of Stanley Livingston, this "synchrotron" allowed proton bunches to travel in fixed orbits inside a circular vacuum pipe. Cyclotrons are limited by the size of the magnet pole faces one can manufacture, but synchrotrons are limited only by the available real estate and finances. The chosen site was an abandoned army training center, Camp Upton; the newly organized Atomic Energy Commission provided the funds. In 1952, the Brookhaven National Laboratory and its 3000 MeV "Cosmotron" finally rose from the sands of Long Island.

So why the pressing need for ever higher energies? Simply that the heavier the particle you are trying to produce, the more energy you must supply—according to Einstein's $E = mc^2$. A proton sitting around in a laboratory can violate energy conservation very briefly by emitting and then reabsorbing a virtual pion. But if you want to actually *see* the pion in a detector (rather than in an equation), you must supply at least its "rest energy"—the energy equivalent of its rest mass—to create it, plus a bit more to carry it away. Cosmic rays have enough energy to blast pions, kaons, and other, far heavier particles into existence, but only in a haphazard manner that makes further study difficult.

In the late forties and early fifties, the Berkeley accelerator could produce pions but not the heavier V-particles. A cyclotron built at the University of Chicago under the direction of Enrico Fermi, the immigrant Italian genius, fared no better. But the new Cosmotron could deliver negative pions at 1.5 billion electron volts (1500 MeV or 1.5 GeV, for "giga" electron volts), giving experimenters plenty of extra energy to play around with. When a group of physicists fired the pion beam into a cloud chamber in 1953, they found V-particles aplenty. What's more, these particles were always produced *in pairs,* never singly. When a pion encountered a proton in the chamber, they often produced a neutral kaon and a lambda:

$$\pi^- + p \rightarrow K^\circ + \Lambda^\circ$$

After traveling a short distance unseen, the two neutral products of the collision decayed into their characteristic V's. But one never saw, for example, a single kaon being produced along with a recoiling proton.

This phenomenon, called "associated production," had been actually predicted a few years earlier to explain why the strange V-particles could be produced so easily yet decayed so slowly. It gradually dawned on physicists that they were witnessing the effects of *two* forces, not just one. The V-particles were produced copiously by the strong force acting between particles like the pion and proton; they would then meander through space until disintegrating at the hands of the same weak force that caused the slow radioactive decays witnessed in nuclear physics. The confirmation of associated production at the Cosmotron was the final clincher that bore out this interpretation. There were now *four* fundamental forces in Nature: gravity, electromagnetism, and the strong and weak nuclear forces.

But one additional puzzle remained. Why did pion-proton collisions *never* yield a kaon plus a recoiling proton? The energy to do this was certainly available.

An answer, of sorts, came from theorists Murray Gell-Mann at Chicago and Kazuhiko Nishijima in Japan. In 1953, they independently proposed that these strange V-particles carried a new property of matter Gell-Mann dubbed, appropriately enough, "strangeness." This property had something to do with the fact that kaons occurred in pairs, not a threesome like the pions—and that the lambda was a single particle, not a pair like the proton and neutron. "Normal" particles like the proton, neutron, and pion had zero strangeness, while the K° was arbitrarily assigned a strangeness of +1 and the lambda a strangeness of −1.

Strangeness is conserved whenever two particles interact by the electromagnetic and strong forces but can change whenever the weak force is involved. In a strong interaction, the total strangeness before the collision had to equal the sum afterward:

	$\pi^- + p \rightarrow K^\circ + \Lambda^\circ$
Strangeness:	$0 + 0 = +1 + -1$
Electric charge:	$-1 + +1 = 0 + 0$

Conservation of strangeness is a lot like conservation of electric charge, which holds true in *all* subatomic particle collisions. Interactions like $\pi + p \rightarrow K + p$ that fail to conserve strangeness can proceed only far more slowly through the weak force; thus, they are essentially never seen.

Because strange particles decay only through the weak force, they survive long enough—about a billionth of a second, give or take a factor of ten—for their tracks to be seen in cloud chambers or photographic emulsions. Neutral strange particles leave no tracks, but their existence and properties can be inferred from the characteristic "V" emanating from apparently empty space. A billionth of a second may not seem very long to the casual observer, but it is an eon to high-energy physicists. Subatomic particles are generally moved at close to the speed of light—300,000 kilometers per second—and can cover visible distances in such a brief time, leaving fairly long footprints.

Two other families of strange particles surfaced during the 1950s, both closely related to the proton, neutron, and Λ particle. The first included the three Σ (the Greek letter *sigma*) particles at masses clustering around 1190 MeV. Then there were the two "cascade particles"—so-called because they decayed into a cascade of lighter fragments—or Ξ's (the Greek letter *xi*) at about 1320 MeV. All these particles are produced rather easily by the strong force but decay slowly through the weak or electromagnetic forces. All three sigmas—positive, negative, and neutral—have the same strangeness as the lambda, or -1. But the cascade particles, one negatively charged and the other neutral, proved to be even stranger. In order that strangeness be conserved in the strong collisions producing them, the cascade particles had to be assigned a strangeness -2.

If all this talk of "strange" particles and "strangeness" sounds like so much conjury and leg-pulling, please be patient. High-energy physicists are acknowledged masters of the art of sorcery—able to extract hard concepts from nothingness just as readily as charged particles out of the void. "Strangeness" is a term, a name, for a property of matter seen only in violent collisions of subatomic particles; it has no corresponding meaning in the everyday world of our senses that language was originally developed to describe. Physicists have had to create a new nomenclature in order to begin *talking* with each other about the weird subatomic landscape. One way they do so is to borrow words from everyday language and use them in entirely new ways.

A name like strangeness *acquires* meaning, along with texture and depth, through repeated usage by a community. Such was the case with electric "charge," a concept scientists introduced more than two centuries ago to "explain" the attraction and repulsion of electrified matter. In a similar fashion, strangeness has become one of the central concepts used by particle physicists today.

To cope with this bewildering proliferation of supposedly "elementary" particles, physicists began to focus upon what are called quantum numbers. The quantum numbers of a particle embody the magnitude of its intrinsic properties—its very identity. (You have already encountered two of these properties: charge and strangeness.) These numbers gave scientists a convenient way to classify the growing horde of particles into affinity groups or families. In lieu of a truly fundamental theory of elementary particles, this "zoological" approach made good sense. It was like biology before Darwin: animals had to be grouped into phyla, genera, and species before an evolutionary paradigm could begin to unify the entire kingdom.

One extremely important quantum number is angular momentum, a property of matter in motion about some axis. A rock tied to a string and swung round and round above your head has angular momentum about you; increase the rock's mass, the length of the string, or the rotary speed and the angular momentum increases accordingly. During the 1920s, physicists realized that the angular momentum of a subatomic particle is not a smooth, continuous entity; it occurs only in discrete amounts that are whole-number multiples of Planck's constant h divided by 2π, or $\hbar$ (called "h-bar"). An electron orbiting a nucleus can only have an angular momentum of $0, \hbar, 2\hbar \ldots$ and so forth.

In addition to orbiting a nucleus, electrons rotate about their own internal axis—as do all subatomic particles. This *intrinsic* angular momentum, or "spin," can be compared to the motion of a spinning top (although here I am stretching the analogy a little). Through a quirk of quantum mechanics, the spin of a subatomic particle can be either a whole-number or a half-number multiple of $\hbar$. Thus the permitted values of spin are $0, \hbar/2, \hbar, 3\hbar/2, 2\hbar \ldots$ and so forth.

Now a subatomic particle has a *definite* spin quantum number that never changes (unless the particle disappears). Its spin is part of its very identity, its badge of membership in a select society. Protons, neutrons, electrons, positrons, and muons all have "spin-½," by which we mean that their spin is $\hbar/2$. Photons have spin-1, while pions and kaons have spin-0. The massive strange particles Λ, Σ, and Ξ (cousins of the neutron and proton) all were eventually shown to have spin-½.

Subatomic particles belong to two very broad classes depending on whether they have half-number or whole-number spin. Those with half-number spin we call "fermions" after Enrico Fermi; those with whole-number spin, "bosons." Loosely speaking, fermions comprise the "stuff of matter" while bosons are the "force particles"—like the photon and pion—holding things together. We cannot create or destroy a single fermion, at least not all by itself, but bosons can appear and disappear willy-nilly.

Another important distinction is that between particles and their "antiparticles." Recall that the positron first surfaced as a negative-energy solution of the Dirac Equation, while the electron is the positive solution. They are identical in every respect except charge; physicists say the positron is the antiparticle of the electron. Because of relativity, every subatomic particle has such an antiparticle. Thus, the positive muon is the antiparticle of the negative muon, and the positive pion and kaon are antiparticles of their negatively charged twins. A neutral particle can be its own antiparticle, as happens with the photon and neutral pion; or its antiparticle can be a distinct particle, as is the case for the two neutral kaons.

In the early 1950s, this division of the particle kingdom had not yet become dogma, although the evidence was becoming more and more persuasive every year. Still missing was the antiparticle of the proton, the "antiproton" $\bar{p}$. Although there had been hints seen in cosmic-ray experiments, few physicists were convinced. And Brookhaven's Cosmotron did not have enough energy to produce an antiproton artificially. So people could doubt, and some prominent scientists did.

Meanwhile, accelerator physicists at Berkeley's Radiation Laboratory had been building their own synchrotron, one that would surpass the Cosmotron by boosting protons to energies of 6 GeV. The "Bevatron," as it was named, would have just enough energy to produce antiprotons—if they indeed existed. When it began operation in 1955, one of the earliest experiments to be attempted was a search for the antiproton. In photographic emulsions taken with the new machine, a team led by Owen Chamberlain and Emilio Segrè found convincing evidence for a negatively charged particle with exactly the proton mass. The doubters could doubt no more. Here, at last, was the long-sought antiproton.

The division of the subatomic world into particles and antiparticles can be said to parallel the division of the animal kingdom into two sexes. Most animal species—especially the more complex ones—have male and female members that are otherwise virtually identical. But some simple species, like snails and sea anemones, combine male and female traits in one and the same individual. And what happens when particle and antiparticle meet roughly resembles the sexual interaction of the male and female of the same species: they produce a state of pure energy that can give birth to other particles and antiparticles.

One more particle would show its face before the fifties ended. The "neutrino" was detected in 1956 by Clyde Cowan and Frederick Reines of the Los Alamos Laboratory.

This ghostly particle had been anticipated since the early 1930s, when Wolfgang Pauli proposed a neutral, invisible mote to resolve certain anomalies witnessed in the radioactive decays of nuclei. Energy seemed to evaporate during these so-called beta decays, whereby a nucleus spontaneously emitted an electron through what we now recognize as the weak interaction. Niels Bohr was even ready to abandon the sacrosanct concept of energy conservation, at least in the subatomic realm. But Pauli could not bring himself to make such a sacrifice. He suggested instead that the missing energy was actually spirited off by a spin-½ particle that emerged hand-in-hand with the electron, but had virtually no interaction with matter and hence remained undetected. Emboldened by James Chadwick's discovery of the neutron, Pauli finally announced his prediction at the Solvay Congress of 1933.

The following year, Enrico Fermi published a detailed theory of beta decay based on this idea. He called Pauli's new particle the "neutrino"—Italian for "little neutral one"—and denoted it by ν (the Greek letter *nu*). The success of this theory helped convince many scientists of the neutrino's existence. The very same particle was also believed to be responsible for the energy missing when a pion decayed into a muon.

But there was still no direct evidence for the neutrino before 1956. Sensitive measurements of the energy and momentum of a recoiling nucleus had shown that *something* was indeed eloping with the missing energy. The same observation held true for pion decays. However, this evidence was only circumstantial: the bed was empty and there was a ladder at the window sill, but just one set of footprints in the dew. Hardly enough to get a conviction. It would be necessary to catch the guilty particles in the act.

What was needed to detect the neutrino was a very intense source of these elusive particles plus a huge detector. In the 1950s, large nuclear reactors emerged to provide a suitable source; Cowan and Reines supplied the detector. They buried over three thousand gallons of cadmium chloride solution between liquid scintillators next to the giant Savannah River nuclear power plant in South Carolina. Of the many trillions of neutrinos speeding through the detector, a few occasionally collided in the solution, each producing two successive light flashes in the scintillator, a sure sign of a neutrino.

These ghostly particles might not leave any footprints in the dew, but every once in awhile one of them would trip over a string stretched in its way, setting off the alarm! Here, at long last, was unmistakable evidence for the neutrino. Reines immediately sent a telegram to Pauli, who received it in Geneva at a conference. The aging theorist held the telegram up for all to see and exclaimed, joyfully, "The neutrino exists!"

Actually, the particle Cowan and Reines detected was the "antineutrino," or $\bar{\nu}$, the antiparticle of the neutrino. In further experiments they showed that there were two distinctly different particles with all the same properties but one: their spins were oppositely aligned. With zero charge and spin-½, both have a mass at least twenty thousand times smaller than the electron mass; they were assumed (until about 1980) to have exactly zero mass. Because they interact with matter only through the weak force, both particles can easily pass through the entire universe without ever interacting. This is fortunate, because neutrinos are everywhere—as common as light. Over a billion just passed through your head while you were reading this sentence.

* * *

By the late 1950s, particle physicists were beginning to feel as if they were drowning in a Greek alphabet soup. "If I could remember the names of all these particles," Fermi once complained, "I'd have become a botanist instead of a physicist." Where, three decades before, there had been just three elementary particles—the electron, proton, and photon—by 1959 there were thirty, if you included all the antiparticles. In their attempts to understand the forces holding nuclei together and tearing them apart, these scientists had broken open a veritable Pandora's Box.

Fortunately, patterns were starting to appear within this morass of new particles. An astute observer could begin to discern the vague outlines of an orderly mosaic. The situation at the decade's close is summarized in the table on p. 70, taken from a 1959 lecture given at Princeton by the Nobel laureate Chen Ning Yang. Here, particles are grouped into three broad classes according to their quantum numbers.

The baryons and leptons require further explanation. Baryons (and antibaryons) are heavy fermions (spin-½ particles) with masses equal to or greater than the proton mass; their name comes from the Greek word *barys* for "heavy." The leptons and their antiparticles include all the lighter fermions; their name derives from the Greek *leptos* for "small." These two families differ in more than just their weight. Baryons feel the effects of the strong force while leptons do not.

The badge of membership in either class is yet another quantum number. The "baryon number" of a particle, denoted in Yang's table by N but more commonly by B, is +1 for baryons, −1 for their antiparticles, and 0 for everything else. The "lepton number" l is +1 for leptons, −1 for their antiparticles, and 0 for everything else. Thus a proton has $N = B = +1$ and $l = 0$, while an electron has $N = B = 0$ and $l = +1$; like all bosons, the pion has $N = B = 0$ and $l = 0$.

Like the quantum numbers for charge and strangeness, the baryon and lepton numbers are more than idle games played by particle physicists. They are *names* for actual physical properties of matter in the world of the very small. Like charge and strangeness, these names acquire meaning through their use by a community of scientists. In particular, the baryon and lepton numbers are *conserved* quantities that do not change in any

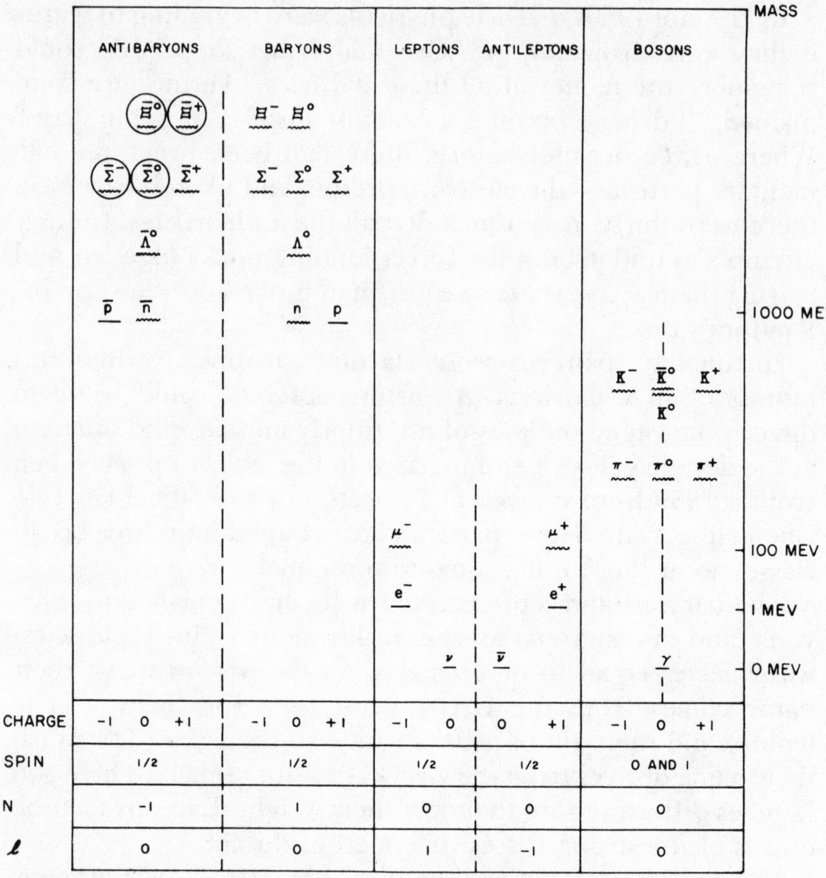

	ANTIBARYONS	BARYONS	LEPTONS	ANTILEPTONS	BOSONS
MASS	$\bar{\Xi}^0$ $\bar{\Xi}^+$ $\bar{\Sigma}^-$ $\bar{\Sigma}^0$ $\bar{\Sigma}^+$ $\bar{\Lambda}^0$ $\bar{p}$ $\bar{n}$	Ξ^- Ξ^0 Σ^- Σ^0 Σ^+ Λ^0 n p	μ^- e^- ν	$\bar{\nu}$ μ^+ e^+	K^- $\bar{K}^0$ K^+ K^0 π^- π^0 π^+ γ
CHARGE	-1 0 +1	-1 0 +1	-1 0	0 +1	-1 0 +1
SPIN	1/2	1/2	1/2	1/2	0 AND 1
N	-1	1	0	0	0
ℓ	0	0	1	-1	0

1000 MEV — 100 MEV — 1 MEV — 0 MEV

Yang's table of the elementary particles recognized in 1959. Unstable particles (those that decay) are identified by wavy lines; stable particles by solid lines. Those encircled were expected to exist but were not yet confirmed by experiment.

interaction—strong, weak, or electromagnetic.* Their sum stays the same before and after a collision. This indestructibility of baryons and leptons is the characteristic of fermions that makes them the "stuff of matter."

Bosons, the "force particles," can come and go pretty much as they please. There are no conservation rules for "boson number," and thus it receives little attention—even though such a

* Modern-day, "grand-unified" theories permit baryon and lepton numbers to change during an interaction, but there is no experimental evidence (as yet) for such behavior.

quantum number is indeed conceivable. An electron or proton can emit one, two, or thirty-seven photons without ever breaking the rules.

Yang could have made a further subdivision in his table, one that many particle physicists of the day actually made. He could have divided the bosons into two separate groups: the pions and kaons in one group called the mesons, and the photon all by itself in another. Like the distinction between baryons and leptons, mesons participate in strong interactions while photons do not.

Finally, the class of all particles that feel the strong force is called the "hadrons," from the Greek *hadros* for thick or heavy. This is a class that includes the baryons, their antibaryons, and the mesons. Thus the subatomic particle kingdom of 1959

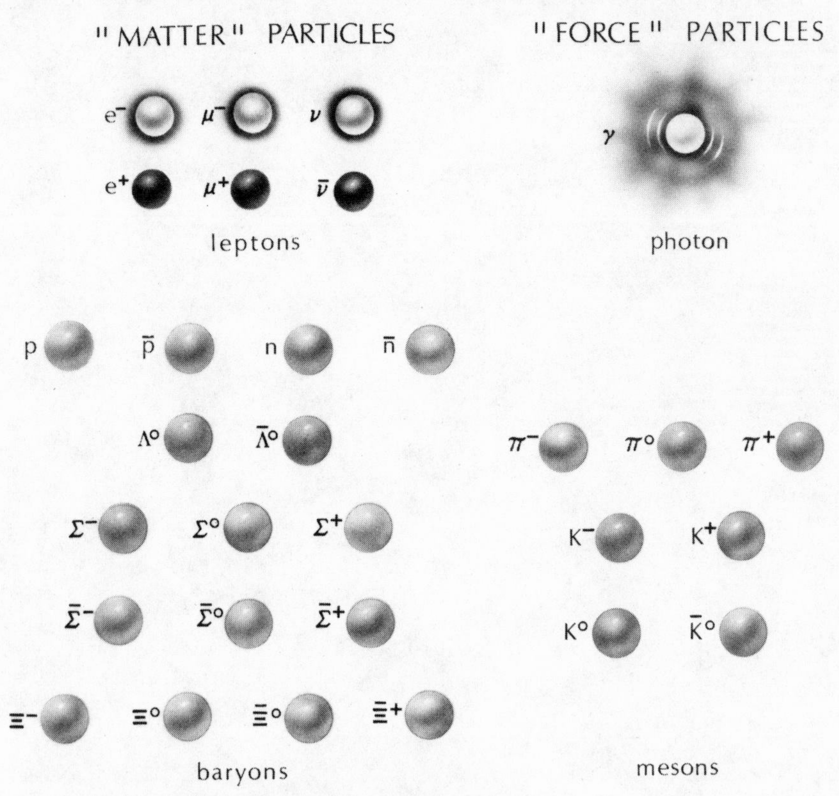

The principal subdivisions of the particle kingdom, circa 1960. The hadrons, which include both mesons and baryons, feel the strong force; leptons and photons do not.

could be grouped into two broad classes, depending upon whether or not a member feels the strong force. Hadrons feel it, while leptons, antileptons, and the photon do not.

But however useful these divisions of the particle kingdom might have been, they were still far from a complete understanding of this unruly zoo. Physicists prefer simple, elegant explanations needing a minimum of arbitrary numbers and assumptions. This is the principle of "Occam's razor," more a prejudice than a scientific judgment. Orderly or not, a table of thirty different elementary particles was deemed highly unsatisfactory. What physicists now sought was a deeper meaning behind this order.

CHAPTER 3

THE TAMING OF THE ZOO

The fact that, at least indirectly, one can actually see a single elementary particle—in a cloud chamber, say, or a bubble chamber—supports the view that the smallest units of matter are real physical objects, existing in the same sense that stones or flowers do.

—Werner Heisenberg, THE PHYSICIST'S CONCEPTION OF NATURE

THE early sixties were years of rapid expansion in American science. We were emerging from a cynical decade that had witnessed the first H-bomb tests and fallout shelters, and Russia had just beaten us into space with her Sputnik. Prodded by a young, vigorous, and popular President, the United States seemed determined to regain its lead in scientific matters.

By then the construction of larger and more powerful accelerators had become an international obsession. In 1957 the Russians took the energy lead by building their 10 GeV Synchro-Phasotron. Western Europe joined the race three years later with its own 28 GeV synchrotron at CERN, the European Center for Nuclear Research, located near Geneva. Not to be outdone, the United States regained the lead in 1961 with the Alternating Gradient Synchrotron, or AGS, a 30 GeV proton machine built at Brookhaven. Both employed the new principle of "strong focusing" developed in the early fifties by Stanley Livingston and his associates to permit use of far smaller magnets and vacuum pipes. Almost a thousand feet in diameter, these two circular accelerators could not fit inside a single large building like earlier devices; they had to be housed instead in circular tunnels built under mounds of earth.

Smaller, more specialized machines were underway at Harvard, Princeton, and the Argonne National Laboratory near Chicago. And funding for the Stanford accelerator, physically the largest and most expensive accelerator then planned, finally came through early in 1961.

But all eyes were turned toward the powerful new proton synchrotrons at CERN and Brookhaven. The beams of protons, antiprotons, pions, and kaons emanating from these accelerators had far more energy than available in earlier machines. Thus they could produce new and much more massive particles—with well over three times the proton mass—than had yet been discovered. Scientists flocked to these two machines like eager schoolchildren to the seashore, searching for rare and exotic new seashells tossed up by the waves.

Physicists were aided in the hunt by a new kind of particle detector developed in the fifties, the "bubble chamber." In essence, this device is a kind of cloud chamber in which the gas inside has been replaced by a liquid—commonly hydrogen, freon, or propane. Kept at a low temperature extremely close to its boiling point, the liquid is exposed to a beam of high-energy particles. Just as the particles pass through, the pressure in the chamber drops suddenly, bringing the liquid immediately above its boiling point. Tiny bubbles of gas begin to form around the many ions left in the wake of a charged particle. By flashing a bright light into the chamber at just the right instant, one can take photographs in which the tracks of such particles appear as lines of bubbles—footprints in the dew.

The bubble chamber was invented in 1952 by Donald Glaser, who had been one of Carl Anderson's Caltech graduate students. At the time, he was a young postdoc at the University of Michigan. According to folklore, Glaser hit upon the idea at a local Ann Arbor pub while staring absentmindedly into a bottle of beer and wondering how the streams of bubbles formed inside.

His first bubble chambers were tiny, hand-held devices with ethyl ether used as the liquid. After demonstrating that clear tracks could be observed in these prototypes, Glaser and coworkers built larger chambers and experimented with other liquids. Hydrogen, for example, gives a target of pure protons and electrons, and propane has roughly equal numbers of protons, neutrons, and electrons. When heavy particles like protons or pions traverse a chamber, their glancing collisions with the atomic electrons can be ignored; one can focus instead on their much stronger interactions with the target nuclei.

Bubble chambers have two great advantages over cloud chambers. For one, the target volume is a thousand times denser, and the number of particle collisions increases accordingly. And because the visible tracks can be seen in much finer detail, physicists can find very short tracks left by particles with exceedingly brief lifetimes. Photographic emulsions have these advantages, too, but their analysis is tedious and painstaking. With bubble chambers, picture after picture can be snapped, the film developed, and then teams of "scanners" can pore over it looking for characteristic footprints.

The first bubble chamber actually used in high-energy physics experiments was built in the midfifties at Brookhaven's Cosmotron. Soon thereafter, they began to appear at particle accelerators around the globe. Any atom-smasher worth its salt had to have one. So crucial was the bubble chamber to particle physics that Glaser received an unshared Nobel prize for his invention.

It was Luis Alvarez and his Berkeley group who lifted bubble-chamber physics to the status of high art. Like Ernest O. Lawrence, Alvarez had a flair for grand ideas and enormous projects. Once he learned of Glaser's invention, he gathered a large group of experts and proceeded to build a series of ever larger chambers for the Bevatron. Even Lawrence had his doubts about the wisdom of these efforts, but Alvarez persisted. By 1959, his group had built a chamber 72 inches long containing 500 liters of liquid hydrogen—by far the largest bubble chamber of its day.

Almost singlehandedly, this immense device kept the Bevatron in the thick of the search for new particles—despite the fact that its top proton energy was now far smaller than that of CERN or Brookhaven. With a powerful electromagnet surrounding the chamber and complicated cryogenics equipment maintaining the liquid hydrogen at exactly the right temperature and pressure, the 72-inch bubble chamber required its own separate building next to the accelerator. But it rewarded its far-sighted builders with millions of sharp, detailed photographs of subatomic collisions.

The stark pictures that poured forth from this device were reminiscent of Abstract Expressionist paintings—especially the dynamic, "picture-plane" imagery of Jackson Pollock or Franz Kline. No two were quite the same. They provided a fascinating new window upon a world in constant turmoil—with gently arching tracks ending suddenly in great explosions of raw exuberance, or the whooping whorls left by dying particles clutching desperately to the last remnants of their energy. Here were brief glimpses of Nature caught under the most violent conditions imaginable. You could appreciate these bubble-chamber photographs for their sheer beauty alone.

To the practiced eye, however, these pictures are also fraught with information, pregnant symbols of a hidden world just beyond our five senses. Different particles leave characteristic tracks that help you distinguish one from another. The mass of a particle can be estimated from its track's density, its lifetime from the length, and its charge and momentum from the curvature of its track in the magnetic field permeating the chamber. Even properties of neutral, uncharged particles—which leave no tracks—can be inferred if they happen to collide or decay within the chamber, producing charged particles whose tracks *are* visible. Working backward with conservation laws, you can deduce the properties of the parent from those of the offspring.

To harvest all the information contained in this wealth of bubble-chamber photographs, Alvarez assembled a small army of scanners and computer programmers led by his cadre of physicists. Aided by a machine dubbed "Franckenstein" after the engineer who built it, the scanners shuffled through thousands of pictures, looking for characteristic signatures. The

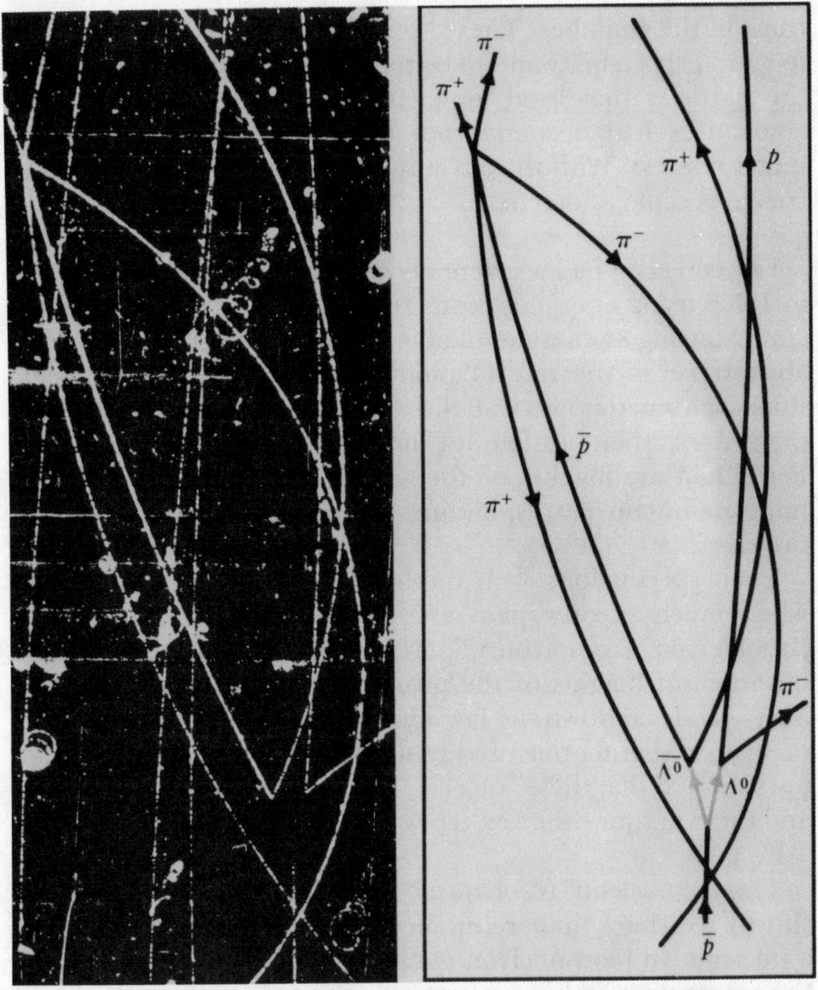

Photograph of an antiproton colliding with a proton inside a bubble chamber. Their annihilation creates a neutral lambda and its antiparticle, which reveal their existence by the characteristic V pattern of their decay products.

selected events were then encoded into a digital form comprehensible to the computer, which calculated the relevant particle properties in the blink of an eye. Scanning provided ample part-time work during the sixties for hundreds of housewives and students, of which Berkeley had a seemingly endless supply.

Working backward, physicists soon began searching for particles whose lifetimes were far too short to leave a visible track in the chamber. These "resonances" were just beginning to gain respectability and to be treated on an equal footing with the particles that lived long enough to leave tracks. A few resonances had been noticed in the fifties, without stirring much interest. With the advent of large bubble chambers, this trickle became a torrent.

Enrico Fermi had accidentally discovered the first resonance in 1952, using energetic pions from Chicago's 450 MeV cyclotron. Starting at energies of about 50 MeV, he encountered an abrupt rise in the rate of pion-proton scattering. The steep slope continued upward all the way through 135 MeV—the top pion energy then possible at Chicago. Theorists speculated that Fermi had stumbled upon the leading edge of a broad peak, a huge mountain corresponding to an excited relative of the proton.

Their speculations were amply confirmed the following year when much greater pion energies became available at the Brookhaven Cosmotron. There experimenters finally scaled the topmost heights of the peak—at an energy of about 190 MeV—and slid down the far side. The huge peak loomed large no matter whether one used positive or negative pions. Further tests showed that these "objects" had spin-3/2 and strangeness 0, and came in four different types with charges of +2, +1, 0, and −1.

These "nucleon resonances" have proved to be excited, short-lived states that are momentary flirtations of a pion with a nucleon. In pion-nucleon scattering, a pion can be caught in the momentary embrace of a nucleon; the two orbit each other briefly before separating once again. Think of two dancers coming together in a square dance, embracing warmly, and swinging fondly before parting again and going their way.

Most of the intrinsic properties of these resonances derive from those of their component pion and nucleon. The spin of 3/2 results from adding the nucleon's spin-1/2 with the pair's one unit of orbital motion about each other. A charge of +2, for example, comes from adding the charges of a positive pion (+1) and a proton (+1), while −1 results from a negative pion (−1) and neutron (0) combination. Pion-nucleon resonances have strangeness 0 because pions and nucleons have strangeness 0. And the resonance mass of 1232 MeV is merely the sum of the pion mass of 140 MeV, the nucleon mass of 938 MeV, and the mass-energy of their orbital motion.

The brief lifetime of a resonance can be estimated by invoking the Heisenberg Uncertainty Principle, which requires that the uncertainty in the energy of a resonance times its average lifetime be greater than *h*. The uncertainty in its energy is just the width of the broad peak in pion-nucleon scattering, or about 100 MeV. When you divide *h* by this quantity, you get an average resonance lifetime of about 10^{-23} second, or ten trillionths of a trillionth of a second—about the time it takes light to flash across a proton. The romance of a pion and nucleon is fleeting indeed! Maybe "flirtation" is a better word, after all.

Initially, physicists denoted these resonances by the label N*—as if they were mere excited states of a flirtatious nucleon and merited nothing more than an asterisk. After all, they lasted only 10^{-23} second, left no tracks in any detector, and showed up only as a bump in some obscure graph. Resonances were generally omitted from tables of elementary particles, as witness Yang's 1959 table. But theoretical advances of the late 1950s and early 1960s put them on an equal footing with the more "stable" elementary particles. In recognition of this elevated status, nucleon resonances were finally granted their very own label Δ (the Greek capital letter *delta*).

A good way to explain resonances is to compare them to the notes produced by a violin, flute, or recorder. When you blow into a recorder, you are injecting sound energy of various frequencies into the tube. But the instrument has a will of its own and "resonates" only at specific notes determined by its internal geometry and by the holes you have covered. These we call its "natural frequencies" of vibration. Such frequencies occur throughout physics, almost wherever matter and energy meet. If you plot a graph of sound intensity versus frequency, you discover a series of narrow peaks or bumps, each centered at one of the recorder's natural, resonant frequencies.

These peaks are curiously reminiscent of those that occur in graphs of pion-nucleon scattering. Here, the incoming pion may have a whole range of energies, but the pion-nucleon system has only a few specific resonant energies at which it can

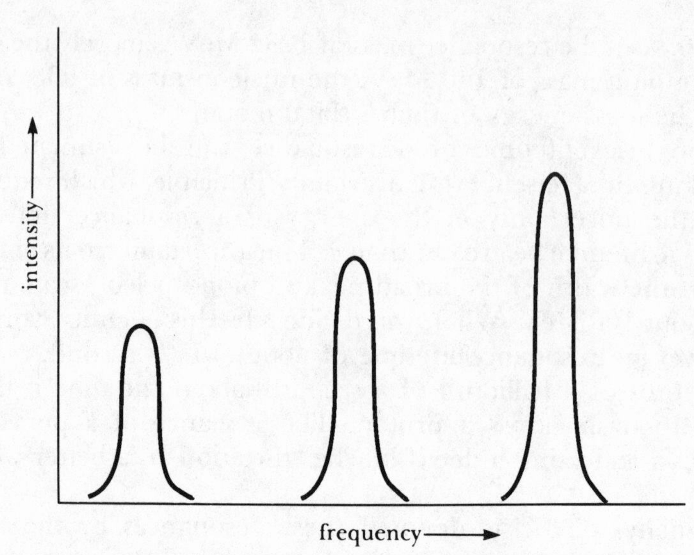

Graph of the resonant frequencies, or resonances, encountered in playing a recorder.

get excited. These are the energies at which the pair prefers to "vibrate." The recorder analogy is more obvious if you recall that, in quantum mechanics, energy and frequency are closely tied to each other by Planck's constant. The recorder vibrates at specific frequencies, while the pion-nucleon system resonates at specific energies.

In searching for resonances, physicists could never hope to find their tracks directly. With a lifetime of 10^{-23} second, resonances travel but a trillionth of an inch before decaying into two or three other particles. Bubble-chamber photographs show only a spray of tracks emanating from the point where the resonance was formed. These sprays were the key. Physicists and scanners sorted through thousands upon thousands of pictures looking for particular spray patterns. Aided by a large computer, they calculated the momentum and energy of each visible remnant, combined them in various ways, and plotted graphs of the results. Resonances appeared as peaks or bumps in these graphs—some obvious and unmistakable, others vague and obscure.

This sport of "bump-hunting," as it is called, became a favorite pastime during the 1960s at Berkeley, Brookhaven, and CERN. Bubble-chamber physicists found a huge number of nucleon resonances using such methods. But interest was beginning to turn to meson resonances, too. These flirtations of two or three mesons—a pion and a kaon, for example, or even three pions—are usually called "vector" mesons because they are spin-1 objects like the photon, not spin-0 like "normal" mesons. Unit spin is a "vector" quantity, one that has a preferred direction in space. Hence the name vector meson, because the spin gives it direction in life.

By the late fifties and early sixties, theorists were clamoring for vector mesons. The experimenters did their best to oblige them—or to confound them by discovering other, as yet unpredicted resonances. With the 72-inch bubble chamber at its disposal, the Alvarez group had a decided advantage in the quest. The first vector meson, the K^*, was discovered in Berkeley late in 1960. This pion-kaon resonance has spin-1 and a mass of about 890 MeV, almost that of a proton. By the next summer a pion-pion resonance, the ρ (Greek *rho*), and a three-pion resonance, the ω (Greek *omega*), had shown up at Brookhaven and Berkeley, both with spin-1 and masses of about 760 MeV. The floodgates had burst.

In a desperate attempt to reduce the growing complexity to a few fundamental particles or principles, theorists developed a progression of theoretical models. But the particle onslaught repeatedly foiled their plans.

In a visionary article, "Are Mesons Elementary Particles?," written a decade earlier, shortly after pions were first identified, Fermi and Yang had proposed that the three pions were not really "elementary" but instead were different combinations of a nucleon—a proton or neutron—and an antinucleon. This was a daring idea in 1949, because antinucleons had not yet been seen, although there were good reasons to believe they actually *did* exist. The proposal foundered, however, because it could not explain the purported forces binding nucleons to antinucleons. And it sank almost completely from sight because it failed to explain the proliferation of strange particles in the early 1950s.

Later in the decade a group of Japanese theorists tried to solve these problems. To include strangeness in the picture, Shoichi Sakata and followers added a lambda and an antilambda to the list of supposedly "fundamental" particles. They could build all the pions and kaons by pairing the members of a basic triplet—the proton, the neutron, and the lambda—with their three antiparticles. A proton (charge +1) paired with an antineutron (0) yielded a positive pion (+1), for example, while a lambda (charge 0, strangeness −1) combined with an antiproton (−1, 0) gave a negative kaon (−1, −1).

Sakata could even derive the properties of the sigmas and xis from threefold combinations of the fundamental triplet. But the fact that these heavy baryons had to be treated very differently from the lambda itself—supposedly a close cousin—offended many aesthetic sensibilities. Why should these two baryons be composites while the lambda baryon itself was "elementary"?

With the explosion of new resonances in the early 1960s, constituent models plunged into disfavor. It became increasingly difficult to imagine *any* member of this unruly zoo being somehow more "fundamental" than the others. Quantum field theories of the strong force were also in decline. The time was ripe indeed for new and radical approaches. Out of this ferment came two theories that were to dominate particle physics for the rest of the decade—the "bootstrap model" and the "Eightfold Way."

The bootstrap model began by rejecting the whole idea of elementary particles—at least as far as the hadrons were concerned. The greatest proliferation was occurring among these heavy particles, which all felt the effects of the strong force.

At Stanford, Robert Hofstadter and collaborators were showing that the proton and neutron had a definite *size,* unlike the electron, which seems to be concentrated at a mathematical point. There were good theoretical reasons to believe that *all* the hadrons had a similar size, too. Any particle that felt the strong force—pions, kaons, protons, sigmas, the resonances, etc.—was not a pure state but instead a busy cloud of virtual mesons, baryons, and their antiparticles. This cloud supposedly gave hadrons their apparent size.

How could something with size be elementary? On a gut level, the property of "elementarity"—of being indivisible and irreducible—seemed the exclusive birthright of particles with a pointlike nature. If a particle had a *size,* then we should be able to cut it into smaller pieces with a sufficiently sharp knife. By this common-sense criterion, the electron might indeed be elementary; baryons and mesons were decidedly not.

It was Professor Geoffrey Chew of the University of California who nurtured these ideas into full flowering with his bootstrap model of the hadrons. Charismatic and outspoken, Chew thrust this model into world prominence largely through the force of his own personality. "The notion that certain particles are fundamental while others are complex," he wrote in a seminal 1961 article, "is becoming less and less palatable for baryons and mesons as the number of candidates for elementary status continues to increase."

The bootstrap model was more a long-range program than a full-fledged particle theory. It accepted, on faith more than anything else, the proposition that there were no elementary hadrons and considered these particles instead as being built entirely from one another. Thus a proton, for example, was a complex combination of a "bare" proton, a proton and a pion, a neutron and a pion, a lambda and a kaon, a xi and two kaons . . . *ad infinitum.* Any combination of hadrons that had the right total charge, strangeness, spin, and other quantum numbers made at least a small contribution to the physical proton that we could see leaving a bright streak in bubble chambers. The same held true for any other hadron. By virtue of their involvement in the strong force, hadrons were all "mutual composites" of each other that hauled themselves up into existence "by their bootstraps."

The proton and neutron had no special significance in the bootstrap framework. They were no more "fundamental" than the other, heavier baryons like the lambda or the sigmas. They were just the lightest and most stable baryons encountered.

In the bootstrap model, intrinsic properties of hadrons like masses and lifetimes were to be calculated not by appeal to some limited group of simpler, "fundamental" building blocks. Instead, they would result from a "self-consistent" solution to a set of equations linking each hadron to all the others. In principle, it was a beautiful idea. One need not insert any arbitrary, unexplained numbers—like the masses of supposedly fundamental particles—into the theory. This was Occam's razor at its sharpest. As Chew himself put it in 1963:

> Each particle helps to generate other particles, which in turn generate it. In this circular and violently nonlinear situation, it is possible to imagine that no free parameters appear and that the only self-consistent set of particles is the one we find in Nature.

For Chew and the bootstrap school, Nature at her deepest levels was a delicately woven fabric. Any crass attempts to shred this beautiful cloth were ultimately doomed to failure.

In practice, the equations of the bootstrap were so horrendously complex that they defied all efforts to solve them. At best, the bootstrap approach was able to provide a qualitative understanding of only the lightest resonances. During the late 1950s, Chew's own success in these predictions had been an important catalyst for his conversion to this new faith. By 1963, he was its acknowledged Messiah. "Twentieth century physics already has undergone two breathtaking revolutions—in relativity and quantum mechanics," he exulted at a Cambridge University lecture. "We [now] may be standing on the threshold of a third."

Part and parcel of the bootstrap approach was the so-called "S-matrix" theory of hadron collisions. Like the bootstrap model, it was more an ambitious program than a full-fledged theory. The S-matrix traced its lineage to a proposal Heisenberg had made in 1943. In essence, any collision of two hadrons can be summarized in a two-dimensional array of complex numbers called the S-matrix—a vast, largely unknown table of numerical probabilities for transitions from one hadron state to another. Physicists of the sixties believed they could eventually map out the S-matrix by a systematic study of hadron collisions. Hand in hand, theory and experiment would find their way through this uncharted territory.

Intrinsic to S-matrix thinking was the belief that quantum field theories like quantum electrodynamics, which was so successful in the arena of electromagnetic forces, were hopeless when it came to strong interactions. According to this viewpoint, all efforts to reduce hadron collisions to the exchange of a few quanta at a few space-time points were doomed to failure.

Emission of a virtual "force particle" at one point and its absorption at another could never describe these vast complexities. The very existence of these unseen processes was seriously questioned. A realistic theory of the strong force supposedly had to be written only in terms of the *observable* parameters—the masses, energies, spins, charges, and other physically measurable quantities of the subatomic particles entering and leaving a collision.

From the S-matrix viewpoint, a potentially infinite series of particles were swapped when two hadrons collided. As long as the total energy, momentum, charge, and certain intrinsic quantum numbers remained the same after the collision as before, any hadron could join the melee.

This observation, together with discoveries made by the Italian physicist Tullio Regge, gave S-matrix advocates a useful way to organize the hadrons. From parallels in low-energy atomic physics, Regge proposed that hadron spins and masses were related to each other in very specific ways. The higher the spin, the greater the mass. Baryons of the same strangeness and other quantum numbers, for example, would lie on a straight line in graphs of their mass versus spin. For mesons, similar straight lines would occur in plots of their mass-squared versus spin. Dubbed "Regge trajectories," these oblique lines extended upward and outward to infinite mass and spin. Observable baryons (or mesons) were expected to occur only at those values of mass (or mass-squared) where the trajectories crossed lines of half-number (or whole-number) spin—the only values allowed by quantum mechanics. To a great extent, these expectations proved correct.

An entire Regge trajectory—not just a few virtual particles—was thought to be traded in the collision of two hadrons. A whole string of beads, not just one. Differing only in their masses and spins, the members of a trajectory were all in some sense equivalent to one another, equal partners in a classless society of subatomic particles. And no trajectory was in any way "superior" to any other. One did not need to invent a lower caste of elementary, somehow more "fundamental" building blocks to explain hadronic behavior. "If there is no need for aristocracy among strongly interacting particles," asked Chew in 1964, "may there not be democracy?" This idea of a "nuclear

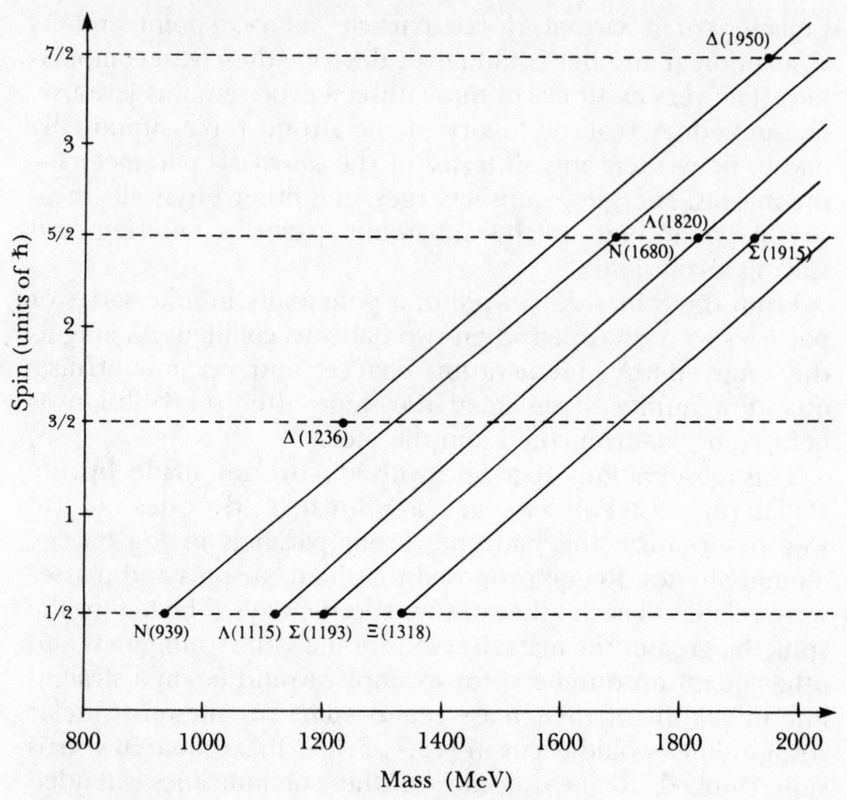

Organization of the baryons according to their Regge trajectories, circa 1963.

democracy," a key tenet in the bootstrap faith, gained a wide following among particle physicists of the 1960s. Converts to the new faith looked for guidance to Berkeley, where other voices calling for democracy were growing increasingly louder as the decade progressed.

Simultaneous with the bootstrap model, the "Eightfold Way" was gaining its own adherents during the early sixties. This idea, which stressed the importance of symmetry in the strong interactions, was developed independently in 1961 by Murray Gell-Mann at Caltech and Yuval Ne'eman at Imperial College in London. Gell-Mann was a leading theorist by this time; Ne'eman, on the other hand, was almost completely unknown, having only a year earlier resigned his position as colonel in the Israeli army and as a military attaché in London to pursue a doctorate in theoretical physics at age thirty-five. Ne'eman's work was actually the first to be published (in the European journal *Nuclear Physics*), but Gell-Mann's reputation gained him a wider audience.

Born in New York, Gell-Mann entered Yale at fifteen, eventually deciding to study physics instead of archaeology. MIT awarded him a Ph.D. in 1951, when he was only twenty-one—an age when most physics students are still grappling with their first course in quantum mechanics. After a stint at Chicago, he moved on to Caltech, collaborating there with Richard Feynman in 1957 on a highly regarded theory of the weak force.

By all accounts, Gell-Mann has a prodigious intellect. Not content with his vast accomplishments in theoretical physics,

Murray Gell-Mann and Richard Feynman in 1959.

this urbane and literate polymath has taught himself several languages, including Swahili, and is conversant in many other sciences. He often tries to overpower people with his brilliance, as if he were not quite sure whether he is the brightest man alive and still had to prove it.

In the early 1960s, Gell-Mann produced a long series of publications elaborating the Eightfold Way, including a pivotal 1962 *Physical Review* paper, "Symmetries of Baryons and Mesons." The account that follows hews closely to his approach.

Mesons and baryons appear in closely knit families or "multiplets"—a threesome of pions, two pairs of kaons, a pair of nucleons, a single lambda, three sigmas, two xis—whose members differ only by mass and charge. The slight mass differences (only a few MeV) within each family could easily result from the different charges of the various members. So you can think of each family or multiplet as a single particle with a new intrinsic property—its "multiplicity."

The strong force is insensitive to electric charge. It does not distinguish between positive, negative, and neutral pions, for example. To the strong force these particles appear as the three sides of an equilateral triangle seem to a casual observer: they might have different orientations in space, but they are otherwise *identical.* The same holds true for any other family of mesons or baryons; the strong force treats each member exactly alike. The slightly different masses of the family members arise because the much weaker electromagnetic force interacts differently with their different charges.

For a number of years, Gell-Mann had been trying to find a general symmetry principle that could incorporate strange mesons and baryons into a broad classification scheme including all the hadrons. To him, the strangeness of a particular hadron family was closely related to its multiplicity. Kaons were "strange" mesons *because* they came in two pairs, not a triplet, as one might have expected by naive analogy with the pions. The same was true of the three sigmas and the lambda; these heavy baryons were strange because they did not come in pairs like the two nucleons—the proton and neutron. Were these just curious coincidences? Or evidence of a still deeper symmetry? Gell-Mann was convinced of the latter.

Late in 1960, he finally found an appropriate mathematical

framework for his ideas in the abstract formalism called "group theory." That year a number of physicists were rediscovering and applying the work of Sophus Lie, an obscure Norwegian mathematician who had elaborated a peculiar class of these theories during the nineteenth century. Lie groups are collections of a finite number of abstract objects, all related to one another by specific transformations. Gell-Mann realized that one particular Lie group, labeled SU(3) for *S*pecial *U*nitary group in *3* dimensions, had just the properties he needed to classify *both* the mesons and baryons. At about the same time, Ne'eman was having a similar epiphany.

Using the group SU(3), Gell-Mann arranged the known mesons and baryons into hexagonal patterns according to their charge and strangeness. (Actually, he used related quantities—"isospin" and "hypercharge"—that are more abstract and difficult to describe.) Patterns for the spin-0 mesons and the spin-½ baryons are shown on p. 90; the spin-1 vector mesons, then mostly a cherished hypothesis of many different theorists, supposedly fit into a similar hexagonal pattern.

Now note that the meson pattern has seven members while the baryon has eight—exactly the number required by the SU(3) group. Based on this discrepancy, Gell-Mann predicted an *eighth* spin-0 meson with zero charge and strangeness, which he called the χ (the Greek letter *chi*). Within months, this eighth meson was discovered by the Alvarez group at Berkeley and dubbed the η (from the Greek letter *eta*), with a mass of 549 MeV, slightly heavier than a kaon. Both the baryons and mesons now fell into separate eight-member groups, or "octets."

Like Dmitri Mendeleev almost a century earlier, Gell-Mann had organized an unruly zoo of supposedly elementary objects into a simple geometrical pattern, a kind of Periodic Table of the mesons and baryons. From a suspicious gap in that pattern, he predicted the existence of yet another member—in analogy with Mendeleev's prediction of three new elements. When experiment bore out this bold prediction, physicists figured they might finally be on the right track.

The acknowledged master at *naming* the subnuclear world, Gell-Mann reached into Chinese literature to christen this classification scheme the "Eightfold Way," a name that soon

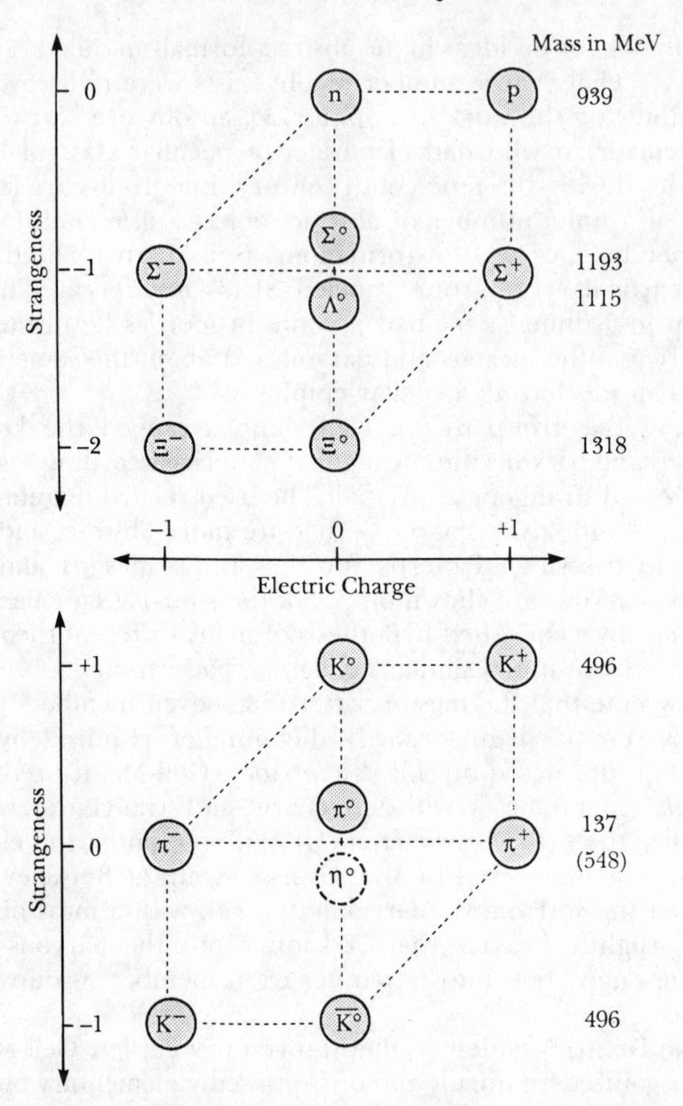

The hexagonal patterns predicted by SU(3) symmetry for the spin-0 mesons (bottom) and spin-½ baryons (top). Based on the discrepancy between the observed patterns, Gell-Mann predicted an eighth meson, eventually named the eta (dotted circle), which was discovered within a year.

captured the imagination of physicists around the world. The phrase comes from an aphorism attributed to Buddha, about the appropriate path to Nirvana:

> Now this, O monks, is noble truth that leads to the cessation of pain; this is the noble Eightfold Way: namely, right views, right

> intention, right speech, right action, right living, right effort, right mindfulness, right concentration.

Within a year, the octet of spin-1 vector mesons required by the Eightfold Way had been largely verified in bubble-chamber experiments at Berkeley, Brookhaven, and CERN. In the subnuclear realm, Nature seemed to favor such extended families, or "supermultiplets," of eight members. The particle zoo was well on its way to being tamed.

The concept of SU(3) symmetry at the heart of the Eightfold Way can be difficult to grasp because it is highly abstract. A concrete way to visualize this symmetry is by analogy with crystals, whose apparent beauty lies mainly in their regular geometrical features. In the growth of crystals we can witness Nature, as if by magic, taking some of the simplest shapes of mathematics. The "symmetry" of these gems means that we can perform certain geometric operations on them, and they end up looking *exactly* the same. Rotate a cube, for example, about an imaginary axis drawn through the centers of two opposite faces. After turning it 90 degrees, you notice it looks just the way it did at the start—no matter what its original orientation.

We can compare SU(3) symmetry to the symmetries of an octahedron—the shape assumed by common alum salts upon crystallizing (see diagram, p. 93). Consider the baryon octet and imagine each of the eight baryons inhabiting one of the eight triangular faces of a regular octahedron (like the cube above, one of the five Platonic solids). On the face immediately opposite you is a specific baryon, say a proton. It can be changed into any other baryon, the lambda for example, by an appropriate rotation about one of the major axes (the lines drawn through two opposite corners). Similarly, we might instead have populated the eight faces with members of the meson octet; the very same rotations would change one meson into any other.

The essence of the Eightfold Way is that all the members of a particular octet are related to one another by a group of transformations very much like the rotations of this regular octahedron, only these transformations actually occur in an abstract realm, not the tangible, three-dimensional space where geometric rotations occur. The "eight" in Eightfold Way, strictly speaking, refers to eight quantum numbers in this realm (including isospin and hypercharge) that are either conserved or altered by these transformations, just as geometric rotations can change the identity of the particle on the crystal face opposite you. The member particles of a given octet, however, are viewed as being just the different states of a single entity; each corresponds to a different "orientation" of something called its "unitary spin"—an abstract property that lies at the heart of the Eightfold Way. Geometric rotations turn one side of a triangle into another, or one crystal face into another; unitary rotations turn one baryon into another, or one meson into another. Symmetry is the key.

Though it seems obvious now, in the early sixties the Eightfold Way was no small leap of the imagination. The spins of the strange mesons and baryons were still ambiguous then, and it took considerable insight and guesswork to group the mesons and baryons into octets at all. Often one had to ignore misleading experimental data. Then, too, the masses of these particles had *widely* differing values. At 938 and 940 MeV, it was easy to think of a proton and neutron being closely related, but how could one group a 135 MeV pion together with a 494 MeV kaon?

To Gell-Mann, the symmetry was right there for all to see; it had merely been "broken" by some unknown mechanism that gave individual members their widely different masses. To his credit he saw beyond all the apparent chaos to the underlying symmetry within, SU(3) symmetry. This was Gell-Mann's particular genius. Like Plato, he could look at a pile of coal and see only diamonds.

The Eightfold Way does not require that *every* hadron belong to an octet. There can be other groupings of one, three, ten, twenty-seven, or more. In the same vein, the octahedron is not the only regular, Platonic solid. There are also the tetrahedron with four sides, the cube with six, the dodecahedron with twelve, and the icosahedron with twenty. In both cases, the underlying symmetry has a number of concrete manifestations. What is most intriguing is how both the Eightfold Way and the Platonic solids each single out certain specific pure numbers—and none other.

It was the discovery of the ten-member group, called the

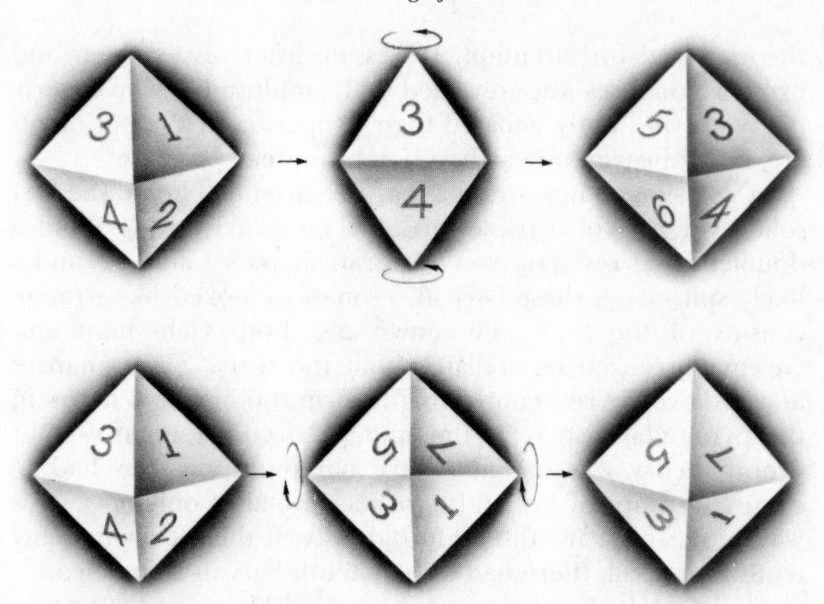

Symmetry operations on a regular octahedron.

"decimet," that really iced matters for the Eightfold Way. The members of this group had to be ten baryons with spin-3/2. Four members were already well known: they were just the four nucleon resonances—the Δ^{++}, Δ^{+}, Δ^{0}, and Δ^{-}—that had been discovered by Fermi and others in the 1950s. According to SU(3) symmetry, these four particles formed the base of a pyramid in a graphical display of this baryon decimet.

The next two layers were revealed at the 1962 International Conference on High-Energy Physics held at CERN. Known as the "Rochester Conferences" since they originated at the University of Rochester, these biannual gatherings are the premier event on the calendar of high-energy physics. Invitations to the Rochester Conferences are eagerly sought by particle physicists throughout the entire world. For experimenters, this is *the* marketplace where you put your new results on the auction block and shop for the latest in theoretical fashions.

Legions of bubble-chamber physicists from Berkeley, Brookhaven, and CERN itself—as well as smaller groups from the French and Russian accelerators at Saclay and Dubna—converged on Geneva in July of 1962, their carts laden with new baubles, their pockets bulging with cash. The Eightfold Way, the Sakata model, the bootstrap model, and a host of other theories vied for attention. In session after session, new and exotic resonances were revealed to the multitudes, who eagerly discussed and hotly debated their properties. To experimenters, it was heaven on earth, a bump-hunter's paradise.

In a session on new strange particles, scientists heard the first solid evidence for a triplet of Σ^* (then called the Y_1^*) and a doublet of Ξ^* resonances. With strangeness −1 and −2, and a likely spin of 3/2, these baryon resonances looked like strange cousins of the four well-known Δ's. Both Gell-Mann and Ne'eman realized immediately that the three Σ^* resonances and the two Ξ^* resonances might form the next two layers in the pyramidal display of the spin-3/2 baryon resonances. But group theory was unambiguous: baryon resonances had to come in groups of *ten*, and here was a total of only *nine*. This was an ideal test for the Eightfold Way: if the theory had any truth to it at all, there had to be a tenth baryon resonance.

At the end of the session, both Gell-Mann and Ne'eman

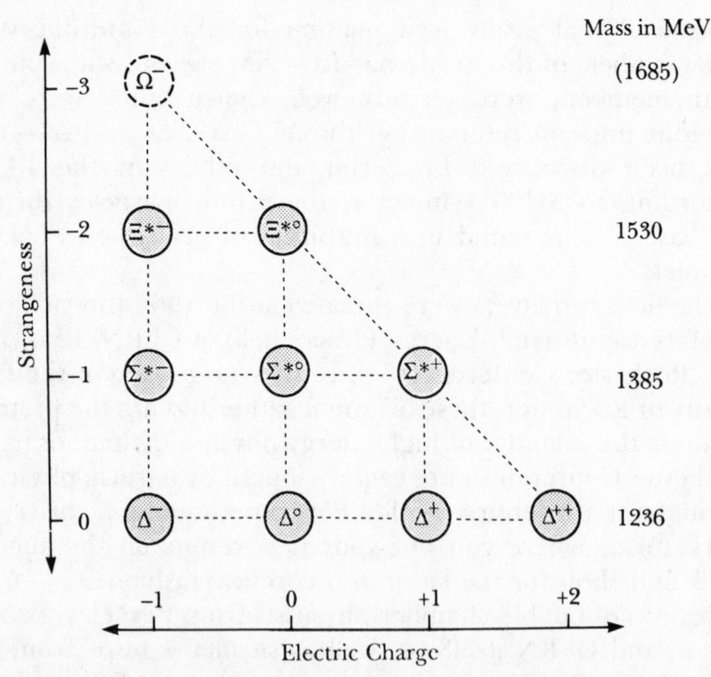

The decimet of spin-3/2 baryons required by the Eightfold Way. After the second and third rows had been confirmed in 1962, Gell-Mann predicted a tenth and final member, the omega-minus (dotted circle).

raised their hands to speak. Better known and sitting at the front, Gell-Mann was recognized first. He strode to the blackboard to suggest that "we should look for the last particle called, say Ω^-, with strangeness equals -3" at a mass of 1685 MeV, almost *twice* the proton mass. The Greek letter *omega*—the last letter in the Greek alphabet—seemed an appropriate choice for this final member of the baryon decimet. In the audience, Ne'eman could only sit and listen while Gell-Mann predicted and named the very same particle he was about to suggest himself.

To return to the crystal analogy, the Eightfold Way also permitted a ten-sided crystal structure and experimenters had discovered particles for nine of the faces. The remaining empty face cried out for a tenth. In the two-dimensional display of this group, the tenth particle sat at the pyramid's apex. What's more, the "omega-minus" had to have a negative charge and a relatively long lifetime. Thus it should leave a visible *track* in bubble chambers, a real footprint for once, something that could actually be *seen*.

Among the experimenters present when Gell-Mann predicted the omega-minus was a young Brookhaven physicist, Nicholas Samios. Earlier, he had presented his group's results on the discovery of the Ξ^* resonance, whose existence had spurred Gell-Mann's prediction. Samios recognized that a search for this very strange baryon was an important goal for the new 80-inch bubble chamber then under construction at the Long Island laboratory. The two shared lunch at the Conference, and Gell-Mann obligingly wrote Samios a short note on a paper napkin urging the Brookhaven director, Maurice Goldhaber, to authorize a search for the omega-minus.

With 900 liters of liquid hydrogen, the 80-inch device gave Brookhaven a bubble chamber every bit the equal of Berkeley's 72-inch chamber. Until then, the Alvarez group had dominated bubble-chamber physics only because of the sheer volume of target liquid it could present to beams of high-energy particles. What the Berkeley physicists lacked in particle energy they easily recouped through the extensive collision details their chamber could reveal. But Brookhaven's new chamber was soon to end their long reign.

Samios had been working with bubble chambers since their infancy in the midfifties, when he was a graduate student at Columbia. With jet-black hair swept back from his forehead, this short, wiry, fast-talking son of Greek immigrant parents seems as if he'd be equally at home hustling bets in the pool halls of his native Manhattan as on the experimental floor of a particle accelerator. An intuitive, no-nonsense scientist, Samios recognized that Brookhaven would have the edge in the search for the omega-minus once its enormous bubble chamber came on line. The other key was a high-energy beam of negative kaons to shoot at the chamber and smash into the protons inside. Just such a beam, with energies up to 5 GeV, was already under construction there.

Finally, by mid-December of 1963, all the complex parts of this massive enterprise were ready to go—beam, chamber, and physicists. Working around the clock, a team of more than thirty scientists logged photographs and took them immediately to the scanning tables to search for telltale tracks that might give evidence for the omega-minus. They were watching in particular for the characteristic V's left by the charged remnants of neutral strange particles. The V's provided arrows pointing back toward the actual collisions, where the experimenters hoped to find a very short omega-minus footprint.

Things were fairly uneventful for the first week, until an ominous problem began to reveal itself just before Christmas. Under the intense cold of the liquid hydrogen, narrow segments of a mirror inside the chamber were suddenly cracking away from their moorings and falling up against one of the windows through which the cameras photographed events. Under normal conditions this would have been only a minor nuisance. But at temperatures close to absolute zero it held a real potential for catastrophe. If the brittle window itself ever cracked and the liquid hydrogen escaped the chamber, an ensuing fireball could easily level the entire building and kill everybody inside.

Shaking his head as he recalled this tense moment, Samios remembered his great relief when the decision about whether to proceed fell to Ralph Shutt, then leader of the Brookhaven group and a man even more experienced than he in bubble-chamber work. Shutt studied the problem gravely for a long while and finally announced on Christmas Eve that the experiment could continue in spite of the apparent danger. As if by magic, the problem soon disappeared. No more mirror segments cracked away, and the rest of the run went smoothly.

By the end of January the group had obtained 50,000 good photographs without any evidence for an omega-minus. But on the evening of January 31, 1964, Samios spotted a photograph with just the right signatures: a few V's pointing back to the collision region, where the very short track of a negatively charged particle could be easily discerned. He was elated. Here at last was a "gold-plated event" exactly like what they were seeking.

There followed several feverish days of analysis, as group members identified each particle in the photograph and calculated their energies and momenta. By working backward, they soon concluded that the short track had been made by a negative baryon with strangeness −3 and a mass of 1686 MeV, give or take 12 MeV—almost exactly what Gell-Mann had predicted!

The activity now turned toward publication of this discovery.

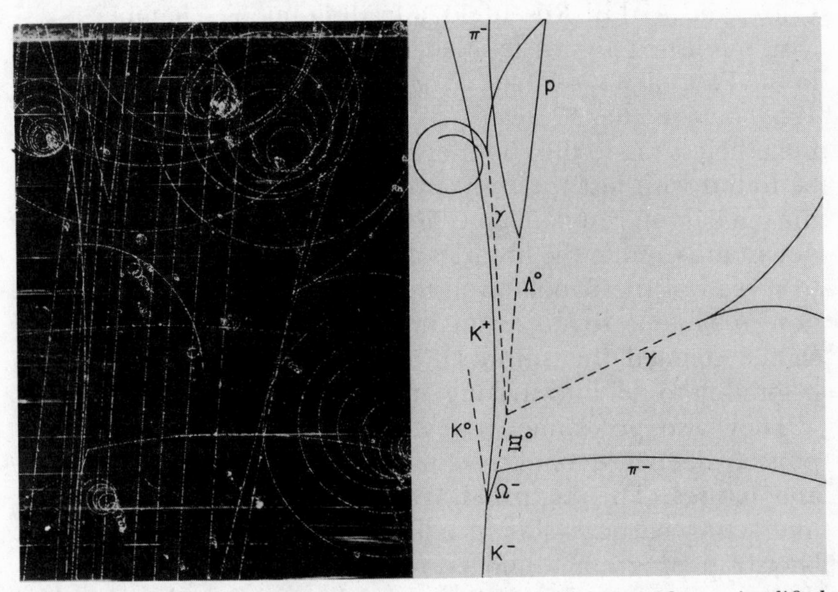

Bubble-chamber photograph of the first omega-minus event, plus a simplified sketch of the particle trajectories involved. The short track at lower left is the footprint of this very strange baryon.

On February 11 the finished paper reached the *Physical Review Letters.* Two weeks later, this journal published the paper under the unassuming title, "Observation of a Hyperon with Strangeness Minus Three." At the head of the article were the names of the thirty-three physicists, including Samios and Shutt, who had participated in the experiment. To the casual observer it might have seemed a bit obsessive: *thirty-three* names on the paper and only *one* event. But oh, what an event!

After the omega-minus discovery, the Eightfold Way emerged as the orthodox theory of the strong interactions. The Sakata model and other "constituent" models withered away because they had no easy explanation for the properties of the baryon resonances. S-matrix theory and the bootstrap program remained in favor, however, because of their promise to derive all the baryon and meson masses from the sole requirement of self-consistency. And with the help of Regge trajectories, they could explain the behavior of hadron collisions at ever higher energy.

As witness to the growing orthodoxy, Gell-Mann, Geoffrey Chew, and Arthur Rosenfeld, a leading bubble-chamber physicist, published an article titled "Strongly Interacting Particles" in the February 1964 issue of *Scientific American.* Written late in 1963 before the discovery of the omega-minus, the article nonetheless made the confident prediction it would eventually be found with just the right properties to cap the pyramid of the ten baryon resonances. Coincidentally, this issue was on the newsstands and in the libraries at the very moment word began to spread of the Brookhaven find. "The great unifying invention analogous to quantum mechanics is still not clearly in sight," claimed the authors, "but the experimental data are beginning to fall into striking and partly predictable patterns."

They also proclaimed the entire bootstrap program as a possibly deeper, dynamical explanation for hadron symmetries and masses. The Eightfold Way explained *how* the strongly interacting particles should fall into regular patterns, and the bootstrap program would eventually explain *why.* Or at least that was the fond hope. Promoted by eminent theorists like Chew and Gell-Mann, S-matrix theory and the bootstrap program had a rosy future indeed.

Two books appeared in 1964 that helped harden the growing orthodoxy: *The Eightfold Way,* by Gell-Mann and Ne'eman, and *The Analytic S-Matrix,* by Chew. They were little more than hastily edited lecture notes or annotated collections of journal articles, but both books became virtual bibles that dominated particle physics in the mid-to-late 1960s. Library copies were usually dog-eared from overuse, if their pages didn't fall out first.

At the heart of the new orthodoxy was a firm faith that Nature was *mathematical* at her most fundamental. Reality could be reduced to pure number, nothing more. This belief is the modern expression of an idealist philosophy over 2500 years old, dating back at least to Pythagoras and Plato. Appeals to a materialist conception of reality, in which the great diversity of phenomena could be explained by invoking a small collection of "fundamental" building blocks, were supposedly doomed to failure. Symmetry and self-consistency were all one really needed—or so the vast majority of particle physicists thought at the time.

CHAPTER 4

A QUIRK OF IMAGINATION

Well! I've often seen a cat without a grin, thought Alice; but a grin without a cat! It's the most curious thing I ever saw in all my life!

—Lewis Carroll, ALICE IN WONDERLAND

ALMOST a year before the omega-minus discovery, Murray Gell-Mann had paid a fateful visit to Columbia University. Arriving there on Monday, March 25, 1963, he stayed through Friday and delivered three well-attended lectures on the strong force, the Eightfold Way, and the bootstrap model. Physicists from throughout the New York area crammed into the lecture hall in Columbia's Pupin Labs to hear the very latest about these topics straight from the horse's mouth. Early that same week Gell-Mann had a luncheon conversation with Columbia theorist Robert Serber. The subject of their discussion eventually transformed the entire field of particle physics.

Serber had been working in physics since 1934; a slight, soft-spoken, unassuming scientist who generally avoided the limelight, he had been a postdoc with Robert Oppenheimer at Berkeley and had served as one of his trusted lieutenants at Los Alamos. Now working at Columbia, Serber had encountered the Eightfold Way only a few weeks before Gell-Mann's scheduled lectures. He began to wonder why a fundamental threefold group of particles—the triplet—did not seem to appear in Nature like the octet and decimet. The triplet, after all, was the most obvious grouping predicted by the three-dimensional SU(3) symmetry. Serber began to work out this idea on paper, soon discovering a way to build octets and a decimet by making various different combinations of three fundamental objects.

The day before Gell-Mann's second lecture, Serber and other Columbia theorists took him to lunch at the faculty club. While they were eating, Serber mentioned his work and asked why a fundamental triplet did not appear in Nature. According to Serber, Gell-Mann's immediate reply was, "That would be a funny quirk!" Tsung Dao ("T. D.") Lee, a Nobel laureate at Columbia, chimed in, calling it "a terrible idea." Pulling out pen and napkin, Gell-Mann showed Serber why. Such a triplet, if it existed at all, had to have *fractional* charges of $2/3$, $-1/3$, and $-1/3$. Serber had to agree that these were odd birds, indeed; ever since Robert Millikan's famous oil-drop experiment, nothing but whole-number charges had ever been observed.

While a visiting lecturer at MIT that year, Gell-Mann had been seeking a simple basis for the Eightfold Way. He knew about triplets but always passed them over without much thought because of the fractional charges involved. Serber's offhand query now catalyzed his thinking. Later that evening and the following morning, Gell-Mann began to realize that fractional charges were *not* completely absurd—as long as they could never appear in Nature. "If the bootstrap approach were correct," he reasoned, "then any fundamental hadrons would have to be unobservable, incapable of coming out of the baryons and mesons to be seen individually." By some unknown, paradoxical twist of fate, these oddities might remain forever *trapped* within the physical protons, neutrons, pions, and other hadrons that did appear in Nature. Nuclear democracy could still hold sway.

Gell-Mann began calling these whimsical motes "quorks," a nonsense word he had used previously, in the spirit of Lewis Carroll, to mean "those funny little things." Serber thought it was a play on the word "quirk" used at lunch. Whatever the case, Gell-Mann mentioned these "quorks" in his next lecture, giving Serber credit for proposing the fundamental triplet. Although he quickly moved on to other topics, "quorks" were a favorite subject of discussion at the subsequent coffee hour.

Gell-Mann was in no great hurry to publish the "quork" idea. There were a number of puzzles to resolve, and he had other commitments demanding his attention. Back at Caltech that fall, he began working out the details in earnest. In a phone call to his old MIT thesis adviser Victor Weisskopf, then Director of CERN, Gell-Mann mentioned he was working on a new and exciting idea: that baryons and mesons were composed of three fractionally charged entities. "Please, Murray, let's be serious," came Weisskopf's reply. "This is an international call."

By then the widely read theorist had found a telltale passage in James Joyce's enigmatic novel *Finnegans Wake:*

> *Three quarks for Muster Mark!*
> *Sure he hasn't got much of a bark,*
> *And sure any he has it's all beside the mark.*
> *But O, Wreneagle Almighty, wouldn't un be a sky of a lark*
> *To see that old buzzard whooping about for uns shirt in the dark*
> *And he hunting round for uns speckled trousers around by Palmerston Park?*

This poem is the drunken dream of Humphrey Chimpden Earwicker as he lies passed out on the floor of his Dublin tavern. Gell-Mann happily borrowed the "quark" spelling for his imaginary triplet. No doubt he also enjoyed the irony of the last few lines. There is an obvious parallel between the trials of Muster Mark and the seemingly hopeless plight of eager experimenters who would inevitably go searching for motes they could never possibly find.

Late in 1963, Gell-Mann finally wrote a short article and sent it to the new European journal, *Physics Letters.* When asked why he didn't send it to the more prestigious (his own adjective was "pompous") American journal *Physical Review Letters,* he replied that it "probably would have been rejected" there. The paper was published on February 1, 1964, the day after Samios had found the "gold-plated" omega-minus event at Brookhaven.

Gell-Mann's two-page article, "A Schematic Model of Baryons and Mesons," is brief and to the point. After a few preliminary remarks about other possibilities, he observes that "a simpler, more elegant scheme can be constructed if we allow non-integral values for the charges. . . . We then refer to the members $u(2/3)$, $d(-1/3)$ and $s(-1/3)$ of the triplet as 'quarks'." The *u, d,* and *s* are his shorthand for "up," "down," and "strange" quark varieties, and the fractions are their electric charges. In the very last sentence, he thanks Robert Serber for stimulating these ideas.

As often happens in theoretical physics, the idea of quarks had a second, completely independent birth. The midwife this time was George Zweig, a young CERN postdoc who had just received his Ph.D. from Caltech. Gell-Mann had been his first thesis adviser there, and Feynman took over when Gell-Mann left for his visiting lectureship at MIT in early 1963.

What spurred Zweig into thinking about quarks were the odd properties of a ninth vector meson, the ϕ (the Greek letter *phi*). With a mass of 1020 MeV (slightly heavier than the proton) and no electric charge, the phi found a ready berth within the Eightfold Way, which required a neutral "singlet" meson (a one-member group) at about the same mass as the other eight vector mesons. During 1962 and 1963, bubble-chamber experiments had given solid proof for the ϕ, which appeared in mass plots as a very sharp, narrow peak.

When he first read the publications of these experiments in the spring of 1963, while still at Caltech, Zweig was deeply perplexed. What puzzled him most was the observation that this phi meson preferred to decay into two kaons instead of other, energetically more favorable combinations. The phi itself had barely enough mass-energy to produce the two kaons, which had a combined mass of 990 to 995 MeV; that meant only 25 to 30 MeV remained to carry them away. The main reason the phi showed up as a very narrow peak was the fact that it took a long time for the reluctant corpuscle to summon enough courage and go through with this arduous process. Why, thought Zweig, did this particle elect to decay by the most difficult path available?

He at first guessed that a conservation law was at work: the phi chose the most difficult path because this was the only one allowed. "Feynman taught that in strong interaction physics everything that possibly can happen does, and with maximum strength," Zweig explained years later. "Only conservation laws suppress reactions."

This line of reasoning led the young theorist back to the Sakata model—to a *constituent* picture of the mesons and baryons. If the phi meson were composed of two constituents, one with strangeness +1 and the other −1, then its great preference for kaon decays could be explained. These two strange objects would find it difficult "to eat one another," so two individual kaons (one with strangeness +1 and the other with −1) were needed to carry both of them away.

But the Sakata model had its own problems, namely that it just could not reproduce the baryon octet and decimet found in Nature. While fiddling with group theories, however, Zweig inadvertently discovered that he *could* get these collections if baryons were built from three constituents with *fractional* charges. At first, it seemed an artificial solution:

> Fractional charges bothered me because I wanted a correspondence between leptons and the constituents of hadrons. To have one set of these particles integrally charged and the other set fractionally charged was ugly, but at this point there seemed to be no choice.

Now things began to fall into place. Zweig was elated:

> This was a fantastically exciting time. It was impossible to finish even the simplest calculation without jumping up, pacing back and forth for a few minutes, and rushing back to see if things were still working after all.

Fractionally charged building blocks proved the key to a new view of matter.

At CERN that fall, Zweig wrote up his discoveries for publication, calling his fractionally charged particles "aces." Mesons, which were built from pairs of these aces, formed the "deuces" and baryons the "treys" in his deck of cards. Just cut the deck, shuffle the cards, and you could deal all kinds of poker hands. His discovery first appeared as an internal CERN document, or "preprint," in mid-January 1964; meanwhile, Gell-Mann's quark paper was at *Physics Letters* awaiting publication, and Samios was scanning bubble-chamber photographs for an omega-minus footprint. It was a busy month for particle physics.

While Gell-Mann and Zweig were talking about essentially the same physics, stylistically they were poles apart. To the former, quarks had only a "mathematical" existence, if any; they were not respectable corpuscles in the same sense as leptons, mesons, and baryons, which left footprints or other clear evidence in detectors. To Zweig, however, aces were real, concrete fragments. Form versus substance. Along these two lines of reasoning, the ongoing debate between the ideal and the material continued.

To the casual reader, Gell-Mann's brief article seems ambiguous. He begins with a paean to the bootstrap and only begins mentioning quarks about halfway through the text. As regards the actual "existence" of these curiosities, he muses that "It is fun to speculate about the way quarks would behave if they were physical particles of finite mass (instead of purely mathematical entities as they would be in the limit of infinite mass)." To him, the question was one of "mathematical" versus "real" or "physical" quarks. The former could never be seen in Nature, while the latter would leave obvious tracks in detectors.

In fairness to Gell-Mann, he had been working with triplets long before Serber or Zweig stumbled across them—at least since his first (1961) paper on the Eightfold Way. There he built up the baryons by appeal to two triplets of "particles" (his quotes) with integer charges. But his intentions then were pedagogical; the triplets were little more than a convenient way to introduce the underlying symmetries. Later, he used quarks to develop abstract algebraic relations called "current algebra" without ever granting them any material existence. As he put it in May 1964, the whole process had a strong parallel in French cookery:

> We construct a mathematical theory of the strongly interacting particles, which may or may not have anything to do with reality, find suitable algebraic relations that hold in the model, postulate their validity, and then throw away the model. We may compare this process to a method sometimes employed in French cuisine:

a piece of pheasant meat is cooked between two slices of veal, which are then discarded.

From beginning to end, by contrast, Zweig's long paper interprets these triplets as physical, material objects. Though liberally sprinkled with the caveats expected of a young academic, it is obvious throughout where his bets are placed: he takes his aces for real. "We will work with these aces as fundamental units," he states early on, "from which all mesons and baryons are to be constructed." In great detail, Zweig then imbues his aces with normal physical properties like masses and spins, and shows how the observed properties of the mesons and baryons can result from various combinations of these subunits.

The most telling aspect of this paper is the graphical, *visual* means he uses to portray these building blocks. The three different aces Zweig represents by a solid circle, triangle, and square; their antiparticles are open versions of the same symbols. Baryons are combinations of three solid symbols joined by lines representing the forces between the aces; mesons are just the different pairs you can make, each with a solid symbol and an open symbol—an ace and an anti-ace. These are no abstract mathematical symmetries requiring a knowledge of group theory to understand them. These are real material objects you can almost *see;* you can combine them as easily as a child building with blocks.

Gell-Mann and Ne'eman had shown that Nature prefers certain specific clusterings—the singlet, the octet, and the decimet—for mesons and baryons we actually observe. These orderly, symmetrical patterns can however be explained by appeal to a deeper level of existence, whether quarks or aces. The same kind of reasoning is used to explain symmetrical crystal shapes: symmetry is the result of orderly arrangements of a few basic building blocks—the atoms or molecules filling a regular crystal lattice. For Gell-Mann, however, the symmetry patterns of the Eightfold Way were *themselves* the fundamental reality. Zweig took the additional step from form into substance; to him, the aces were real bits of material—the "stuff" of matter.

In his efforts to publish his paper, Zweig encountered

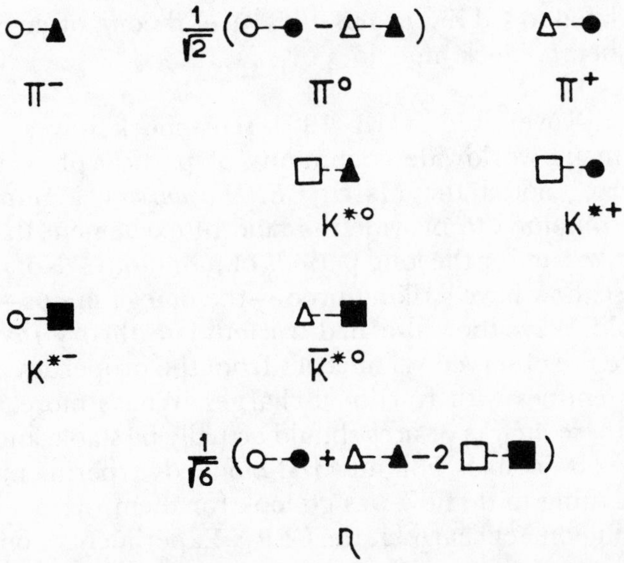

○—▲ π^- $\frac{1}{\sqrt{2}}$(○—● − △—▲) π^0 △—● π^+

□—▲ K^{*0} □—● K^{*+}

○—■ K^{*-} △—■ $\bar{K}^{*0}$

$\frac{1}{\sqrt{6}}$(○—● + △—▲ − 2 □—■) η

The various combinations of three "aces" and their antiparticles to make up the octet of spin-0 mesons, as given by Zweig in his 1964 paper. A solid symbol is an ace, and the same open symbol is its antiparticle.

extreme difficulty. Given the theoretical climate of the day, with the Eightfold Way and the bootstrap model ascendant and constituent models in overwhelming disfavor, his was a radical idea notably lacking in support. To make matters worse, he wanted to publish in the American journal, *Physical Review,* but Leon Van Hove, the director of CERN's theory group, insisted he use a European journal instead. Going over Van Hove's head to Weisskopf, Zweig finally got his way but was soon swamped by complaints from referees—other physicists asked to judge the paper's scientific merit. He eventually abandoned trying altogether. The paper languished unpublished until it was reprinted in a historical collection almost two decades later.

What physicists knew of Zweig's work usually came in over the grapevine or through circulation of his CERN preprint in a kind of underground communications network. To his credit, Gell-Mann himself was unstinting in his efforts to call attention to the paper; he repeatedly cited his young colleague for making an independent, equivalent derivation of the same basic idea. After Zweig joined the Caltech faculty in late 1964, Gell-Mann was perhaps his strongest advocate for tenure. But he still liked to kid Zweig about his "ace" theory, often calling it "the concrete block model."

The discoveries of early 1964 sent shock waves rippling through the worldwide community of particle physicists. "In February," noted the March *CERN Courier*, "a number of events combined to provide the kind of excitement that more than makes up for the long periods of monotony." Not only did physicists now have striking proof—the omega-minus—for the Eightfold Way; they also had a candidate theory that could produce the observed symmetries from the properties of three curious entities with fractional charges. What's more, at least one of these quarks or aces should actually be stable and hence directly observable. For any red-blooded experimenter, the obvious thing to do now was go look for them.

Meeting on February 11, the CERN Experiments Committee decided that aces (as they were originally known at CERN) should be taken seriously. That afternoon, Zweig explained his theory at the scheduled Experimental Physics seminar, and the scientists there debated the best way to search for them. With charges of ⅓ and ⅔, aces should leave obvious footprints in bubble-chamber photos. At high energy, the density of a track (defined as the number of bubbles per centimeter) is proportional to the square of the particle's charge. Aces should thus leave tracks with only ⅑ or 4⁄9 the density of a high-energy pion. A bubble-chamber experiment, the CERN physicists agreed, was the best way to search for these fractionally charged particles.

In fact, such an odd particle might have already been photographed in previous bubble-chamber experiments. One of the physicists present, Douglas Morrison, realized that his group had already taken some appropriate pictures back in 1960. He quietly slipped out of the room, grabbed some of his colleagues, and went back to reexamine these photos. In a single night they looked through 10,000 pictures, but found no fractionally charged particles. Another CERN group soon scanned through 100,000 old photos and got the same null result.

In March, a hastily scheduled experiment with the CERN 81-centimeter bubble chamber and a particle beam optimized to produce aces also failed to show any evidence. Similar ad hoc experiments at Brookhaven came away empty-handed, too. If aces or quarks existed, they were produced far less frequently than normal particles like pions. Or else they had masses above 3000 MeV, more than three times the proton mass, and could not be produced in existing particle beams.

To confuse matters still further, well-known theorists weighed in with fundamental triplets in which the members had *integer* charges. For more than half a century, experiments had revealed nothing but whole-number multiples of the electron charge *e;* integer charges were not going to be abandoned lightly. Leading theorists including Van Hove and T. D. Lee published models with two or three fundamental triplets whose members had charges 1, 0, and −1 (times *e*). With only one triplet, the original quark model was still the simplest theory, but you could not convict these other theories on grounds of aesthetics alone.

But the original quark model was having problems, too. If you took the Δ^{++} resonance to be composed of three *u*, or up, quarks each with charge +⅔ and strangeness 0, then you could reproduce its observed charge of +2 and strangeness 0. This meant, however, that you had identical, spin-½ quarks in the exact same quantum state. That was expressly forbidden. Such a configuration violated the Pauli Exclusion Principle, developed by Wolfgang Pauli way back in 1925 to "explain" why two electrons never occupied the exact same turf. This law requires that no two identical fermions (spin-½ particles) can ever be in the same quantum state. Without it, all matter would collapse upon itself. But if the Δ^{++} and other baryons were made of three spin-½ quarks, then they apparently violated this principle.

A way around this dilemma was discovered and published late in 1964 by O. W. Greenberg of the University of Maryland. He proposed that there were not one but *three* triplets of fractionally charged entities—"paraquarks" he called them—possessing some unknown physical property that distinguished one triplet from the next. Thus he could build up all the baryons from three paraquarks without violating Pauli's principle, as long as they were all somehow different. No longer *identical* fermions, two paraquarks could occupy the same quantum state.

Although it was a workable solution, Greenberg's idea did

not arouse much enthusiasm. It was just one further degree of complexity we had to accept in order to believe in quarks at all. What was this new physical property? Who needed it, anyway? At least with "strangeness" one could point to a track in a bubble-chamber photograph and say, "Now there's a strange particle!" But paraquarks just did not seem to appear in Nature, any more than quarks. The whole idea reeked of artifice. Taken together with the fact that nobody had yet *seen* a quark in a detector, this quandary helped to put their actual existence in serious doubt by early 1965.

If quarks existed at all, it seemed, they had to be awfully massive. The first experiments had seen nothing, but these were sensitive only to masses up to about 3000 MeV, or 3 GeV. Even though the CERN and Brookhaven accelerators could deliver 30 GeV protons, only a fifth of this, or 6 GeV, was available to produce new particles, and that small amount had to produce both a quark *and* an antiquark. The remaining energy carries the debris away from the point of collision. If a trailer truck smashes into an idle Volkswagen, the lion's share of the collision energy scatters the fragments forward—truck included.

There were good theoretical reasons to believe in massive quarks. The quark theory could explain the observed masses and other properties of the mesons and baryons only if the forces between two quarks did not depend very much on which way their spin vectors were pointing. This meant they had to be moving very slowly—much less than the speed of light. Because of the uncertainty principle, once again, the only way there could be slow-moving quarks inside such a tiny space as a proton was to make them very ponderous. Masses of 4 to 10 GeV, or even more, were expected.

But wait a minute. If there are three quarks in a proton, each with a mass of 4 to 10 GeV, then you expect—by simple addition—a total mass of 12 to 30 GeV. And the proton mass is slightly less than 1 GeV. Why the great discrepancy?

The answer lay in the possibility that quarks might be very tightly bound together inside protons; to pry them apart you would have to supply a tremendous amount of energy. This "binding energy" makes a *negative* contribution to the total because it represents energy you have to *put in* if you ever want to see quarks in their free, unbound condition. It's like a bucket of water down a deep well: before you can slake your thirst, you have to supply a lot of your own hard work (energy, that is) to haul it up.

In the case of the proton, the quarks had to be down an incredibly deep well. To make the proton mass come out right, the total binding energy had to be somewhere between −11 and −29 GeV—assuming the quarks themselves weighed in at 4 to 10 GeV. Thus, a proton needed to be clobbered awfully hard before it would disintegrate into these hypothetical fragments. Small wonder nobody had yet seen a single solitary quark.

An ingenious detour around this roadblock was taken by a team of experimenters at Columbia University led by Leon Lederman. These scientists realized they could produce much more massive particles if they took advantage of the internal motion—known as "Fermi motion"—of protons and neutrons buzzing around within nuclei. Especially in heavier nuclei like those of copper and iron, Fermi motion can occasionally be quite large, the equivalent of hundreds of MeV for a single nucleon. Protons fired at these nuclei usually hit a slow-moving nucleon, or one moving at some random angle. But, in the rare instances where the proton strikes a fast-moving nucleon head-on, there is a lot of extra energy available to produce new particles—with masses upward of 5 GeV, in fact. Far more damage is possible when a trailer truck hits a Volkswagen in a head-on collision.

Early in 1965, the Columbia team set up their experiment at the Brookhaven synchrotron. They slammed the proton beam into a copper target and studied the debris reeling off at an angle of 5 degrees. But rather than look for fragments with fractional charges, as most earlier experimenters had done, these scientists watched for very heavy particles, especially those with masses between 3 and 5 GeV. By using *mass* as the criterion, they could also search for quarks with whole-number charges—which were expected to be heavy, too.

But once again the experimenters came away empty-handed, without a single "naked" quark to show for all their hard work. If quarks had a material existence at all, they had to be heavier than 5 GeV. Or they were still too heavily clothed to appear unclad in Nature, despite the clever ploys experimenters might concoct to find them.

Nature, however, had a consolation prize. Among the debris they found evidence for the "antideuteron," the antiparticle of the heavy hydrogen nucleus. Composed of an antiproton and an antineutron, this particle showed up as a peak in the mass plots that "stood out like Mont Blanc on a clear day," Lederman recalled. To physicists it was no great surprise, but when the press showed up eager for a story, it found a good one. "The existence of antideuterons," Lederman told one reporter in a swashbuckling moment, "is the final proof that antipeople could possibly exist." After his unguarded comments made it to the front page of *The New York Times,* the letters began to pour in from people who needed antipeople.

Having failed to find any quarks at atom-smashers, experimenters turned to the heavens for an answer. Cosmic rays had been the source of many particles discovered before 1960, so why not look for quarks in the hail of radiation raining down from the skies? If quarks were indeed very massive, as now seemed likely, cosmic rays would be an ideal place to search. To produce truly massive quarks extremely high energies were needed, well beyond those available at Brookhaven or CERN. The primary cosmic rays hitting the upper atmosphere sometimes had energies of a *trillion* electron volts, or 1 TeV, enough to produce quarks with masses above 20 GeV.

The possibility of discovering quarks brought a number of old cosmic-ray physicists out of the woodwork. Their science had been in the doldrums since the midfifties, eclipsed by the new accelerators with their uniform particle beams and their amazing bubble chambers. Most of these physicists had drifted off to find work at the big laboratories, with only a few old diehards keeping the flames alive.

Late in 1965, a group of five scientists at CERN led by Antonino Zichichi decided to search for fractionally charged quarks in cosmic rays. A Sicilian by birth, Zichichi had studied cosmic-ray physics in Britain during the 1950s before returning to Italy, where he became Professor of Advanced Physics at the University of Bologna. A short, energetic, highly expressive man, he is perhaps best known as Director of the Ettore Majorana Centre for Scientific Culture at Erice, perched on a scenic mountaintop overlooking the Mediterranean near the western tip of Sicily.

To detect quarks, the CERN experimenters built a simple device called a "cosmic-ray telescope" from spare parts. It was essentially a club sandwich with six horizontal slabs of plastic "scintillator" material, each slab viewed by its own photocell. When a charged particle passed through any layer, its trail of ions generated a small light flash, or "scintillation," that the photocell then translated into an electronic pulse. Like the track density in bubble chambers, the size of this pulse was roughly proportional to the square of the particle's charge. Thus, a quark with a charge of $-1/3$ or $2/3$ would generate pulses in each of the six layers that were about $1/9$ or $4/9$ the pulses expected from a normal cosmic-ray muon with unit charge.

The CERN experimenters ran their cosmic-ray telescope for almost three months in 1966 without finding any truly convincing evidence for quarks. Because the cosmic-ray energies were largely unknown, they could not even put a lower limit on the quark mass, as accelerator experiments had done. The most these physicists could say was that quarks, if they existed at all, were coming down at a rate of less than one quark per square foot per day—far, far less than normal cosmic rays.

There *was*, however, one particular event, Zichichi later admitted, that was a "solid candidate" for a quark. Occurring early in the experiment, this morsel showed up with abnormally low pulses in all six layers of the sandwich. Skeptics all, the CERN scientists tried hard to explain away its occurrence without having to rely on fractional charges. If all the power supply voltages had momentarily dropped, Zichichi admitted, they would get uniformly low pulses from a normal cosmic-ray muon. Unable to come up with a convincing explanation, however, they continued to run the experiment, hoping to encounter a few more odd events like this one. But after three months of anxious waiting, they had seen no more.

They finally ended the experiment, Zichichi recalled, "more out of exhaustion than anything else." When the results were published later that year, no mention was made of the odd event. Given the theoretical climate of the day, they needed much better evidence than that to claim the discovery of a quark.

Yet another way to search for quarks was to look for them in normal, garden-variety matter. Cosmic rays have been raining down on the earth ever since it coalesced about 4.5 billion years ago; even if the percentage of quarks in this radiation is minuscule, some appreciable number of them must have accumulated in the earth's crust—or perhaps in the atmosphere and oceans. Then, too, single quarks might have been created in the Big Bang; perhaps a small number were drifting around naked and lonely in ordinary matter, unable to find the mates they needed to become respectable protons or neutrons. Doomed to the solitary life, these supposedly "liberated" quarks would wander from encounter to brief encounter until a fortuitous meeting with other quarks resulted in the long-sought union. Or so went the speculations.

Once again, fractional charge provided the telltale "signature" for this search. If a tiny sample of matter had a quark in it, its fractional charge should be measurable by variations of the classic Millikan oil-drop experiment that had established the electron charge back in 1910. Experimental techniques had come a long way in fifty years, and highly sensitive methods could easily spot the difference between fractional and integer charge. Physicists were fond of citing a telling sentence in Millikan's 1910 paper: "I have discarded one uncertain and unduplicated observation apparently upon a single charged drop, which gave a value of the charge on the drop some 30% lower than the final value of *e*." Could he have casually discarded the one measurement that disagreed with the hypothesis of integer charge—thereby missing the discovery of a naked quark?

From astrophysical calculations it seemed there might be somewhere between one and one hundred quarks per gram of matter near the earth's surface—well within detection capabilities. But the first search for quarks in ordinary matter, made by three scientists at the Argonne National Laboratory near Chicago, found none. They examined meteorites, seawater, air, and even the dust gathering in their ventilation shafts, all in a vain search for quarks. They passed huge volumes of air and water through electric fields, trying to concentrate whatever quarks might be there, all to no avail. Reporting their work in the summer of 1966, these scientists only could speculate that there was less than one quark per *kilogram* of air or seawater. Naked quarks were rare indeed.

One of the leading figures in the early quark searches was an Italian theorist at the University of Genoa, Giacomo Morpurgo. Early in 1965, he had written a key paper interpreting quarks as real, material—albeit very massive—entities and proposing a simple way they might be bound into hadrons. "If the present ideas are valid, the quarks should exist," he concluded; "they should not be only mathematical entities." Having thus published his ideas about quarks, Morpurgo then took an unusual step for a theorist: he decided to do an experiment to search for them. "I simply thought that a realistic model should imply real heavy quarks," he reminisced, "and I thought it would be fun to look for them."

Born in Florence in 1927, Morpurgo had come to the United States for a year in the early 1950s to work as a postdoc at Chicago with Enrico Fermi, perhaps the last particle physicist thoroughly comfortable and accomplished in both theory and experiment. Some of this influence had obviously rubbed off on Morpurgo, who said he was much impressed by his mentor's insistence on simple physical explanations. This slight, gentlemanly scientist had little patience with the complicated theories being advocated in the 1960s to explain the apparent symmetries of subnuclear matter. For him, the answer lay in a few very heavy building blocks with charges a fraction of that of the electron.

With the help of able coworkers, Morpurgo built a small device in his Genoa laboratory that could "levitate" tiny granules of graphite using only magnetic fields. These scientists then measured the charge on a grain by observing its movement in an electric field. The amount of deflection told them the size of the grain's net charge; its direction, left or right, told them whether it was positive or negative. By shining ultraviolet light upon the grain, they could expel electrons and change its total charge at will. If a granule started with integer charge, it eventually would reach zero charge and show no deflection at all, left or right. But if a granule began with a fractional charge, it could never hit zero exactly. The minimal deflection would provide a measure of its fractional charge—and solid proof for the existence of quarks.

During the first part of 1966, Morpurgo and his team constructed their magnetic levitation device and began testing the first few granules. Most grains had to be rejected because they contained far too many extra electrons, but the Italians eventually found five to their liking. None of these, however, showed any evidence of fractional charge. As the total amount of graphite examined was exceedingly small, about ten billionths of a gram, they could not reach any strong conclusions from this first, preliminary experiment. But Morpurgo and his collaborators spent another decade perfecting their device and measuring larger and larger volumes of material—all without ever finding a quark.

If you had invested in quarks, your stock had fallen to new lows by the summer of 1966. Almost twenty experiments had already searched for them, and not a single naked quark had yet been seen. One could of course argue that they were far too

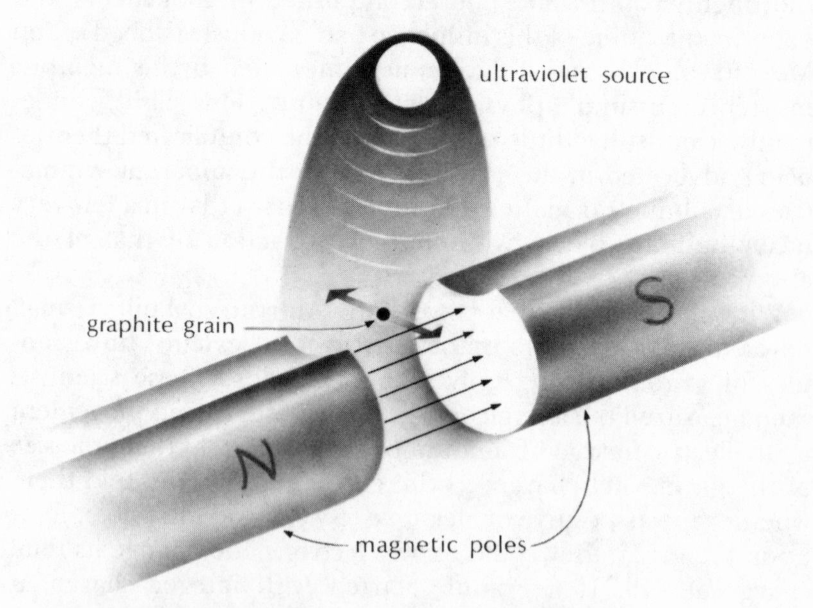

A simplified sketch of Morpurgo's apparatus. Graphite grains suspended in a magnetic field were deflected left or right by an electric field.

massive to emerge, that they were produced only very rarely, or that experimenters had simply been looking in the wrong places. These arguments, however, were by now getting pretty far-fetched.

The simplest interpretation was that quarks just did not exist as normal, material particles. If they had any "existence" at all, it was only mathematical—as Gell-Mann had claimed from the start. In a speech that May to London's Royal Society, he told his audience an anecdote:

> One atomic spectroscopist friend of mine rings me up, sometimes at midnight, to report his progress in a search for quarks in sea water. He has electrolyzed a huge amount of sea water to look for characteristic atomic levels of quark atoms. He thought he found one once, but it turned out to be an unknown line of tungsten. Since then, he has decided that the chemical properties of real quark atoms—if they exist—would be very strange indeed. And since most things with curious chemical behavior in the ocean eventually are eaten by oysters, he is grinding up oysters and looking for quarks in them. He has not yet seen any, nor have any been found at very high energies in cosmic rays.

"So we must face the likelihood," Gell-Mann concluded, "that quarks are not real." They might appear in mathematical equations—but not in particle detectors.

Such mathematical entities without a separate, material existence were hardly new to physics. In crystals, for example, the transmission of heat and sound can be readily understood by introducing corpuscles called "phonons" that carry these forms of energy throughout the crystal lattice. Like photons, each phonon carries energy equal to Planck's constant times its frequency of vibration. Many properties of crystals can be explained using these phonons, but they have absolutely no "meaning" outside the lattice. If you smash a crystal, you won't find any phonons lying around on your table. They "exist" only within its confines.

Like phonons, quarks could explain an impressive array of physical properties by 1966. All the known baryons could be built from combinations of three quarks; all the known mesons could be built from the quarks and their antiquarks. The quark model also placed stringent limits on the baryons and mesons that might actually exist. Whereas the Eightfold Way allowed a cluster of twenty-seven baryons, for example, the quark model permitted families of only one, eight, and ten. Experiments bore out the latter. Using the quark model, too, physicists could calculate magnetic attributes of the proton and neutron simply by adding up the corresponding properties of their three constituent quarks. And the decays of baryons and mesons were readily explained by assuming them to be built from quarks. This is what led George Zweig to his prediction in the first place.

Late in the summer of 1966, the biannual Rochester Conference took place, this time at the Lawrence Radiation Laboratory in Berkeley, California, site of the Bevatron and Mecca to S-matrix and bootstrap theorists around the globe. These theories were back in fashion now, after they had fallen briefly out of favor at the 1964 conference held in the Soviet Union. For Luis Alvarez, Geoffrey Chew, and the rest of the Berkeley scientists, this week-long exercise affirmed a style of doing particle physics that had flowered largely on their own campus.

Among the hundreds of physicists who arrived from all over the world were CERN experimenters Douglas Morrison and Antonino Zichichi. Giacomo Morpurgo came, too, and submitted *nine* papers to the sessions on quarks. Robert Serber, T. D. Lee, and Leon Lederman arrived from Columbia, as did Brookhaven's Nicholas Samios. From Caltech came Richard Feynman, Murray Gell-Mann, and George Zweig.

At the request of the organizers, Gell-Mann gave an introductory talk at the opening session to summarize the current state of particle physics and set the tone for the whole affair. After relating some recent successes with S-matrix theory, he turned to the quark model. "Now what is going on?" he asked his audience. "What are these quarks?" About their possible existence in Nature, he was none too hopeful:

> The idea that mesons and baryons are made primarily of quarks is difficult to believe, since we know that . . . they are mostly, if not entirely, made up out of one another. The probability that a meson consists of a real quark pair rather than two mesons or a baryon and antibaryon must be quite small. Thus it seems to me that whether or not real quarks exist, the q and $\bar{q}$ we have been talking about are mathematical. . . .

Here was the leading theorist of his day, a man soon to receive the Nobel prize for developing the Eightfold Way, warning the entire Conference not to take his very own quark hypothesis too literally. One had to be brave indeed to continue believing in quarks.

But despite his admonitions, a small group of physicists there *did* continue believing quarks were real. One of their leaders, in fact, was the scheduled rapporteur for the session on symmetries of the strong interactions. A theorist at Oxford University, Richard Dalitz embraced the quark model in 1965, shortly after Morpurgo had written his first paper about quarks. He had entered particle physics from nuclear physics, where it was natural to explain complex objects (the atomic nuclei) by breaking them down into their constituents—the protons and neutrons. At a summer institute that year in Les Houches, France, he stayed in his room almost the entire time, reading everything he could find about quarks and working out his own, "nonrelativistic" quark model, leaving only to eat his meals and teach his course. Together with Morpurgo, he is widely regarded as one of the two fathers of this model.

A balding, bespectacled man of medium height and average build, Dalitz was supposed to discuss abstract symmetries in his Berkeley talk. But he concentrated almost exclusively on his quark model, much to the chagrin of many theorists present. After a summary of the basics and a short survey of the unsuccessful quark searches, Dalitz launched into a detailed account of how all the known mesons and baryons could be built up from heavy quarks and antiquarks—explaining their masses, spins, decays, and other observable properties. To many present, this was like the "messy" nuclear physics they had left behind to work in particle physics. Yawns appeared on many faces and more than a few physicists began to nod off. Gell-Mann left abruptly during the first hour of the talk. After a coffee break, only a third of the original audience returned to endure the final hour.

That same year Stephen Gasiorowicz, a University of Minnesota physicist who had been present at Berkeley, wrote a textbook titled *Elementary Particle Physics.* A far-reaching introduction to the field, his book eventually went through several printings and was used as the main text in many graduate courses in particle physics, including the one I took at MIT in

1969. It was an influential book from which many young particle physicists of the late sixties learned their trade. In a total of over 600 pages, however, Gasiorowicz devoted just one paragraph to quarks, and even there he was not too sanguine:

> If such quark particles existed, at least one of them would have to be stable, because baryon conservation would forbid the decay into "ordinary" particles. A search for new particles of this type has so far proved unsuccessful. In view of this failure and the difficulty of inventing a mechanism which would bind a quark and an antiquark, or three quarks, . . . we will not discuss the quark model further.

The next 200 pages, encompassing eleven chapters in all, is devoted to discussions of the Eightfold Way, the S-matrix theory of the strong force, and the bootstrap model. Quarks are never mentioned again.

In the *Timaeus,* Plato gives us his fullest account of the origins of the universe and the phenomena of Nature. With Socrates for a mouthpiece, he explains the creation of the world according to a divine plan, the motions of the heavenly bodies, and the internal structure of matter. "It is clear to everyone that fire, earth, water, and air are bodies, and all bodies are solids," he observes. "All solids again are bounded by surfaces, and all rectilinear surfaces are composed of triangles." Here, in a nutshell, we glimpse Plato's theory of matter: the four elements reduce to four of the five regular solids (the Platonic solids), which can themselves be reduced to triangles. At its very deepest level, "matter" is form, not substance. Everything reduces to pure geometry, to mathematics.

There are striking parallels between Plato's theory of matter and the dominant midsixties theories of particle physics. At the time there was a growing conviction that physicists had at last found the deepest level of matter, the innermost layer of the cosmic onion, where everything would eventually be explained by appeals to pure mathematics—to the harmonies and symmetries of "pure," abstract thought. To try to explain these regularities by calling upon a still deeper level of existence was a base and ultimately fruitless program. To ascribe any material essence to quarks was to flirt with heresy.

The parallel between Plato's triangles and Gell-Mann's imaginary quarks is more striking yet. Not only is the same whole number 3 dominant in both theories, but their fundamental entities are both abstract products of the mind, not substantial objects of our senses. Much more than mere stylistic preferences, these are profound and far-reaching statements about the locus of physical reality. True reality is "in here," not "out there"—contemplative, not material. Plato's idealistic world view, which dominated Western thought for more than two thousand years, obviously exerts a strong influence even today.

By 1966 the quark hypothesis was in danger of becoming another "grin without a cat," a good idea without any substance behind it. Lewis Carroll's memorable phrase has been interpreted to mean the abstract world of pure mathematics, while his "cat without a grin" connotes the material world of our five senses. Although they explained many known facts about mesons and baryons, quarks had won only a grudging acceptance, if any, as abstract mathematical entities.

Particle physicists need more tangible evidence than that. We need to see, feel, and taste a particle before we can believe it exists. For quarks, such a proof was still completely lacking.

For more than a decade, Murray Gell-Mann had been a central figure—perhaps *the* central figure—in particle physics. He had elaborated the idea of strangeness to the point where it was crucial to a thorough understanding of the strong and weak forces. With the Eightfold Way he had tamed the growing proliferation of elementary particle species and organized them into a few structured families with simple, regular symmetries. And he had introduced quarks as a possible deeper explanation for these symmetries. In 1969, Gell-Mann finally received the Nobel prize he so richly deserved for all these impressive contributions.

But leadership was about to pass to a new generation of particle physicists. Here Gell-Mann brings to mind another ancient figure. Like Moses had done with the Israelites, he had led particle physicists out of the desert to the gates of Jericho. Other, younger physicists would now take over and lead us into the Promised Land.

CHAPTER 5

THE BIRTH OF A MONSTER

But to try to apply an experimental test would be to show ignorance of the difference between human nature and divine; for god alone has the knowledge and the power that makes him able to blend many constituents into one and to resolve the resulting unity again into its constituents, but no man can or will ever be able to do either.

—Plato, TIMAEUS

IN 1963 a peculiar 2-mile gash appeared in the rolling foothills behind Stanford University. Soon workmen had laid a 4-inch copper tube along the entire length and encased this tube in concrete. After filling the ditch again, they erected a long, narrow structure above it to house the more than 200 klystrons (microwave oscillators) supplying power to the tube. Meanwhile, two tremendous concrete buildings, each as big as an airplane hangar, began to emerge at the eastern extremity. The whole ensemble resembled an enormous freight train chugging relentlessly toward the station.

Hardly noticed amidst the exciting discoveries being made almost every month at Berkeley, Brookhaven, and CERN, the Stanford Linear Accelerator was coming to life. Then the biggest and costliest atom-smasher in the world, it was largely ignored by most particle physicists, who preferred instead to work with the high-energy beams of mesons and baryons available from proton synchrotrons. Few of these scientists understood the potential of the 20 GeV electrons soon to emerge from this odd new machine.

Completed in 1966, this accelerator was the culmination of an evolutionary process that had begun at Stanford over thirty years before. This work occurred in parallel to, but largely separate from, the cyclotron and synchrotron development happening first at nearby Berkeley and then elsewhere around the world. For most of these years, linear accelerators labored in the shadows cast by their circular cousins, which could push heavy subatomic particles to far higher energies.

The pioneering work on linear accelerators had been done by William W. Hansen, a young assistant professor at Stanford who first realized their potential in 1935. With the help of Russell and Sigurd Varian, he developed "klystron oscillators" as a source of microwave power to accelerate electrons. As with so many other scientists, however, Hansen's work was interrupted by World War II. His experience with microwave oscillators was sorely needed in the development of radar devices for the war effort.

After the war, Hansen returned to Stanford and resumed his accelerator work. By 1947, he had built and tested his first full-fledged machine, the Mark I, which was twelve feet long and accelerated electrons to 6 MeV. Encouraged by its success, Hansen envisioned a far larger machine delivering electrons with energies of up to 1000 MeV or 1 GeV. Unfortunately, he died of a chronic lung disease in 1949, well before his dream could be realized.

Hansen's visionary project was carried forth by other Stanford physicists under the general direction of Professor Edward L. Gintzon, who supervised the construction of the Mark II and Mark III accelerators. By early 1952, the Mark III had reached 80 feet in length and delivered electrons with energies up to 200 MeV. Construction of additional accelerator sections was limited by the available funds, some of which had been diverted into the ongoing research program. Eventually, by 1960, the Mark III stretched over 300 feet in length and delivered 1 GeV electrons to the experimental area. After more than a decade, Hansen's dream had finally become reality.

The Mark III made several important contributions to physics, among them the definitive studies of nuclear structure by Robert Hofstadter and his collaborators. Beginning in 1952 and throughout the rest of the decade, these physicists fired electrons at a wide variety of atoms—from hydrogen to lead—and used magnetic spectrometers to look for projectiles scattered at large angles. In experiments that were similar to Rutherford's alpha-particle scattering, they soon established that the proton and neutron were not point particles but instead had a very definite *size,* however small. They were tiny spheres about 10^{-13} cm in diameter—a hundred thousand times smaller than the atom itself.

Hofstadter, who won the 1961 Nobel prize for this revealing work, understood only too well the relationship between the size of his equipment and that of the subatomic features he could examine with it. The bigger the equipment, the smaller the features. So he was always building larger and larger spectrometers—especially when accelerator improvements boosted the Mark III's peak energy. His colleagues remember him proudly displaying a watercolor painting of a colossal spectrometer towering upward into the sky, its topmost parts piercing the clouds above.

When the Stanford physicists began to discuss what would succeed the Mark III, Hofstadter naturally urged them to build the largest possible machine. "Why not build one a mile long?" he suggested in 1954. Skeptical at first, the others soon realized that such a monstrous machine was a practical, if expensive, idea. Dubbed "The Monster" or Project M, a 2-mile accelerator was eventually proposed to the federal government in April 1957. As far as anyone can remember, 2 miles was chosen simply because it was the longest possible straight line one could reasonably draw on the sprawling Stanford campus.

Edward Gintzon, who originally served as the project director, recalled the planning process of 1956–7:

> Many of the committee meetings took place at Rosotti's Beer Parlor on Alpine Road. It is hard now to convey the spirit of

those meetings. There was a sense of excitement, of promise, and of personal pride in participating in a project of great importance. Good beer didn't hurt any!

Negotiating the proposed accelerator through the thickets and ambushes of Congress took five years. Some legislators were completely opposed to the Monster and others wanted the plum for their own state. But the Stanford scientists stuck to their guns. During a congressional hearing on the proposal, Gintzon was asked why he wanted to build such a colossus. "Senator," he snapped, "if I knew the answer to that question, we would not be proposing to build it."

In 1961, Congress finally authorized the project. Stanford University signed a contract with the Atomic Energy Commission in April 1962, with the total estimated cost put at $114 million. This may not seem like much today, when new machines cost billions, but at the time it was a *phenomenal* amount of money—more than had ever been spent by any nation (or group of nations) on an atom smasher. The five-year wait had probably been inevitable.

When groundbreaking for the Stanford Linear Accelerator Center (SLAC) finally occurred in 1962, Hofstadter was not a party to the ceremonies. He had opposed the idea of the Monster's becoming a national laboratory, wanting it instead to be a university facility reserved for the exclusive use of Stanford physicists. But most of the others felt that the U.S. physics community would never support a machine so tightly controlled by Stanford, especially not when they were asking the government to foot a $100 million bill. When it became apparent that the accelerator had to be a national facility open to any reputable scientist with a good idea, Hofstadter bowed out. Thus began a deep division in the Stanford physics community, which has begun to heal only recently.

Late in 1961, after the Monster received final congressional authorization, Gintzon resigned as director of the project to devote full time to being Chairman of the Board at the nearby Varian Corporation. His shoes were ably filled by Wolfgang K. H. Panofsky. "Pief," as he is known throughout the physics community, is a tiny, balding man hardly over five feet tall.

Shuffling through the SLAC corridors in his everyday wardrobe of loose-fitting, rumpled clothes, he has more than once been confused with the janitor. But if he is small in physical stature, Panofsky is a giant in the world of ideas and intellectual accomplishments. Under his leadership SLAC became the Cavendish Labs of our time.

Born in Weimar Germany, Panofsky had come to the United States shortly after Hitler's rise in 1933. Following college years at Princeton and Caltech, he worked on the Manhattan Project in Los Alamos. After the war he joined the University of California at Berkeley, where he helped isolate the neutral pion in 1949—the first subatomic particle to be discovered at an accelerator. Resigning abruptly during a 1951 loyalty oath dispute that decimated the ranks of Berkeley physicists, Panofsky moved across the Bay to Stanford. Aside from his broad physics interests, he is a leading authority on nuclear weaponry, having served as a consultant to the U.S. Arms Control and Disarmament Agency since its inception.

Through his many Washington contacts, Panofsky was in-

Wolfgang Panofsky presenting plans for SLAC to a 1962 meeting of Stanford trustees.

strumental in getting the Eisenhower Administration to support Stanford's accelerator proposal. Unfortunately, this support triggered efforts by the Democrat-controlled Joint Committee on Atomic Energy to stall the project in Congress. Only when Kennedy came to power was it possible to free the bottleneck.

During the final negotiations of the contract between Stanford and the Atomic Energy Commission, Panofsky discovered an obscure but pernicious clause that could, at a later date, limit free publication of the laboratory's results. He objected to it. The AEC negotiators insisted it was only a standard piece of "boilerplate" included in all such contracts. But Panofsky refused, effectively pushing the $114 million back across the bargaining table. Fortunately, the AEC backed down and dropped the clause, opening the way for the freedom of expression so characteristic of SLAC.

If SLAC were the Cavendish Labs of our time, then this intense little man was surely its Lord Rutherford. From the very beginning, Panofsky was intimately involved with every aspect of the accelerator and its experimental facilities. He also advised a group of staff scientists and physicists from MIT and Caltech that designed and built the massive spectrometers in End Station A. These alone cost another $6 million, a figure that rivaled the entire pricetag of a few contemporary accelerators. But despite his heavy commitments, Panofsky was readily available to everybody at SLAC; the door of his office was always open to anyone who had a problem or just wanted to talk physics. The democratic spirit he encouraged brought out the best from everybody there—staff and visiting scientists alike.

To most particle physicists the big new accelerator seemed destined for obscurity, a relatively unimportant machine yielding interesting but second-rate experiments. Few scientists wanted to leave Berkeley, Brookhaven, and CERN, where powerful hadron beams were unlocking the secrets of the strong force, to toil in the listless backwaters of electron scattering. To staff SLAC and build its detectors, therefore, Panofsky drew heavily upon what he termed the "HEPL alumni"—younger physicists he had worked with as Director of Stanford's High Energy Physics Lab. With fond memories of their HEPL (pronounced "hepple") years, they responded eagerly to his call.

HEPL alumni Jerome Friedman and Henry Kendall had left Stanford for faculty positions at MIT, where they did electron scattering experiments using the nearby Cambridge Electron Accelerator. But in 1962 they joined forces with SLAC and Caltech to build the spectrometers for End Station A. SLAC staff under Richard Taylor, another alumnus, built the magnets and support structures for these devices. Friedman and Kendall contributed the particle detectors sitting inside the shielded caves—the retinas for these huge "eyes"—at the back of each spectrometer.

Even though they had built almost all this equipment, there was no absolute guarantee that these young scientists would get to do the important experiments when the accelerator finally came on line. They feared that Robert Hofstadter might try to preempt the collaboration and use this very same equipment to do the experiments himself. A Nobel laureate, he wielded enormous power and influence in the physics community. Earlier there had been a bitter struggle between the two factions over the spectrometer design, with Hofstadter advocating a scaled-up version of the vertical devices he had used so successfully at HEPL. To handle the much higher SLAC energies, however, his colossus would have stretched almost *300 feet* into the sky. "He abandoned that idea," quipped Taylor, "when he realized you'd be able to see it from Berkeley!"

But Hofstadter did not challenge them further, and the SLAC-MIT-Caltech collaboration performed the first major experiment after the machine began full operation in early 1967. This experiment was an extension of electron-proton scattering to the ultra-high energies that had just become available. A team of physicists—which included Panofsky and most of the men who had built the spectrometers—fired electrons with energies as high as 18 GeV into a liquid hydrogen target. Like Rutherford and Hofstadter before them, they detected scattered particles at a variety of angles, using the smaller of the two spectrometers available.

These scientists were looking only for those electrons that scattered "elastically" from the proton nuclei inside the hydrogen atoms. That is, the electron and proton collided like two hard billiard balls, without any of the electron's energy being used to break up the proton. But it received a strong kick (called a "momentum transfer") and bounced away in another direction like the object ball in a game of pool. Only the ricocheting electron was detected in these experiments; from its energy and the angle at which it shot out, scientists knew that the proton was still intact.

This elastic scattering experiment provided no great surprises. The data proved to be a smooth continuation of earlier data measured at lower energies and momentum transfers (smaller kicks) using the Cambridge Electron Accelerator (CEA) at Harvard and the Deutsches Electronen Synchrotron (DESY) in Hamburg, Germany. All these experiments measured a quantity called the "cross section" for elastic electron-proton (e-p) scattering.

The cross section, a concept used throughout high-energy physics, represents the effective "size" of whatever object one is trying to hit. Imagine you are throwing a rock at a window; your chances of hitting it are proportional to its total area, say 10 square feet. This is the *total* cross section it presents to your projectile. Assuming it hits the window at a certain velocity, there is some probability the rock will glance off harmlessly without breaking it, say 30 percent. A high-energy physicist analyzing your experiment would conclude that the cross section for *elastic* rock-window scattering is 3 square feet. The cross section for any particular process is a measure of the probability it will occur, and of the frequency at which it occurs when the experiment is repeated many times.

The high-energy electrons available at SLAC allowed scientists to study features far smaller than a proton diameter. The characteristic de Broglie wavelength of their probe was now far smaller than a proton, enabling them to examine its internal structure in much finer detail than ever before. If the proton were a diffuse snowball of matter, as was commonly surmised in those days, then the cross section should drop sharply when large momentum transfers (hard kicks) were imparted. Actually the cross section was declining even more rapidly than had been anticipated by theory—a fact that puzzled almost everybody.

Compared to other experiments of the sixties, electron-proton scattering had an extremely simple interpretation. Although there were ambiguities due to quantum and relativistic effects, it still offered a strikingly "visual" mental image of the proton—its floor plan, so to speak. This experiment allowed us to "see" how matter was spread out inside protons.

These protons seemed to have even more empty space than any theorist had previously anticipated. At such ultra-high energies, too many electrons seemed to charge straight through undeflected or to glance off at very small angles. Or, as a few theorists were saying at the time, the proton had a lot of trouble sticking together after such a vicious pummeling.

Some of the more fashionable theories of the day, like "Regge poles" and "vector meson dominance," had a difficult time explaining these observations. They could accommodate the data, but only after an unsavory dose of mathematical gimmickry. Physicists generally distrust explanations that are not fundamentally simple and beautiful. These were most definitely *not.*

After completing this first experiment, the original collaboration of spectrometer builders began to break up. The Caltech physicists left for what they thought were greener pastures, and Panofsky returned to the pressing duties of running a national laboratory. The leadership of the SLAC contingent now fell to Richard Taylor, a tall, ferocious bear of a man whose booming, angry voice is often heard echoing through the corridors there. Those who know him well, however, realize that he just likes to challenge people and shake them up—to see if they really believe what they're saying.

Only the MIT and SLAC scientists now remained to take the next step—to begin searching for electrons that collided "inelastically" with protons. In such a collision, the proton fractures. It absorbs more energy from the onrushing electron than it gets during an elastic collision, and the extra energy excites it or creates additional subatomic debris. In the billiard-ball analogy, an inelastic collision between cue and object balls is one in which either ball *shatters* under the impact. At SLAC the electron had to pony up the extra energy needed to smash the proton, and thus rebounded with *less* energy than an electron

Richard Taylor relaxing during the first End Station A experiment as elastic electron-proton scattering events accumulate in the computer display behind him.

leaving an elastic collision. So the way to search for these kinds of smash-ups was to tune the spectrometer magnets to accept recoiling electrons of lower energy.

Two things can happen during inelastic collisions. For one, the proton can jump into a higher energy level—as the hydrogen atom does when its electron absorbs a photon and hops into a higher orbit. In the same way, the "parts" of the proton may stick together after such a clout, but they *do* grab extra energy from the electron and jump briefly into higher levels or "resonances." Such resonances are produced only when the proton absorbs relatively small amounts of energy from the electron. When it takes a really hard bashing, however, all kinds of mayhem can follow. Additional subatomic particles are created, often in abundance. When a hydrogen atom is clobbered very hard by an X ray, by analogy, it disintegrates. In the

MIT-SLAC experiments, such truly violent collisions took place only when the poor proton absorbed *lots* of energy and momentum from the electron. Hence the name "deep inelastic" scattering.

Earlier inelastic scattering experiments at CEA, DESY, and HEPL had concentrated on production of resonances; Panofsky himself had performed the first of these experiments in 1957. As the energy setting of the spectrometer dropped, the measured data appeared as a succession of tall peaks and steep valleys, when plotted versus the scattered electron energy. Each peak meant a specific resonance was appearing; it showed up in these graphs at a precisely determined value of this energy.

Deep inelastic scattering was largely uncharted territory in 1967. Harvard scientists had glanced briefly into the terrain at CEA, but found nothing interesting. The MIT-SLAC experiment itself was designed to examine resonances produced at high electron energies and momentum transfers, with only a passing look at the deep inelastic region. One theorist even thought these measurements a complete waste of time. "Why do you want to measure deep inelastic scattering," he chided David Coward of SLAC, "when CEA has already shown there's nothing there?"

At first the MIT-SLAC physicists encountered the familiar landscape of tall peaks and steep valleys in the raw "on-line"

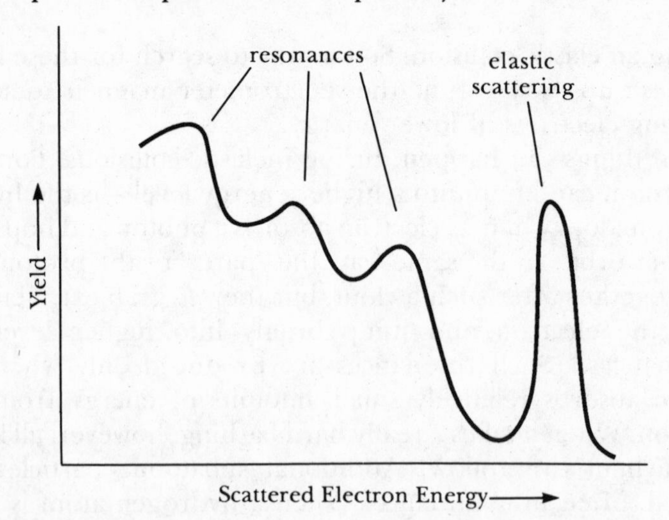

A simplified graph of the elastic peak and resonance bumps encountered in electron-proton scattering at CEA, DESY, and HEPL.

data available during the experiment. But as they continued to lower the spectrometer setting and moved out into the uncharted deep inelastic region, they experienced a surprisingly large and unanticipated number of events "about an order of magnitude" (ten times) larger than expected. At first most of them figured this was "noise," not physics. This was, after all, a region with an extremely high percentage of spurious, background events—both electrons and other subatomic particles—that could be rattling up the spectrometer and masquerading as real events. Better to delay any firm conclusions until after the experiment was over, when a more careful "off-line" analysis could allow better judgment.

But a thorough check of the detectors and a more detailed data analysis failed to eliminate very many of the puzzling events. By early 1968, the MIT and SLAC scientists began to realize that they were witnessing a *real* effect—not some spurious background noise.

Their computer plots of the inelastic data revealed a succession of sharp peaks (corresponding to resonance production) followed by a broad, smooth plateau one could interpret as the disintegration of the proton. A closer inspection showed that these peaks withered away at higher momentum transfers. But the broad plateau in the deep inelastic region refused to disappear. At the highest energies it became the dominant feature of the entire spectrum.

These first deep inelastic measurements were made with the larger, 20 GeV spectrometer set at a few relatively small angles—4, 6, and 8 degrees—to the electron beam. The broad plateau continued to persist, however, when this spectrometer, and later the 8 GeV spectrometer, crept slowly out to larger and larger angles, but the sharp peaks and steep valleys disappeared almost completely.

Here was a result strikingly similar to what Rutherford and his assistants had witnessed in alpha-particle scattering half a century earlier. An unexpectedly large number of particles were caroming off the target at large angles (or large momentum transfers)—far more than expected if the target were just a featureless blob of diffuse matter. At Stanford, the analogous interpretation was not long in coming: the electrons were striking something hard and tiny inside the proton!

There followed several months of hectic activity as the

MIT-SLAC team refined and sifted their data in preparation for a report to the International Conference on High-Energy Physics, to be held that summer in Vienna. After much debate, they finally agreed to present only the cross sections measured at an angle of 6 degrees, where the data were most complete, and to play down the deep inelastic results. Promising careers have been ruined by overzealous presentation of preliminary data later found to be wrong. They were all relatively young scientists, with many possibly productive years ahead of them, and wanted to be careful. And besides, there might well be other acceptable interpretations that were not so earth-shaking.

So the report carried to Vienna late in August 1968 by Richard Taylor and Jerome Friedman was a highly guarded document. Friedman presented this paper before an obscure parallel session, where it attracted little attention. He showed just a few samples of the inelastic data and tried to interpret the resonance behavior they had found. Only two pages and three graphs had anything to reveal about the far more interesting deep inelastic results. And even those bets were hedged with qualifications.

But Panofsky, an old hand at these affairs, delivered the plenary session talk that eventually became the surprise of the Conference. As one of the "rapporteurs," he was responsible for summarizing all the experimental work on electromagnetic interactions that had occurred during the previous year. He had spent a good portion of his time at Vienna cooped up in his hotel room, poring over the many publications submitted for his review. Finally it came time for his summary talk to more than 800 scientists from forty countries assembled in the chandeliered conference hall at the Hofburg, the grand Imperial Palace of the Hapsburgs. Hardly visible behind the lecturn, his head just peering over the top, his voice barely audible, Panofsky dutifully recounted all the year's salient results—coming finally to the MIT-SLAC paper. Turning at long last to the deep inelastic scattering, he announced:

> The qualitatively striking fact is that these cross sections . . . are very large and decrease much more slowly with momentum transfer than the elastic scattering cross sections and the cross sections of the specific resonant states. . . . Therefore theoretical

speculations are focused on the possibility that these data might give evidence of point-like, charged structures within the proton.

Although his audience did not realize it then, this was the dawn of a new age in particle physics. Fitting it was that Panofsky should be the one to usher it in.

Only a few physicists present at Vienna got very excited about Panofsky's revelations. Intrigued by all the new resonance discoveries which had dominated the Conference, most of the others there paid little heed to developments in the unfashionable subject of electron scattering. Raw experimental data need to be interpreted within a common theoretical framework—familiar to the majority of workers in a particular field—before they can have a major impact. Such a common framework was largely absent at Vienna. But stimulated by this MIT-SLAC experiment, it had already begun to take shape back in California.

CHAPTER 6
A POINT OF DEPARTURE

It is also a good rule not to put too much confidence in observational results until they are confirmed by theory.

—Arthur Eddington, NEW PATHWAYS IN SCIENCE

WHILE the Stanford Linear Accelerator was still being built, a young theoretical physicist returned to Palo Alto from the Niels Bohr Institute in Copenhagen. At first glance, James Bjorken, known as "BJ," seemed just another gangly midwesterner who had wandered in off the farm. Mild-mannered to a fault, he was the Clark Kent of SLAC. You might never have guessed that this self-effacing man was one of the most brilliant theorists ever to work there. But when he began talking physics, all your illusions faded. One Stanford physicist compared the experience to "standing in front of a huge fan and getting blown away."

In 1965, Bjorken began working in a branch of particle theory known as "current algebra," which had its origins in Gell-Mann's "veal-and-pheasant" cookery of the early sixties. Instead of working with specific, concrete models of the hadrons, theorists dealt with abstract properties that should hold true universally. Developed largely at Caltech, this approach gave quantum field theorists, then largely out of favor, a safe haven to ply their trade without attracting too much attention. Current algebra had made some obscure predictions called "sum rules" for the collisions of neutrinos with protons or neutrons. But neutrino scattering was then in its infancy, the first neutrino having been detected by Cowan and Reines hardly a decade before, so there was no easy way yet to test these sum rules.

In 1966, just before SLAC was completed, Bjorken discovered a way to carry these sum rules over to electron-proton scattering, where they *could* be tested by accurate experiments. The electron and neutrino are both leptons, and both were thought to be "point" particles—at least to the limits of possible measurement. So similar rules should hold for both kinds of probes. The electron has a cloud of virtual particles swarming about it, complicating the interaction, but quantum electrodynamics allows a correction for this swarm. Thus the electron can be treated *as if* its charge were indeed concentrated at a mathematical point.

When an electron strikes a proton, three things can happen. It can rebound elastically; it can kick the proton into a higher energy state; or it can disintegrate the proton completely. The Feynman diagrams for these processes on p. 138 are all very similar. Feynman had introduced such diagrams, you recall, as symbolic representations of particle encounters that help one calculate the chances of colliding. In each case the electron and proton interact by swapping a virtual photon that carries the electromagnetic force between them; the only real difference is what happens to the proton *afterward.*

Since the work of Hofstadter and Panofsky during the 1950s, experimenters had concentrated mostly on measuring the recoiling electron, not paying too much attention to what the proton was doing. From the electron's energy and angle of recoil, they could tell a lot about what had happened to the proton anyway. In 1966, a year before the first SLAC experiments (described in Chapter 5) were due to begin, Bjorken was merely trying to anticipate what might happen to the proton at the higher SLAC energies, particularly in the unexplored "deep inelastic" region.

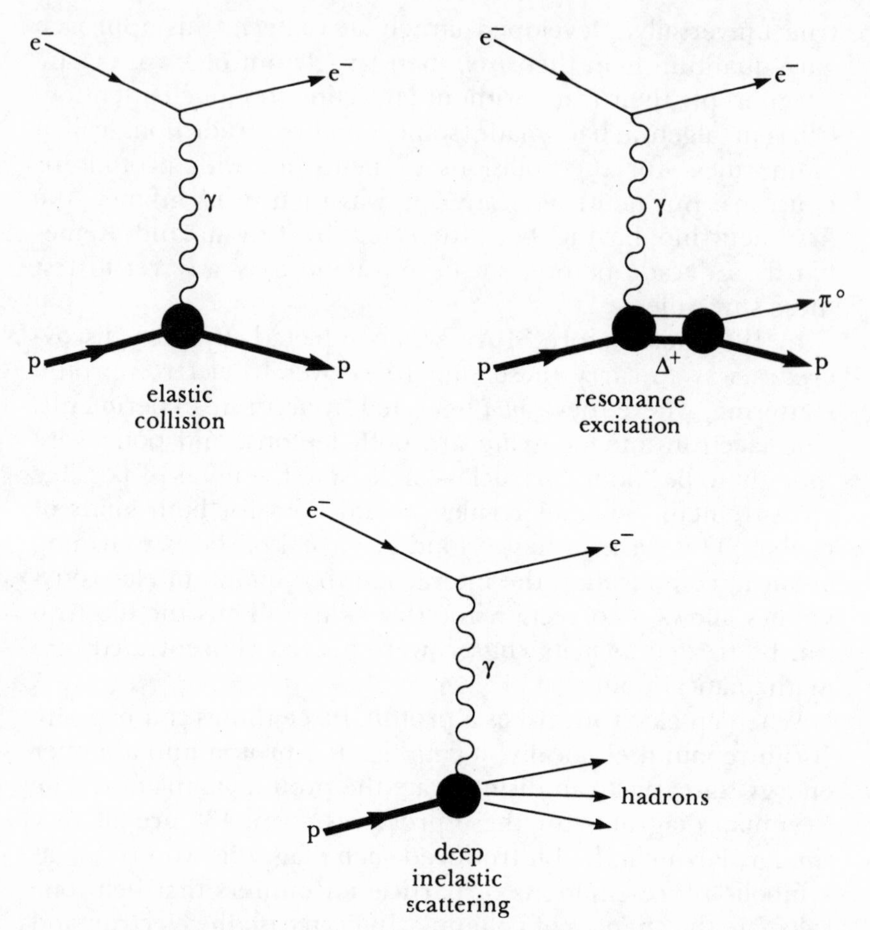

Feynman diagrams for elastic and inelastic collisions of electrons with protons.

If true, Bjorken's results meant that electrons might rebound at large angles far more often than previously thought possible. One had only to imagine that a proton's charge was concentrated at a few points (instead of being spread out evenly, like the snow inside a snowball) for this to happen. An electron could then "meet" a proton only at a few specific points—especially when the exchanged photon's mass-energy was very large. Think back to the basketball player analogy. If two men are passing a light basketball back and forth, they can wander all over the court. But to transfer a cannonball, they'd better meet at a specific, prearranged *point* or forget it.

Bjorken's predictions, however, just did not percolate through to experimenters. Current algebra was so esoteric that its predictions had little impact on day-to-day work. A few of the MIT and SLAC experimenters knew about these sum rules but simply did not understand their significance.

Such a dearth of experimental interest might not have bothered a university-based theorist caught in the throes of abstract mathematics. But an accelerator-based theorist, who wants to anticipate and interpret experimental results, cannot afford to be so blasé. So in early 1967 Bjorken began seeking better ways to communicate. However dimly, he recognized we might be facing a similar choice as had Rutherford and Thomson in the early 1900s: a "nuclear" versus a "raisin-pudding" model of the atom. Only this time it was the nucleon, not the atom, under scrutiny.

In one crude model Bjorken examined the case where nucleons were composed of three light, "pointlike" quarks. If so, he figured, then inelastic electron scattering experiments should reveal a large, very broad peak in the energy spectrum. Based more on "an appeal to history" than on any deductive logic, his intuitive arguments would probably have brought only muffled laughter had they been voiced in the company of other theorists. In May, Bjorken started to write a paper on the subject but soon had a change of heart and shelved it half-finished.

He finally went public in July 1967, at an Italian summer school, and again in September at the International Symposium on Electron and Photon Interactions. Held in the off years between the biannual Rochester Conferences, these increasingly popular gatherings provided a forum for physicists doing experiments with electron or photon beams. The 1967 event was at Stanford, giving participants a chance to see its powerful new accelerator in action. Scheduled to review recent theoretical work on inelastic electron scattering, Bjorken found himself with little to report:

> It is an indication of the state of the subject I am going to talk about that only one paper was contributed to this conference on the theory of inelastic electron scattering. It cannot be stressed too much how ignorant we are of what goes on in this area.

So he devoted most of the talk to his own speculations. "We assume that the nucleon is built out of some kind of point-like constituents, which could be seen if you could really look at it instantaneously in time." He then postulated three light quarks and predicted a broad peak in the inelastic spectrum that should result if electrons caromed off them. Grilled by T. D. Lee about his use of quarks, Bjorken backed off. "I would also like to disassociate myself from this as a test of the quark model," he quavered; "I brought it in mainly as a desperate attempt to interpret the rather striking phenomenon of point-like behavior."

In the very next talk Louis Hand, who had been Panofsky's graduate student at Stanford before becoming a postdoc at Cornell, reviewed the existing experimental data on inelastic electron scattering. He searched for such a broad peak in four of the spectra DESY experimenters had just measured, but could find no evidence for it. All one could see in each graph was a tall, narrow peak corresponding to elastic scattering and a few prominent bumps identified with well-known resonances.

So Bjorken's trial balloons had been shot down rather summarily in the summer of 1967. Because his current algebra was far too esoteric for experimenters, his sum rules failed to gain many supporters among them. And when he turned to quarks to help them visualize current algebra, he ran afoul of the majority of theorists, among whom quarks were then in general disrepute. Typical of these sentiments were the remarks voiced by CERN's Leon Van Hove. "What can we do," he asked Richard Taylor, "to keep BJ from making such a fool of himself?"

In fact, Bjorken was highly ambivalent about these ideas himself. There were four or five different ways to look at inelastic electron scattering, and he had turned to current algebra by a process of elimination as the most reasonable. "The kind of logic I was using was a house of cards," he confessed much later. "It was not at all deductive." Moreover, in 1967 Bjorken was not yet the major theoretical force he soon became. He was just another bright young scientist reluctant to declare that *this* is the way things will turn out.

Along with Bjorken, MIT's Jerome Friedman and Henry Kendall were deeply interested in the inelastic electron scattering experiments. They had done this kind of work at CEA—at lower energies—and viewed the forthcoming SLAC experiment as a natural extension of their ongoing program. So when they visited Stanford in the midsixties, taking a few weeks off from teaching responsibilities at MIT, they usually found time to chat with Bjorken about how best to interpret the data they would soon be collecting.

The resident staff physicists involved in the electron-scattering experiments were organized as SLAC Group A under the leadership of Richard Taylor. These men had the primary responsibility for making all the complex equipment—the two spectrometers, liquid hydrogen target, beam monitors, and minicomputer—work properly. These state-of-the-art devices needed constant attention to detail. For most of the Group A physicists, interpretation of the data could wait until *after* the experiments had been completed.

MIT's Friedman and Kendall are a study in contrast. Both are tall—six feet or better—and both are experimental physicists, but there the similarity ends. A warm, almost grandfatherly scientist, Friedman is constantly concerned about everybody's welfare, on and off the job. Another of Fermi's protegés, he is conversant in both theory and experiment. Round-shouldered, soft-waisted, and bespectacled, his short curly hair gradually graying and thinning, he was the kindly intellectual of the MIT-SLAC collaboration.

Kendall was carved from the same block of New England granite as the Kennedys and Lodges. The scion of one of Boston's wealthiest families and heir to the Kendall Mills fortune, he spends as much time in the mountains as he does on the experimental floor. During the fifties and sixties, this rugged, broad-shouldered scientist established himself as a world-class rock climber, helping to pioneer the new sport. Fond of rubbing shoulders with politicians and government officials, Kendall was frequently called to Washington to give expert testimony. A driving force behind the Union of Concerned Scientists, he helped raise the issue of nuclear reactor safety in the 1970s, long before it had become a major cause. But he always seems cold and aloof—difficult to approach with serious questions.

These two scientists had met in the late 1950s, while working at Stanford as postdocs under Hofstadter. While Friedman understood the theoretical nuances of particle physics better, Kendall was a whiz at electronics. Almost singlehandedly, he designed and built the vast electronic network for End Station A—the sprawling nervous system that extracted and processed signals from the various detectors and devices.

Whenever he visited Palo Alto in the midsixties, Kendall roomed with Bjorken and Herbert DeStaebler, another HEPL alumnus who had joined SLAC Group A at its inception. They shared a cottage in College Terrace, a small peninsula of bungalows jutting into the rambling Stanford campus diagonally opposite SLAC. Weekends often found the three bachelor physicists roped together, scaling the sheer granite cliffs of Yosemite or the Pinnacles further east.

Inelastic electron scattering was a frequent topic of conversation at the College Terrace home. What kind of terrain to expect in this uncharted territory? More peaks and valleys? Often Friedman came by, too, usually bursting with questions about the "radiative corrections" that had to be applied to the measured data to correct for the effects of the virtual particles swarming around an electron.

Because of this swarm, a number of complex processes can occur when an electron and a proton meet, not just the single-photon exchanges pictured on p. 138. What's worse, just before or after colliding the electron can emit other photons that fly off into thin air, spiriting away some of its precious energy. These "radiative" processes occur because the electron is actually a complex object—a "bare" point charge surrounded by a swarm of virtual photons and electron-positron pairs. If its motion stops abruptly, one or more photons can be knocked free. Think of it as a Chevy convertible loaded with carefree teenagers gaily jumping around, from front seat to back and vice versa, as it careens down the highway. If the driver swerves sharply or slams on the brakes to avoid an oncoming truck, one or two unlucky teenagers may go airborne, with disastrous results.

In principle, quantum electrodynamics allows one to make corrections for these radiative processes. In elastic electron-proton scattering, for example, it predicts what percentage of the rebounding electrons escape from under the narrow elastic peak and show up elsewhere in the energy spectrum. So the "radiative correction" to elastic scattering is simply a multiplicative factor that adjusts for this loss. For *inelastic* electron scattering, however, the corrections are more formidable. Here you have extra electrons dropping in as well as the usual escapees. In practice, you must solve complex integral equations to get a correction factor.

So difficult were these corrections that many of the original collaborators had little enthusiasm for the inelastic experiment. It was always "the poor sister" of the others, said Barry Barish of Caltech, who decided to avoid an experiment he regarded as "the dredges" and seek more promising physics at Brookhaven. Friedman and Kendall, he recalled, were the ones who really had the greatest hopes and promoted it the hardest. Having worked on similar experiments before, they had developed computer programs to calculate such corrections. Aided by Bjorken, they finally convinced most of the others that a meaningful measurement was possible.

Thus, an experiment that was to prove pivotal in the history of particle physics began rather inauspiciously—as a difficult measurement at a supposedly "second-rate" accelerator. By the summer of 1967, a full third of the original SLAC-MIT-Caltech collaborators had opted for work elsewhere, leaving inelastic scattering to the remaining group of rather young and unheralded physicists. Even these stalwarts saw their experiment as a mere continuation of earlier measurements, looking mainly for resonances and studying their properties. Few had guessed what they might actually discover.

When the experiment began that fall, the MIT and SLAC physicists used two different frameworks to conceptualize electron scattering. One way, developed mainly by Hand and advocated in his Stanford Conference talk the previous summer, viewed electron-proton collisions as testing the proton's appetite for virtual photons. In the other, which hewed more closely to Bjorken's approach, one took an imaginary pointlike proton as the "ideal," and measured electron-proton scattering to determine a real proton's deviation from this ideal.

Each conceptual framework had its own built-in biases—theoretical and philosophical. Hand's approach was much more compatible with the dominant bootstrap tradition: inelastic electron scattering measured a proton's tendency to gobble up the stray photons boiling off the passing electrons. No "structure" need be imputed to it, nor any "elementary constituents," in order to understand this appetite. By taking pointlike behavior as the ideal, in contrast, the second framework was more compatible with a mental image in which there *were* point charges flying around inside the proton.

In the second approach, physicists could extract two "structure functions," W_1 and W_2, from detailed measurements of electron scattering. These functions contain all the information we could hope to learn from these experiments about proton structure. If the proton were indeed a snowball of evenly distributed matter, as most theorists then anticipated, then W_1 and W_2 should drop rapidly as it was hit harder and harder—that is, as the electron bounced away, at larger angles and with bigger energy losses. But if its charge were somehow concentrated at a few points, as Bjorken had suggested that summer, these functions should decrease far more gradually.

In the first inelastic experiment, beginning in September and continuing through October, the 20 GeV spectrometer collected scattered electrons at relatively small angles. For a survey experiment taking the first tentative steps into uncharted territory, it was wise to start at small angles where confusing backgrounds were low. But even here Friedman, Kendall, and Taylor encountered an abnormally large number of rebounding electrons, often ten times more than expected. The prominent resonance peaks began to wither away rapidly, as anticipated. But a broad plateau extending far into the "deep inelastic" region beyond these peaks seemed unduly persistent, refusing to budge.

Having lived and worked closely with the experimenters, Bjorken got an early look at their raw data—even before the radiative corrections had been applied. In early 1968, while they spent a few hectic months making these corrections and trying to find major blunders, he was busily estimating the second structure function, W_2, and comparing it with his expectations. He could tell, even from these rough data, that it was dropping off fairly slowly, like the number 1 divided by the energy carried by the exchanged virtual photon. This was about the behavior his equations had been predicting: the proton was acting like a point charge.

* * *

Just as the inelastic experiment was beginning to show signs of something interesting, however, Kendall was distracted by events occurring in Vietnam. Shortly after the Tet Offensive of January 1968, he flew to Southeast Asia in the company of other prominent scientists who had been studying the feasibility of erecting an "electronic barrier" across the demilitarized zone. Counting many a high-energy physicist among them, this group had been meeting for six weeks every summer to examine complex technical issues related to national security.

After the press caught wind of Kendall's journey, rumors began to circulate that these scientists were actually in Vietnam to help erect this electronic barrier. By the time he returned to MIT in late February, Boston area radicals were in an uproar over such a "misuse" of science. Ever the stoic, he managed to weather the storm surging about him, but it certainly pried his attention away from the intriguing SLAC results.

Late that March, Kendall returned to SLAC to prepare for another experiment. Bjorken, who had been trying to convince Group A physicists to take his structure functions seriously, strode into Kendall's office one morning with a proposal. If W_2 seemed to drop as the virtual photon energy ν rose, he suggested, why not plot the *product* of these two quantities, $F = \nu W_2$, and see if it remained constant? For the x-axis of the graph, he recommended Kendall use the ratio ν/Q^2—the virtual photon's energy divided by the square of its mass.

Now Kendall loved to plot data, and he was good at it. The old logbooks of these electron scattering experiments are replete with his graphs of this or that functional relationship, all done with a distinctive flair. MIT's version of the data analysis had just been finished, yielding more accurate numbers he could begin comparing with theories. Even though the validity of certain corrections was still in doubt, it couldn't hurt to try.

Kendall wrote a short computer program to calculate F and began to plot the results by hand on two sheets of orange graph paper spliced together. What he found pretty much bore out his former roommate's hunch. Except where there were sharp resonance peaks, the values seemed to lie along a single curve; at any selected position along the x-axis, the measurements gave roughly the same result, within the errors of measurement. No

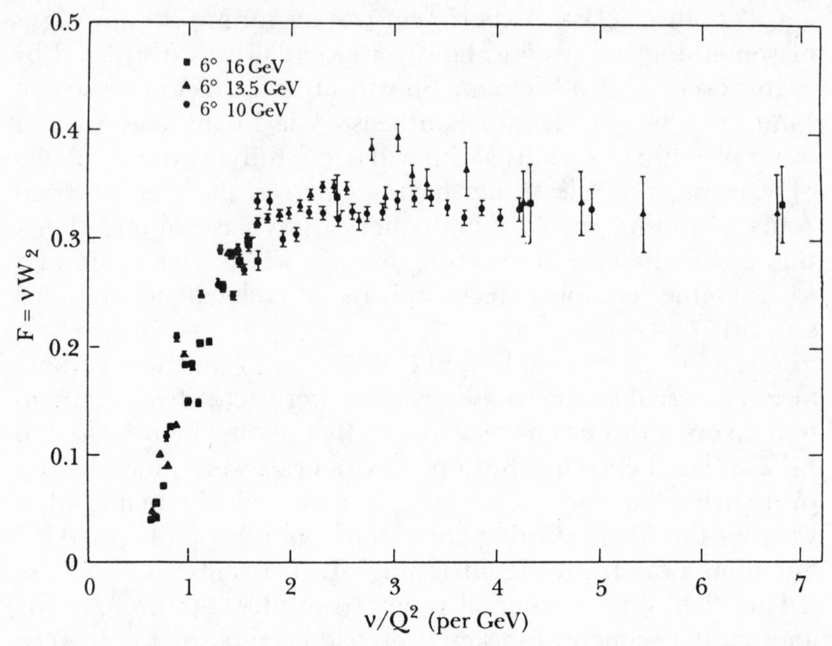

The graph of the proton structure function that Kendall plotted at Bjorken's suggestion.

matter what the electron energy, the values of *F* seemed to cluster about this curve.

In his theoretical studies of "pointlike" scattering, Bjorken had anticipated exactly this kind of behavior, which he dubbed "scaling" in the half-finished paper he had filed away unpublished in mid-1967. Scaling means you get the same answer, independent of the energy scale at which you do the experiment.

But there were other ways to interpret the curious data. Using Hand's conceptual framework for electron-proton scattering, another group of SLAC theorists proposed that the virtual photon traded by the electron and proton became another, *different* particle just before being swallowed up. Quantum mechanics allows such identity changes as long as key properties like charge and spin remain the same before and after the switch. Neutral vector mesons—the rho, omega, and phi—fit the bill perfectly. Thus, the virtual photon became a vector meson that, being itself a hadron, interacted *strongly* with the proton. In effect, the photon occasionally behaved like a hadron; when it did, the proton captured it far more easily. Such a combination should then "dominate" the exchanges between electrons and protons, yielding the excess events then being observed in deep inelastic electron scattering.

During the 1960s, "vector meson dominance" was a popular theory among particle physicists who studied electron and photon collisions. Its origins traced to the late 1950s, when a number of theorists first proposed the spin-1 vector mesons to explain why protons behaved as observed in some of Hofstadter's experiments. By including these meson resonances in the virtual cloud swarming about a proton, you could better understand its short-distance behavior. The theory achieved preeminence under the sponsorship of Chicago's J. J. Sakurai, who had written a long paper in 1960 outlining its major tenets. "Vector dominance," as the theory was called for short, had a ready explanation for just about everything that happened in electron or photon scattering.

Vector dominance was compatible with the bootstrap mentality then permeating high-energy physics. It agreed with the bootstrappers that mesons and baryons were fundamental units of existence, but it went one step further to argue that vector mesons had a special role to play in any nuclear democracy. All hadrons were still equal, but some were more equal than others.

So advocates of pointlike substructure faced a formidable opponent in their interpretation of deep inelastic scattering. Even within the MIT-SLAC collaboration, rival factions began to square off along these lines during the late spring and early summer of 1968. Group A's analysis had just been completed, verifying the earlier MIT approach and bolstering everybody's confidence that the enhancements observed were indeed *real*, not just a spurious effect of the radiative corrections. At stake now was how to present the data and what to emphasize at the Vienna Conference scheduled for late August.

As usual, Kendall parked himself to one extreme of the opinion spectrum; he wanted to emphasize pointlike structure and the graph he had drawn up at Bjorken's suggestion. At the other extreme stood Eliott Bloom, an ambitious young Group A postdoc who was enamored of vector dominance. Kendall and Bloom often found themselves at loggerheads over many issues. Friedman and Taylor ended up in the middle of the feuding pack, trying to act as conciliators.

The arguments over what to report got rather heated. According to Kendall, he and Friedman agreed they'd "absolutely go to the mat and have a big fight with Group A" if necessary to make sure his graph was included. Bjorken's prediction, although based on abstract theory none of them truly understood, seemed to be borne out by the data. Taylor's version of the debate is somewhat different. Bloom was indeed promoting vector dominance, he agreed. "But Eliott was just a postdoc," and nobody paid him much attention. Kendall had a tendency to perceive things in terms of "polar opposites," Taylor felt, and then to gird himself for a fight. He saw himself and Friedman as the mediators who simply wanted to "let the facts speak for themselves."

The actual details of the debate will probably never be known for sure, given the unreliability of memory and the self-serving temptation to reinterpret past actions in light of present knowledge. This is unfortunate, because it was an important confrontation between two radically different views about the ultimate nature of matter. Whatever the case, pointlike substructure clearly won the day. In the paper Friedman presented at Vienna, one finds the graph of F as well as a comparison of the data with sum rules derived from current algebra and the quark model. Nowhere is there any mention of vector dominance.

During the second week of August 1968, Richard Feynman came to SLAC for one of his frequent visits. (He often visited his sister, who lived nearby, so it was natural for him to drop by the new atom smasher and "snoop around," as he put it.) A favorite rendezvous was an oak-studded knoll behind the cafeteria, offering broad panoramas of San Francisco Bay with Stanford's Hoover Tower standing in the foreground and Mount Diablo in the far distance. On warm summer afternoons he often lounged at a picnic table there, happily talking physics or telling humorous anecdotes of his Los Alamos days to a knot of eager listeners. He was one of the most brilliant and articulate men in all science—and great fun to listen to.

Feynman had been instrumental in providing important tools physicists could use to calculate effects of the weak and electromagnetic forces. But the strong force had stumped him. By the early sixties, he had pretty much given up the quest and turned instead to studying gravity.

Having recently taken up the challenge again, he was examining the collisions of two hadrons—a pion and a proton, say—at high energies. When the collision energy was large, he believed, the dynamic details of the strong force between them might become very simple, and hence amenable to calculation. According to Einstein's Theory of Relativity, which says that fast-moving objects are compressed along their direction of motion, the two high-energy hadrons would look like two flattened pancakes flapping together. To illustrate their collision, Feynman lifted his ever-active hands in front of him, thumbs up and palms inward, and clapped them loudly together.

While working at Santa Barbara that summer, he added a new element to this picture. On June 19, as recorded in his notes, Feynman began thinking of each hadron as a collection of smaller parts or entities which, for lack of any better name, he dubbed simply "partons." He thought of them as arbitrary, "bare," ideal particles—the "quanta" of some underlying field. The exact number of partons in any hadron was indeterminate, because they could appear and disappear willy-nilly.

Then the high-energy collision of two hadrons might be viewed as two thin pancakes of these partons flapping swiftly together. Most of the partons in one pancake would pass right through the other unmolested, but two of them—one in each pancake—would occasionally meet. In this picture the chances of the two hadrons colliding was just the sum of the chances of any two of their partons colliding.

For almost another month, Feynman worked out consequences of his "parton model," looking for evidence of its validity in the scanty data then available on high-energy pion-proton and proton-proton collisions. By mid-July, he felt he was on the right track.

When Feynman stopped by at Stanford that August, the MIT-SLAC collaboration was just beginning a second inelastic experiment. The debates over the Vienna paper had subsided, the final version was typed, the graphs were being inked, and it was time to get back to taking data. Friedman had flown out for the occasion, and Kendall was in town, too.

Bjorken was away mountain-climbing that day, but Emmanuel Paschos, his new postdoc, took Feynman aside and told him about the curious results emanating from End Station A. "The experimenters have this puzzling graph of the structure function BJ asked them to make," he confided. "He claims the data should 'scale,' and it does, but nobody seems to understand what this 'scaling' means." Perhaps Feynman could help.

The two of them walked over to the office of Yung-Su Tsai, a theorist who had been helping Group A with its radiative corrections, to look at the data. Taking some computer plots out of a manila folder he kept on his filing cabinet, Tsai showed them the spectra measured at each energy, full of resonance peaks and valleys, getting only a few idle nods. But when he pulled out Kendall's graph of *F* and explained a few details, Feynman had an epiphany. The great Caltech theorist fell suddenly to his knees, clasping his hands prayerfully over his head. "All my life," he exclaimed, "I've looked for an experiment like this, one that can test a field theory of the strong force!"

For the rest of the day Feynman could not take his mind off that graph. He knew that it somehow held the key to his parton model, but he did not yet see exactly *how*. He talked about it repeatedly as Paschos drove him to the Flamingo Motel, a favorite resting place of SLAC visitors, located on a seedy stretch of busy El Camino Real. That evening, Feynman stopped at a nearby bar, one of those bistros distinguished mainly by its scantily clad waitresses. He often visited these joints when he was in a creative mood, lounging at one of the back tables with paper and pen, jotting down ideas as they came to him or just drawing pictures of the girls.

Feynman returned to his motel room still agitated because he had not yet found the solution. But just as he was falling asleep, it suddenly hit him. The graph of *F* was in some way the *momentum distribution* of his partons! Just as the wave function ψ told how a particle was spread out in space, this innocuous-looking curve showed how the partons might be distributed in momentum. Just as the protons and neutrons buzzing around in a nucleus have various momenta called their Fermi motion, so too these partons should have a momentum spread he could figure out by deciphering this graph. He jumped out of bed, grabbed some paper, and worked out the details in an hour.

The next morning, Friedman and Kendall were walking by SLAC's Central Lab when Feynman came running up excitedly. "I've really got something to show youse guys!" he called out in his best Brooklynese; "I figgered it all out in my motel room last night!" They took off immediately in search of a blackboard. A small crowd gathered as he worked out the equations, gesticulating all the while. Friedman offered yet another piece of information supporting the parton idea, and Feynman *really* got revved up. "That's it! That's it!"

Picture him there in front of the blackboard waving one flattened palm through the air to depict the proton chock full of partons, the other hand curled into a fist and smacking the open palm loudly to illustrate an electron caroming off. He could really work the crowd. "Everybody seemed pretty excited at the time," Feynman recalled; "I certainly was!"

When he read Bjorken's papers later that fall, Feynman saw that Bjorken had already had almost all of these ideas. But Bjorken had expressed them in the abstract language of current algebra, instead of with his own far simpler picture in which protons are built of little things flying around inside. The parton idea drew upon a long, hallowed "atomic" tradition familiar to all physicists. Twice already had complex objects been resolved into simpler motes during this century. Atoms are composed of electrons and nuclei, which are themselves made of protons and neutrons. What Feynman added to Bjorken's work, besides his simple picture, was his realization that the graph of F was actually a momentum distribution. When one plotted this function versus his own parameter $x = Q^2/2M\nu$ (essentially the inverse of Bjorken's ν/Q^2), things got clearer still. The height of this curve at any particular value of x gave the chances of finding a parton carrying that fraction of the proton's momentum.

Feynman's happy encounter with the MIT-SLAC data was a complete accident. He knew little about this new accelerator and had just been snooping around, trying to find out more about the place. But he realized almost instantaneously that electron-proton scattering was a powerful way to test his parton ideas. An electron was a well-known point probe that could be used to study a complex object whose structure and interactions one did not really understand. Proton-proton collisions, by

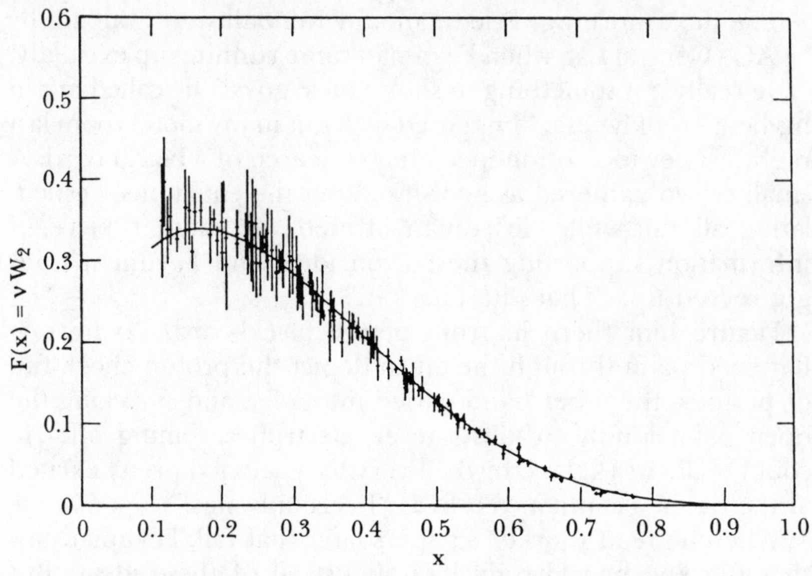

The proton structure function plotted versus Feynman's parameter x.

comparison, were "like smashing two pocket watches together to see how they are put together." SLAC, he realized, was an "incredible new tool" that gave us a far better way to see inside protons.

Feynman returned to SLAC that October to give another presentation of his ideas during a scheduled talk in the auditorium. After that, partons swept through SLAC "like a brushfire," in the memories of most physicists there. Experimenters could finally talk to theorists in a language both understood. Partons were always in the discussion—in offices, seminar rooms, corridors, and on the experimental floor. Bjorken and Paschos soon worked up a detailed parton model based on quarks, while another group of theorists took partons to be "bare" pions and protons, without their virtual clouds. Feynman had again supplied a language, and a strikingly simple mental image, to describe what might be going on in a remote and tiny realm. "I was always delighted," he reminisced, "when something esoteric could be made to look so simple."

It is intriguing to return, with the aid of hindsight, and see how much of the parton model was already implicit in Bjorken's papers before August 1968. Feynman agreed that

Richard Feynman lecturing at SLAC in October 1968. This was his first formal presentation of the parton model, the talk most SLAC physicists recall igniting their interest in partons.

Bjorken had had most of the essential ideas well before his SLAC visit that month; Feynman's only new contribution was his interpretation of the *F* curve as the spread in parton momentum. Why, then, was there such a sweeping reaction there—akin to a religious conversion—to his parton model? The episode, I believe, has much to reveal about the nature of communication within a scientific community.

Bjorken had written originally in the idiom of current algebra—then an esoteric branch of particle theory few experimenters understood or even considered important. Next, he attempted to *communicate* his ideas better by appealing to a specific model with three light, pointlike quarks making up a proton—but his model ran afoul of the accepted dogma of the day. Quarks were themselves out of favor by 1967; who needed *pointlike* quarks? Bjorken himself was not overly enthusiastic about this model, later calling it "the most trivial, simple representation of local current algebra that you could think of." The strong force had been dogging theorists' efforts for two decades; everyone expected the solution to be complex—and the proton to be a very complicated object. Bjorken's efforts to communicate fell largely on deaf ears.

There is another, and perhaps more important, reason for

this failure—one that has much to do with the personalities involved. Physicists like to think they will recognize a good idea, be it voiced by a Nobel laureate or by a graduate student, but they are kidding themselves. It took the enthusiasm of an Einstein, for example, to call attention to the matter waves of de Broglie. A "guru system" operates in particle physics, and Feynman, like it or not, is one of the chief gurus. When he called attention to an idea, physicists listened; he had been right far too often. What's more, he promoted his parton idea quite vigorously, while Bjorken had introduced a similar idea as but one tentative interpretation. The combination of Feynman's vast intellectual weight and his own innate salesmanship finally made physicists sit up and take notice.

The idea of pointlike building blocks may seem obvious today, but it was a radical departure from mainstream thinking in the late 1960s. Almost everybody regarded the proton as a "smear," a "blob," Feynman recalled—like Thomson's raisin-pudding model of the atom. "Nobody thought there might be hard things inside." Even quark partisans thought of quarks as complex, extended objects. "It should be quite clear," declared Giacomo Morpurgo in a major Vienna Conference address a day before Panofsky's, "that like the nucleons in nuclei, the quarks . . . are themselves complicated objects, as are all strongly interacting objects."

But these MIT-SLAC experiments began to suggest instead that the proton's charge was concentrated into a few pinpoints, that could easily be picked off without disturbing the others. Such an observation was tantamount to saying that the strong force was *not* as strong or as complicated as everybody had imagined. It took courage to come out in favor of pointlike constituents. No doubt Feynman's early support helped enormously.

At the Vienna Conference in late August, Friedman refrained from using the word "pointlike" to describe the deep inelastic phenomena when he delivered the MIT-SLAC paper in a parallel session. After the debates of the previous months, the collaborators had finally reached agreement, and he didn't want to subvert that grudging consensus—even though he knew about Feynman's parton model by then. But he and

Taylor coached Panofsky to voice the "pointlike" interpretation in his plenary session talk. This was the appropriate forum to compare theory with experiment, and Panofsky carried more intellectual weight than Friedman.

Still, Panofsky's talk had little impact on the conferees. Many blamed the public address system and the terrible acoustics of the conference hall. Feynman's partons were never officially mentioned at Vienna, although the word was probably used in the corridors. The idea was just too new and radical. Only a few physicists left Austria realizing that something really new and important was happening at SLAC.

By the end of 1968, however, SLAC was ablaze in partons, and small brushfires were breaking out elsewhere, too. The match had been struck by an unheralded group of younger physicists, aided at the last minute by one of the leading theorists of our day. Based on Bjorken's insights, which developed an important conceptual framework for their observations, MIT and SLAC experimenters had begun to peel back the next layer of the cosmic onion. But it would be several more years before the entire physics community became convinced of this fact.

CHAPTER 7

A MATTER OF SCALE

As experimental techniques have grown from the top of a laboratory bench to the large accelerators of today, the basic components have changed vastly in scale but only little in basic function. More important, the motivation of those engaged in this type of experimentation has hardly changed at all.

—Wolfgang Panofsky, Contemporary Physics

Immediately after Vienna, Friedman, Kendall, and Taylor returned to SLAC for further measurements. It was an old confidence game among high-energy physicists: first get approved for a survey experiment, next find some odd results that need further study, then go back and request "more beam." That July, Taylor had appeared before the Program Advisory Committee, data in hand. Panofsky gladly approved another experiment, this time at larger scattering angles.

In August the collaboration had employed the 20 GeV spectrometer at an angle of 10 degrees; now it turned to the smaller and more versatile 8 GeV device. The bigger machine buckled slightly whenever moved, and valuable time was lost while a team of surveyors made sure all its magnets and detectors were lined up correctly. But the smaller device was far more rigid. In a matter of minutes it could pivot around the target on circular rails to begin measuring at another angle. Besides, far fewer electrons bounced off at these large angles, and this colossus had a wider opening to collect them. It was the obvious choice when the experiment edged out to angles of 18, 26, and 34 degrees.

An uncannily similar sequence of events had occurred slightly more than half a century earlier, when Rutherford's "collaboration" discovered the atomic nucleus. After Ernest Marsden noticed a puzzling excess of alpha particles ricocheting at large angles, Rutherford figured they might be hitting something tiny and sent him back to make further tests. Working together with Hans Geiger, Marsden then made a series of more detailed measurements at a dozen different angles ranging from 15 to 150 degrees—forward *and* backward scattering. The detailed *pattern* they finally reported in 1913 was what really established the existence of the atomic nucleus.

The differences between the Rutherford and the MIT-SLAC experiments are largely differences of scale. Whereas Geiger and Marsden pivoted a tiny phosphorescent screen and microscope about a target, Friedman, Kendall, Taylor, and the others pivoted two enormous, complex "eyes" each weighing hundreds of tons and containing many sensitive electronic detectors. Whereas only a closely knit group of three physicists had been involved in the Manchester experiment, more than twenty characters of many different persuasions worked directly on the MIT-SLAC experiments—and many more became involved in their interpretation. And whereas Rutherford's small team perceived features a few billionths of a centimeter across, the big MIT-SLAC collaboration could "see" things almost a thousand times smaller. Pattern recognition at a much larger scale allowed human perception to comprehend a far tinier realm.

By moving around an object like this and "looking" at it from several different perspectives, you can learn more about it than by watching just from a single direction. You know more about the characteristics of a chair, for example, after walking around it and watching the light bounce off at various angles. This is what Geiger and Marsden had done with the atomic nucleus. In 1968, physicists began "walking around" the proton to sharpen their image of its internal structure.

The small-angle measurements made earlier with the 20 GeV

spectrometer had revealed a lot about one structure function, W_2, but next to nothing about the other. To get a better handle on W_1, physicists had to measure the rate of electron scattering at bigger angles. Once they knew *both* functions, they would have a more complete picture of the proton.

By comparing the scattering at different angles, these physicists hoped to resolve one of the principal ambiguities remaining after the earlier experiment. To extract W_2 from the small-angle data, they had to *guess* the relative importance of the two structure functions. Such a guess fixed the value of a key ratio known as R (which we will call L/T). Its true value was completely unknown in 1968, so Kendall had just used the two possible extremes—zero and infinity—in analyzing the small-angle data. Consequently, his graph of $F = \nu W_2$ shown at Vienna actually contained two curves corresponding to these two different guesses. The lack of any solid information about L/T had been a major reason why Friedman had been so equivocal at the Conference. A principal aim of the fall experiments was to *measure* this ratio and clear up this ambiguity.

But it soon appeared that L/T was more than just another idle ratio—that it was interesting all by itself. In November, Curtis Callan and David Gross, theorists at Harvard, circulated a short preprint suggesting that this ratio might give information about the *spin* of whatever it was inside the proton. Like most of Bjorken's earlier papers, this one was expressed in the idiom of current algebra, but it made fairly unambiguous predictions: if the spin was 0, then L/T should be large, possibly infinite; if the spin was ½, then it would be small, possibly zero. Though difficult, the measurement of L/T now assumed added importance.

After Feynman's seminar that October, SLAC blazed with enthusiasm for partons. He provided a remarkably simple, workable framework for thinking about proton structure, one that could be used in many different ways and cried out for further elaboration. Parton advocates there soon broke into two camps: those who took the partons to be bare, pointlike versions of the observable mesons and baryons, and those who imagined them to be the long-sought quarks.

Feynman's genius lay not so much in the content of his parton idea but in the viewpoint he took. He looked at electron scattering not from the familiar perspective of a stationary proton target but from the viewpoint of an onrushing electron probe. Einstein had taken a similar conceptual leap in the early 1900s, when he tried to imagine himself climbing aboard a light ray to view things from its own perspective; what he saw eventually led him to the theory of relativity.

"Riding" a speeding electron beam at relativistic velocities (99.999 . . . percent the speed of light), Feynman could "see" the uprushing protons becoming flattened into thin pancakes by relativistic contractions along his flight path. What's more, the clocks of moving objects slow down; so the virtual partons flitting in and out of existence inside these fast-moving pancakes survived much longer, when seen from the electron's point of view. The fleet-footed electron has plenty of opportunity to single out a drowsy parton and blast it away without molesting any others.

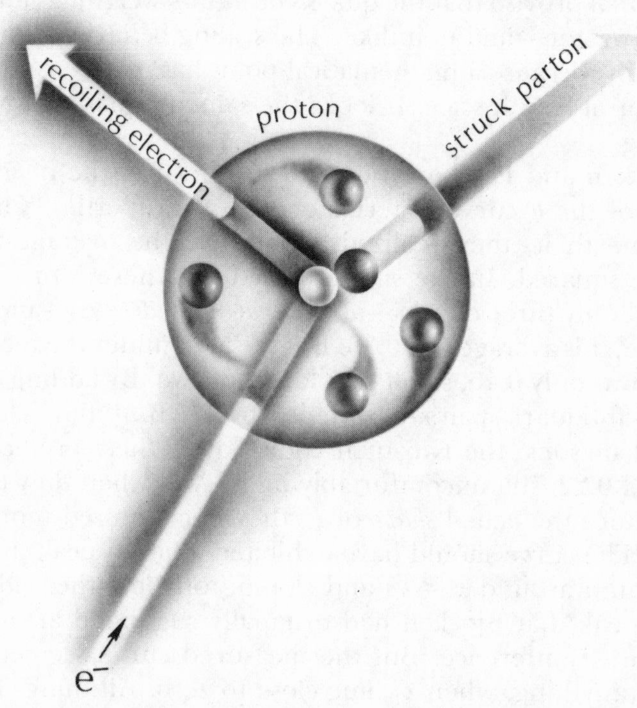

An electron-proton collision, as seen by an observer riding with the electron.

A group of physicists working with Sidney Drell, head of SLAC's theory division, began to elaborate Feynman's ideas in late 1968. A balding, middle-aged theorist with a brusque and businesslike manner, Drell had worked at Stanford since the early 1950s and convinced Bjorken to come there as a graduate student. Together they had written two textbooks on quantum field theory that are widely used in graduate physics courses. Drell's group now picked up where Feynman left off, taking partons to be the bare protons and pions of conventional field theory. They soon discovered how to reproduce Bjorken's "scaling," as the behavior $\nu W_2 = F(x)$ first seen in Kendall's graph was becoming known, by putting constraints on the internal motion of their partons. They also predicted that L/T was small, even zero.

While Drell and company were working out these details, Bjorken teamed up with Emmanuel Paschos to follow an even more venturesome path. They took these partons to be the fractionally charged quarks of Gell-Mann and Zweig, with the additional proviso that the quarks be light—weighing much less than a proton—and pointlike. The scaling behavior then arose naturally, because a mathematical point has no size, and so has no natural length scale. It looks the same no matter how deep you dig.

Bjorken and Paschos also tried to interpret the height and shape of the F curve in terms of parton properties. The area underneath it, they realized, should be the average parton charge squared. In the simplest picture, where a proton was built of only three quarks—$u(+2/3)$, $u(+2/3)$, $d(-1/3)$—and nothing else, this average was 0.33. But the area under the measured curve was only 0.16, small by a factor of two. By adding several quark-antiquark pairs to mimic the surrounding cloud of virtual mesons, the two men could lower their prediction to around 0.22, still uncomfortably high. And when they tried to reproduce the actual *shape* of F, they encountered more difficulty. The curve should have exhibited a broad peak, hitting a maximum around $x = 1/3$ and sloping off on either side—the same peak that Bjorken had originally suggested at the 1967 Stanford Conference. But the measured curve seemed to go flat, if anything, when x came close to zero. All things considered, however, their quark-parton model compared pretty well with the data.

Good theories do more than just fit the available data, however. Should anybody ever measure electron-neutron scattering, claimed Bjorken and Paschos, it would be *smaller* than the proton data—because any hypothetical quark-partons inside a neutron had to have smaller charges, on the average. And they insisted that *L/T* be small or zero, too, because quarks had to have spin ½. In the best traditions of logical positivism, they had made unambiguous predictions and set themselves up for a fall.

On February 3, 1969, Bjorken gave a talk at the midwinter meeting of the American Physical Society in New York. A month later he wrote it up in an unpublished paper with the unassuming title, "Theoretical Ideas on High-Energy Inelastic Electron-Proton Scattering." A far-ranging document, it provides a rare snapshot of a scientific subfield absolutely bursting with energy and about to explode into broader recognition. Twelve of his nineteen references were unpublished preprints or "private communications." The cause of all the recent activity was the single piece of experimental data he showed at New York: the MIT-SLAC graph of *F*.

"The data on inelastic electron-proton scattering represents a study of the proton under conditions of extreme violence," Bjorken began. In a few daring brushstrokes he then described most of the attempts to understand this curious graph, paying particular attention to the parton model as developed by Feynman and elaborated at SLAC. Not content merely to fit the existing data, he compared the various theoretical ideas about *L/T* and some other quantities that can be measured in electron scattering experiments then underway. Then he made educated guesses about related processes that read like brief recipes for much of the important research of the next five years.

By early 1969, partons finally began to percolate beyond their West Coast birthplace. Most of the underground literature Bjorken referenced in his New York APS talk was published that spring. The rest of the physics community began to hear the new word coined by Feynman the previous year.

Meanwhile, theorists working in the bootstrap tradition were quietly mounting a counteroffensive. They struck first at a weakness in the parton lines—the inability to reproduce the exact shape of the *F* curve. Proton constituents should have produced a broad, clear-cut *peak* in this curve. There was nothing of the sort to be found, only a plateau extending to lower values of *x*.

Bootstrap theorists preferred to view electron scattering as determining the proton's tendency to capture or deflect the virtual photon peeling off the recoiling electron. The proton can easily swallow a low-energy photon and use the available energy to jump up into a resonant state. But with high-energy photons like those coming from SLAC electrons, altogether different things might happen. In January of 1969, theorists at Princeton and Israel's Weizmann Institute predicted that the so-called "diffractive" collisions—whereby a photon and its progeny merely glance off the witless proton—should begin to dominate at high energies. Such diffractive encounters led naturally to the flat plateau observed in the *F* curve, without any need for partons.

J. J. Sakurai—the apostle of vector-meson dominance—had his own rival interpretation of the MIT-SLAC experiment. His entire case rested on photons acting like vector mesons whenever they got near a proton. The large enhancements seen at SLAC were compatible with this idea, he announced in a paper circulated in February, if *L/T* were *large* and growing larger as the virtual photon mass increased. The MIT-SLAC physicists might be seeing excess electrons scattered forward—at small angles—but things would return to normal as the 8 GeV spectrometer rolled out to larger angles. "Contrary to widespread beliefs," he wrote, "the observed large cross section in the deep inelastic region does not necessarily force us to the view that we are seeing pointlike structure ('partons', etc.) within the proton." With a little patchwork here and there, vector dominance seemed in fine shape.

Sakurai's paper was finally published in May, just as the SLAC experimenters were beginning to extract *L/T* from the full body of inelastic data measured at both large and small angles. Here was a classic test of competing theories. Parton advocates—especially those who held partons to be spin-½ quarks—needed *L/T* to be small, even zero. By contrast, Sakurai needed a very large ratio, from 1 to 10, for his vector dominance model to explain the large enhancements seen at SLAC.

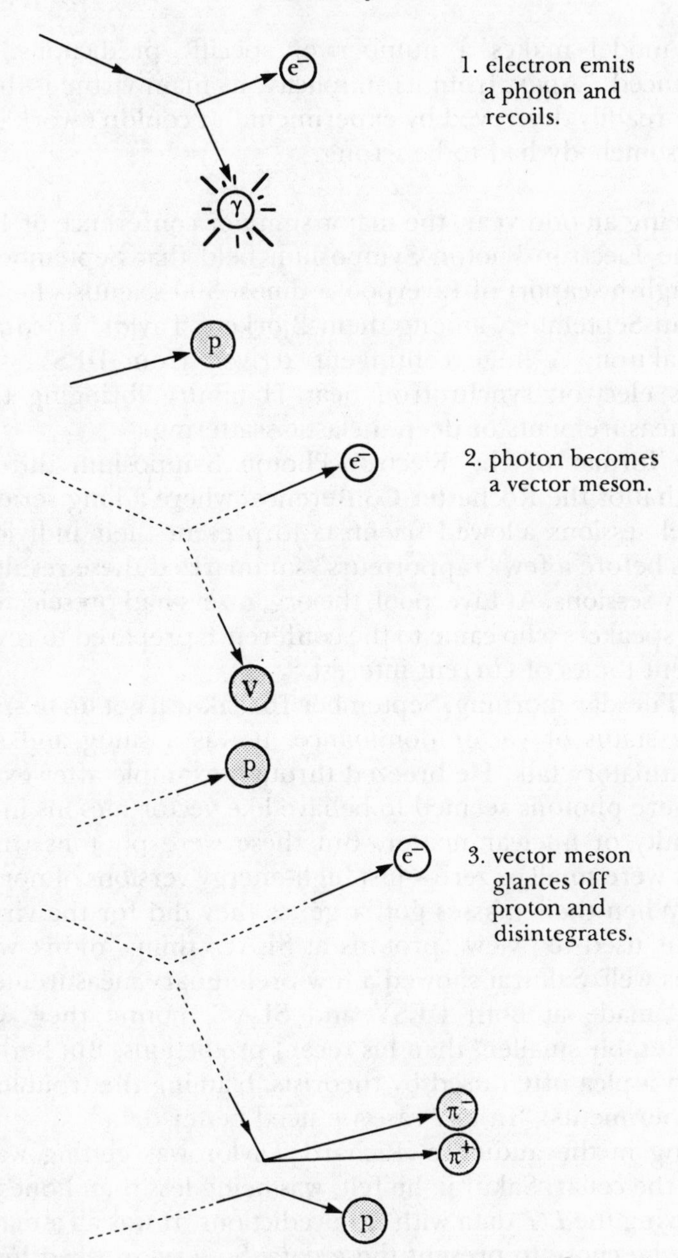

A "diffractive" electron-proton collision. Here the electron senses only the outer reaches of its target, not the inner structural details.

"Our model makes a number of specific predictions," he announced. "Apart from its simplicity, its main virtue is that it can be readily destroyed by experiment." It couldn't work both ways; somebody had to be wrong.

It being an odd year, the major summer conference of 1969 was the Electron-Photon Symposium held that September at the English seaport of Liverpool. Almost 300 scientists headed there in September, among them Bjorken, Taylor, Friedman, and Sakurai. A huge contingent arrived from DESY, Germany's electron synchrotron near Hamburg, bringing their own measurements of deep inelastic scattering.

The format of the Electron-Photon Symposium differed from that of the Rochester Conference, where a long series of parallel sessions allowed scientists to present their individual papers before a few "rapporteurs" summarized these results in plenary sessions. At Liverpool, the organizers had preselected a dozen speakers who came to the conference prepared to review different topics of current interest.

On Tuesday morning, September 16, Sakurai got up to speak on the status of vector dominance. It was a smug and self-congratulatory talk. He breezed through example after example where photons seemed to behave like vector mesons in the proximity of nuclear matter. But these were photons whose masses were small or zero—just high-energy versions of normal light. When their masses got large, as they did for the virtual photons used to "view" protons at SLAC, things didn't work quite as well. Sakurai showed a few preliminary measurements of *L/T* made at both DESY and SLAC, noting they were "considerably smaller" than his recent predictions. But here he copped a plea often used by theorists, blaming the trouble on the experiments: "In any case we need better data."

Sitting in the audience, Richard Taylor was getting warm under the collar. Sakurai, he felt, was being less than honest in comparing the *L/T* data with his predictions. It was all a matter of how he chose to present these data. So Taylor asked him a few penetrating questions at the end of the talk, but Sakurai dodged them skillfully.

Thursday afternoon it was Taylor's turn. He had been added to the schedule at the last minute to give a special talk on the deep inelastic measurements at SLAC, a topic of major interest lately. He began with a detailed discussion of the measurements made with the 20 GeV spectrometer, noting that Bjorken's scaling hypothesis was still in good shape—as long as you took *L/T* to be small. Then he turned to the recent evaluations of this ratio, made using data taken with *both* spectrometers. These were preliminary figures, so Taylor warned his audience against taking the actual values too seriously. But it was evident that the ratio was indeed small, as predicted by the parton camp: the values all seemed to cluster between 0 and ½. With the spectrometers set at larger angles, that is, physicists *continued* to see an excess of scattered electrons.

But Taylor wasn't through with Sakurai quite yet. To the top edge of his transparency showing the *L/T* data, he had scotch-taped several more transparencies showing Sakurai's actual predictions in the same general region. This had been necessary because the data all fell in the range from 0 to 1, while Sakurai's predicted values ranged from 1 to 10. Besides, it helped Taylor dramatize the enormous differences between dream and reality. He gradually unfolded his unique graph, remembered by the physicists present as his "Playboy fold-out," pulling it across the overhead projector. Taylor could be a hard man if you got his dander up. As he unfolded the graph, the auditorium rocked with laughter—from everybody but poor Sakurai, who sat helplessly watching his beloved theory be destroyed by experiment. Utterly destroyed.

Taylor had struck a mortal blow to vector dominance, a blow from which it never recovered. Following Liverpool, Sakurai produced yet another version of his theory, called "generalized vector dominance," that was compatible with the new *L/T* data. But few were listening anymore. Like the Bohr atom in the 1920s, his was a patchwork theory that was rapidly falling apart. The dramatic shoot-out at Liverpool soured relations between Sakurai and the MIT-SLAC collaboration. Before that summer, Friedman recalled, "Sakurai had always been very cordial," but afterward he seemed cold and distant whenever they met.

Sakurai was in some ways a tragic figure of particle physics. He had glimpsed the mountaintop and spent nearly a decade elaborating a beautiful theory that explained a wide range of phenomena about photon-nucleon interactions—as long as the photons were not too massive and just glanced off the outermost reaches. But he became too emotionally involved with his beloved vector dominance and could not let go when high-energy electrons began to probe deep within protons, where his theory could not follow.

Parton theories, by contrast, emerged from Liverpool relatively unscathed, even strengthened by comparison with the most recent MIT-SLAC experiments. A small value of *L/T* was in complete accord with parton thinking: if there were point-like, spin-½ fragments inside protons, then plenty of electrons should ricochet at these larger angles, too. What's more, the fundamental ambiguity in the *F* curve had finally been resolved by evaluating this ratio. The MIT and SLAC physicists now had much more confidence in their results. Virtually all the measured values still clustered about a single universal curve, when plotted versus *x,* as Bjorken had predicted. His scaling hypothesis was rapidly becoming established dogma.

The collaboration finally published the 6- and 10-degree measurements in October of 1969, fully two years after the earliest measurements had been completed. It had taken these experimenters this long to resolve the many ambiguities and uncertainties inherent in the data analysis. Small glimpses of the preliminary results had been shown at Vienna and elsewhere, enough to provoke a few theorists into action. By the time of the Liverpool conference, there were parton models, vector dominance, current algebra, and diffraction models all vying for comparison with the data.

So the collaboration published two consecutive papers in *Physical Review Letters.* The first described the 6- and 10-degree experiments without drawing theoretical conclusions; the second compared the data with predictions of the various theories. These two papers had a far-reaching impact on high-energy physics. Of the ten documents in the field most often cited from 1969 to 1972, these were the only ones about electron scattering; all the rest were squarely in established domains of hadron collisions and bootstrap theories of the strong force. Parton models and electron scattering were still minority topics, but the tide had begun to turn.

Friedman, Taylor, and the others soon hit the lecture circuit with their results. For two years they had remained silent, concentrating on sifting the data and making further measurements. The time had come to spread the word. The fact that they had recently bagged a leading theory, vector dominance, earned them added respect and large audiences.

In December, Feynman published one of his rare papers, "Very High-Energy Collisions of Hadrons," in the *Physical Review Letters*. In this brief document, he applies his parton model to the collisions of mesons and baryons. "I have difficulty in writing this note," begins a quintessentially Feynmanian passage, "because it is not in the nature of a deductive paper, but is the result of an induction. I am more sure of the conclusions than of any single argument which suggested them to me, for they have an internal consistency which surprises me and exceeds the consistency of my deductive arguments which hinted at their existence."

The trouble with particle physics in the sixties, Feynman said over a decade later, was that "everybody was trying to measure everything to a gnat's whisker." By this he meant that most physicists were concentrating on those collisions where only a few well-defined particles, and *nothing else,* came out of a collision. But as the total energy got higher and higher, you could easily create a veritable explosion of debris in any encounter. So collisions that produced only a *few* well-defined particles were an increasing rarity at high energies. Sorely needed at the time was some way to deal with the far more common happenstance where lots of garbage was produced but you examined only one specific particle coming out. This was the motivation that led him to the parton model.

By a happy accident, the MIT-SLAC experimenters had been studying just such a process when Feynman stumbled upon their data in August of 1968. They were looking only at the electrons bounding out of collisions, not caring too much about the details of the hadronic debris emerging from their target on the opposite side. Not only were they measuring just the kind of "inclusive" reactions Feynman was then examining, but they were doing it with a probe whose interactions with matter were well understood. Only by using such a pointlike probe and ignoring the irrelevant details of what happened to the target was there any hope of "seeing" such tiny objects within the proton.

Feynman hit the lecture circuit himself in late 1969, promoting partons and talking up the MIT-SLAC experiments wherever he went. He was a great "public relations man" for the experiments, Taylor observed, and had a tremendous impact on the way the physics community responded to them. "There was an enormous difference between the way I was treated at Liverpool and in Kiev the following year," he recalled. "You could hear a pin drop when I spoke at Kiev."

After Liverpool, partons became respectable topics for discussions and publications. Before that, written information about this renegade idea was hard to find. The smattering of preprints, memos, and other unpublished papers available gave only a partial glimpse of an exciting new theory being passed largely by word of mouth from one oddball physicist to another. But as the sixties ended, partons, scaling, and the MIT-SLAC experiments finally made it into print—and out into the larger consciousness of the entire physics community. What had begun as a tiny brushfire was rapidly burning out of control.

CHAPTER 8
THE ROAD TO OBJECTIVITY

What is a thing? The question is quite old. What remains ever new about it is merely that it must be asked again and again.

—Martin Heidegger, WHAT IS A THING?

DONALD Perkins was one of the few physicists at the Vienna Conference who paid the MIT-SLAC results much heed. A tall, burly, jowly man, he had been pursuing neutrino experiments at CERN for most of the decade, without much success. His colleagues remember him being unusually excited about Panofsky's talk upon his return that autumn to Oxford University, where he was professor of physics.

In January 1969, Perkins was discussing quarks with Frank Close, then a graduate student writing a Ph.D. thesis supervised by Richard Dalitz. When Close questioned the existence of quarks, Perkins grabbed his copy of the Vienna Conference proceedings, tore it open to Panofsky's talk, and pointed to the MIT-SLAC graph of *F*. "If that's not a quark," he declared, "I don't know what is!"

The spark of parton ideas brought back by Perkins found ample tinder at Oxford, where a small circle of graduate students and postdocs under Dalitz had been working out the ramifications of his nonrelativistic quark model. Here was a group of physicists inured to thinking of the proton as built from a small number of fundamental constituents—albeit massive, complex constituents. Partons were soon a favorite topic at Oxford.

At that time, Perkins's group at Oxford was working on neutrino scattering experiments, which had been coming along steadily since the early sixties. In 1962, only six years after Cowan and Reines had first witnessed neutrinos, a *second* kind was discovered at Brookhaven by a team of Columbia physicists. Led by Melvin Schwartz, this group proved that neutrinos given off by decaying pions were in fact *different* from those emitted in the radioactive decays of unstable nuclei. Both of these spooks, and their antiparticles, had no charge and interacted very weakly with matter. But the first kind, called "muon-type" neutrinos, eloped only with muons during various particle decays; "electron-type" neutrinos escaped hand-in-hand with an electron or positron. To make a beam of neutrinos, then, a beam of pions was allowed to decay into muons and muon-type neutrinos. After these decay products passed through many meters of iron or earth, which filtered out the muons, there remained a fairly pure neutrino beam.

In 1963, after one failure, a group of CERN experimenters including Perkins finally made a neutrino beam, which they then shot through a bubble chamber filled with liquid freon, the same cooling agent as used in home refrigerators. In thousands upon thousands of pictures taken, they identified only a few hundred neutrino collisions. Perkins noticed that these rare smash-ups seemed to occur more frequently at high neutrino energies than low. He asked two CERN theorists about the significance of this enhancement, but was told not to worry. The build-up would level off at higher energies. So he dropped the issue and began working on the neutrino beam instead. Until Vienna, that is.

Perkins returned from Vienna in 1968 harboring a vague intuition that the enhancements seen in electron scattering at SLAC and those he had observed five years earlier with neutrinos might be caused by the exact same physics. Both probes were leptons, after all. He worked out the details with a colleague, just before a preprint written by Bjorken arrived in the mail. "Inelastic lepton-nucleon scattering," it declared, "is a very direct means of probing small-distance nucleon structure." Using current algebra, Bjorken predicted that neutrino collisions would occur more often at higher energies. In fact, their frequency should grow in direct proportion to the neutrino energy—roughly what the CERN experimenters had noticed five years before.

Prodded by Perkins, the CERN team reanalyzed their 1963 data, together with some data taken in 1967 using propane in the bubble chamber. All told, there were only 740 events—about what SLAC physicists saw in a typical hour. But the rate of neutrino collisions was indeed rising in direct proportion to the energy, as Bjorken had predicted. Here was a second piece of evidence for pointlike substructure, which the CERN experimenters published in the fall of 1969, just after Liverpool. The paper was noticed at SLAC, but not taken too seriously; with so few events, there were a number of possible interpretations. The CERN data were consistent with a pointlike nucleon substructure, to be sure, but they could not *prove* it existed.

Everybody at SLAC was awaiting the next MIT-SLAC experiment. Electron scattering was becoming a major industry there, and the collaboration had no problem getting more beam. Two months before Liverpool it had asked to go back and study electron-neutron collisions at the same energies and angles used in the earlier proton measurements. The proposal had been quickly approved, and the new experiment began early in 1970.

Electron-neutron collisions were another stringent test of the competing theories. Most parton theories—especially those that took the partons to be quarks—required that electrons collide *less* often with neutrons than with protons. In the model of Bjorken and Paschos, the chances of a smash-up were proportional to the average squared charge of the quark-partons. Because the three principal quarks in a neutron supposedly had charges $-1/3$, $-1/3$, and $2/3$, while those in a proton had $2/3$, $2/3$, and $-1/3$, the ratio N/P of neutron versus proton encounters had to be:

$$N/P = \frac{1/9 + 1/9 + 4/9}{4/9 + 4/9 + 1/9} = 6/9 = 2/3.$$

If there were other contributions that were the *same* for the two targets, then N/P should come out somewhat higher, but still less than 1. Before 1970 the favorite guess was 0.8.

After Liverpool, bootstrap theories of electron-nucleon scattering—in which the exchanged photon behaved like a hadron—were down but not out. Vector dominance was staggering under the blows, but it was not the only alternative. Diffractive models were in fairly good shape because they had not stuck their necks out on L/T and they gave a reasonable explanation for the observed plateau in the F curve. But nuclear democracy insisted that electrons treat protons and neutrons equally. That meant the ratio N/P had to be 1, not 2/3. So an electron-neutron experiment was another classic test; it could not work both ways.

The only trouble was, there were no free neutrons available to use as targets. Unlike the proton, a bachelor neutron is unstable. It has a nasty habit of disintegrating, in a matter of about fifteen minutes, into a proton, electron, and antineutrino. The best alternative is to use a target of deuterium, or heavy hydrogen, whose nuclei consist of a proton and neutron bound loosely together by the strong force between them. Here the neutron is stable: marriage is forever. So you measure the pattern of electrons scattered from deuterium and subtract away the known proton contributions to get the desired pattern of electron-neutron scattering. The MIT-SLAC collaboration now began to fire high-energy electrons at both deuterium and hydrogen targets, detecting the rebounding electrons with the 20 GeV spectrometer set, as before, at angles of 6 and 10 degrees.

With these experiments finished by March 1970, Friedman, Kendall, Taylor, and company had only six months to sift their data before the forthcoming Rochester Conference scheduled for Kiev that September. But they now understood the equipment and analysis programs better, and hoped to present some preliminary results anyway. The physics community would be hungry for new surprises.

Arie Bodek and I had been working under Friedman and Kendall, first as MIT seniors and then as graduate students, since the fall of 1967. We had witnessed their excitement grow as they returned from SLAC every few months with some new piece of data, bringing with them the new parton ideas springing up on the West Coast. I had written my senior thesis on the elastic scattering experiment and Bodek had taken a summer job at SLAC in 1969. In early May 1970, we finally arrived at Stanford to begin preparations for the next inelastic experiment, slated to be the subject of our doctoral dissertations.

Arriving there in the midst of student protests over the Cambodia invasion, we were surprised to find the rest of the MIT group clearing out their comfortable offices in the Central Lab and moving to a shabby old trailer behind the parking lot. Tensions that had been mounting for years were being resolved at last by a complete split between Group A and the MIT physicists. At the heart of the strife were the hostilities between Kendall and Eliott Bloom, who had locked horns during the first MIT-SLAC experiment and never ceased bickering thereafter.

There were good scientific reasons for the split, too. Bloom and Taylor wanted to go back and do more measurements with the 20 GeV spectrometer at small angles, which they felt had been overlooked. Friedman and Kendall preferred to roll the 8 GeV beast out to larger angles and compare electron-proton with electron-deuteron scattering. The time had come for each group to go its own way.

Some of the laboratory staff seemed to be rooting for an MIT failure as we prepared for the experiment largely on our own, aided by a few sympathetic SLAC physicists and technicians led by David Coward. A tremendous esprit de corps among the "in-house" scientists had led to the widespread prejudice that "user groups" like ours could do only second-rate work without heavy staff involvement. Panofsky had deliberately organized the lab around several strong, capable groups that built and operated the big particle detectors. It was indeed difficult for outsiders to set up and do experiments on their own.

The central figure in our efforts was Martin Breidenbach. While an MIT graduate student, he had written his Ph.D. thesis on the very first inelastic experiment. After finishing that in 1969, he stayed on for one more year, working at SLAC as an MIT postdoc before heading off to CERN. A fiery little guy about five foot six, Breidenbach could bark commands like a veteran Marine Corps drill sergeant. His long black hair was thinning fast then, a fact he tried to hide (to little avail) by combing the long remaining strands across a broad and barren forehead. Having worked at SLAC for years, he knew the place inside-out—and had great rapport with most of the scientists and technicians. The spokesman for our experiment, he called the shots as we began to pull the 8 GeV spectrometer out of mothballs.

With only a few months to go and a lot to be done, Breidenbach, Bodek, and I worked well into every evening—and often through the night—trying to get all the detectors, electronics, and computer programs ready. Occasionally we'd work until early morning and try to catch a few hours of fitful sleep in a bunkroom above the counting house. But the steady wham! wham! wham! of an old bubble chamber out in the experimental yard made peaceful dozing impossible.

Midway through those hectic months, Breidenbach and I finally tried to power up the spectrometer magnets. The power supplies for these magnets, able to deliver thousands of amperes of current (household fuses trip out at 30 amps), were kept in a small building just outside End Station A. The "power techs" had finally hooked up these supplies to the magnets and supposedly checked all connections, so we began to turn them on. Standing on tiptoe so he could see the meter better, Breidenbach turned the dial gingerly and sent a few hundred amps trickling through the first circuit. The meter responded as expected with a small offset. Things seemed fine. So he now gave the dial a hearty twirl, sending several thousand amps hurtling magnetward—and watched horror-struck as the pointer wavered menacingly back and forth. A split second later, a deafening thunderclap rocked the entire building, shaking the ground beneath our feet. It had come from the end station, right through four feet of solid concrete!

Had we triggered a spark that somehow touched off a hydrogen explosion? Such a blast had sundered CEA five years before, after a stray spark ignited hydrogen gas escaping from a bubble chamber; an MIT technician had died in that unfortunate disaster. Breidenbach took off toward the end station in a dead run, with me close at his heels.

When we got inside a few moments later, we were relieved to

find the poor technicians still alive and staggering about, their hands on their ears, their faces contorted in pain. No explosion had occurred, luckily. But a quick inspection of the spectrometer revealed that one of the huge brass bolts—thicker than a man's wrist—connecting power-supply cables to magnets had vaporized! The power techs must have missed this one connection. So the great surge of current loosed when Breidenbach twirled the dial had "arced" across the tiny gap like a lightning bolt—causing the tremendous thunderclap and filling the air inside the building with atoms of copper and zinc.

Friedman flew out to SLAC in June with startling news. Analysis of the previous, small-angle experiment, being done back at MIT, showed that the ratio *N/P* was obviously less than 1 and falling sharply. In places, it threatened the quark-parton model limit of ⅔. Electrons treated protons and neutrons very differently, meaning nuclear democracy was dead. Diffractive models simply could not explain these data, which Taylor would be presenting at Kiev shortly, and even partons seemed in trouble. As our own measurements would soon cover the regions where these discrepancies seemed largest, it promised to be an exciting time.

The experiment began in mid-September, just after Taylor returned from Kiev. Right from the start, everything seemed to go wrong. Our electronic logic kept tripping off due to high temperatures. The computer gave absurd answers and had frequent seizures that left it frozen, unable to continue logging data. Then, just as we finally had these problems licked, Breidenbach discovered that the liquid deuterium target was boiling due to overheating by the electron beam. The fan circulating this liquid through the target had frozen stuck and could not be freed without ending the entire experiment. Thus the deuterium density was essentially unknown and our measurements of *N/P* would be meaningless.

In a night of frenzied activity, Breidenbach called on his SLAC friends to help bail us out. They worked until the early morning hours to set up another spectrometer we could use to monitor the deuterium density. By recording the number of protons recoiling out of the target on the other side, we had a measure of the number of deuterons that had gotten in the way of the beam. It wasn't the most pleasing solution (and Taylor thought it a complete waste of time), but at least it allowed us to continue.

After the first few hectic weeks, things calmed down to the point where we could relax a little and just log data. And surprising data it was. The ratio *N/P* continued to drop; it even began to fall well below ½. It seemed as if neutrons were almost *transparent* to electrons. Nobody in the group, not even Friedman, had any idea why the neutron acted so differently. Only a few months before, most of us would have laid bets that *N/P* was around ⅔, and we had planned our experiment accordingly.

By late October, it had become pretty obvious we'd have some exciting results. But these were rudimentary answers being generated by an on-line computer. Still to be applied were a number of important correction factors that could change the values of *N/P* substantially. Kendall began to get nervous that other physicists might come snooping around the counting house, glance at our rough plots, and make the wrong conclusions about what we were seeing. He therefore gathered them up into a three-ring binder deliberately labeled "Administrative and Budget." Nobody, he figured, would ever want to peek inside that.

Taylor's Kiev talk had been extremely well received. These Rochester Conferences attracted the leading particle physicists throughout the world—not just those working in electron and photon scattering. Thus, many scientists from the mainstream fields of meson and baryon physics began hearing about scaling and partons for the very first time.

The September issue of the *CERN Courier* gave the MIT-SLAC experiments top billing as the most exciting results at Kiev. Citing the recent neutron data, this journal of high-energy physics observed that "the measurements come down clearly on the side of the parton model. . . . [But] it still leaves us to get to grips with what partons really are."

After Kiev there was a subtle but important shift in the evolution of the MIT-SLAC experiments. Previously the emphasis had been upon testing various competing theories, including the parton model, to see how well they worked. Now we began trying to elucidate the intrinsic properties of these putative building blocks—which had already passed several trials and were on fairly firm ground. Partons had ceased being mere "possibilities" for us; they were gradually becoming "things." The important question was not so much "Do partons exist?" but "What, in fact, are they?" We still paid lip service to diffraction theories, vector dominance, and other alternatives, but partons clearly had the upper hand. Determining their quantum numbers now became the principal objective of our experiments.

The most startling thing about the data Taylor revealed at Kiev were the low values of *N*/*P* that had been encountered—almost as low as ½. If the partons were three quarks, and all of them were treated equivalently, *N*/*P* had to be about ⅔. This picture soon became known as the "naive" quark-parton model after the early data showed it didn't work too well. More complex models, in which quarks with different charges carried different fractions of the nucleon's momentum, gave values of *N*/*P* as low as ¼, sweeping away the problem.

This is a common process in particle physics. Theorists make a prediction that is not borne out by experiment. But rather than abandon it when proved wrong, they modify their theory to accommodate the latest results and pass the buck back to experimenters. Far easier to change a few parameters than to abandon a cherished idea. Theories are moving targets that are difficult to shoot down permanently. They usually find a way to pop back to life even after the fiercest blows.

Eliott Bloom had a ready explanation for the low values of *N*/*P* observed. An ambitious young scientist brimming over with new ideas, he had captured Taylor's fancy and was rising rapidly within the SLAC hierarchy. Together with theorist Fred Gilman, Bloom had written a paper the previous June that was a curious amalgam of parton and bootstrap thinking. In it, they proposed that the structure functions be plotted versus a new variable x', equivalent to Feynman's x at high energy but different at low. At the end they "predicted" that *N*/*P* would approach ½ when either x or x' got close to 1—essentially what the soon-to-be-reported MIT-SLAC data seemed to indicate.

As a member of Group A, of course, Bloom had the inside

track and already knew the right answers, so this was hardly a good test of the theory. Had Gilman and he published it *before* they knew the answers, it would have been different. But a lot of theory in those days was "phenomenology" just like this: once the experimental results were known, a few parameters could be adjusted to make everything come out right.

From our own coarse, on-line data measured that autumn, however, it appeared we might be able to shoot down *all* these theories. The values of N/P measured at the very highest x seemed to fall well below ½. They would certainly rule out Bloom and Gilman's "prediction" and probably give quark-parton advocates a few sleepless nights, too. But extreme care was necessary here; Fermi motion of neutrons and protons buzzing around within deuterons could skew the ratio substantially.

If there's anything an experimenter likes better than shooting down theories, it is discovering something that theorists had never even anticipated. By October, we thought that maybe, just maybe, we had such a result on our hands. Bodek and I remained at SLAC that fall to do a more careful analysis, while the rest of the group headed off to other work. At SLAC we could use a far more powerful computer and get the work done quickly—in time for next summer's conferences.

Because the ratio N/P was his thesis topic, Bodek began developing a computer program to root out the "smearing" effects of Fermi motion from our data. It was a difficult task that took almost half a year to complete—a long time for him. He is an impetuous physicist, eager to get a job done and published and move on to the next experiment.

By the spring of 1971, he had some early results with the smearing effects removed. The ratio N/P dropped off even faster than before, getting perilously close to the quark-parton limit of ¼. If this trend were to continue, there would be a clear violation of this limit. Even quarks seemed in trouble.

That spring, the MIT-SLAC discoveries finally hit *The New York Times,* catching us completely by surprise. On page one of the April 25, 1971, issue was a headline declaring "Subatomic Tests Suggest a New Layer of Matter," by the highly esteemed science reporter Walter Sullivan.

> A number of physicists believe that, through a variety of atomic experiments, they have begun opening the door to the innermost sanctum of matter.
>
> In the first, and probably most important, of these experiments, conducted at the Stanford Linear Accelerator in Menlo Park, California, evidence has been found of internal components within the proton and neutron—once considered indivisible building blocks of the universe.
>
> Dr. Wolfgang K. H. Panofsky, director of the center, and his staff recently declared jointly that the results "appear to have uncovered another layer of matter."
>
> Specifically these findings suggest the presence, in protons and neutrons, of points of electric charge that, in several respects, resemble the elusive and long-sought quarks.

The names of several other SLAC physicists—Taylor, Bjorken, and Drell—were mentioned in the article, as well as Caltech's Feynman and Gell-Mann. Absent was any mention of MIT's role in the discovery. The one MIT physicist named by Sullivan was Victor Weisskopf, but he was mentioned only as a knowledgeable observer.

Friedman's reaction to the omission, he recalled, was total disbelief. He did not think Panofsky or Taylor could have done such a thing deliberately, and felt it must have happened inadvertently. Kendall, just then writing a *Scientific American* article with Panofsky about the MIT-SLAC experiments, was far more upset, according to Taylor.

What had happened, apparently, was that Weisskopf had tipped off Sullivan, suggesting he write an article about the discovery. Sullivan then phoned Taylor, saying he'd "just been talking to your colleagues at MIT," and now wanted to interview him. Figuring that "colleagues" meant Friedman and Kendall, Taylor obligingly told him the SLAC side of the story. When Sullivan finally wrote the article, unaware of MIT's role, he mentioned only Stanford people.

Follow-up pieces on May 1 and 2 still made no mention of MIT. Weisskopf tried to rectify the deteriorating situation by asking Panofsky to write a letter to *The Times* stressing the importance of MIT's contributions. Published a few weeks later, the letter helped to soothe ruffled feelings a bit, but the damage had been done.

One might question all this concern over a mere newspaper article, but *The New York Times,* as the nation's foremost daily, informs scientists in other fields—and especially Washington policymakers—about new discoveries. In a science so dependent upon government money for its continued progress, Sullivan's front-page article was a valuable trump card in the annual budget scramble. Panofsky understood this fact of political life only too well, far better than Friedman and Kendall. His brief letter only credited them with helping "to design and build the experimental facilities" at SLAC, when in fact their overall contribution was far greater.

Friedman was asked to be the rapporteur summarizing deep inelastic electron scattering at the 1971 Electron-Photon Symposium scheduled for late August at Cornell University. But he declined because of pressing family problems and deferred to Kendall, who had never before served in such a role. About a month before the conference, Kendall flew out to SLAC, where Bodek and I were finishing up the data analysis. In previous summer conferences the MIT-SLAC experiments had always sprung a big surprise, and it looked like we might have another—the plummeting values of *N/P*. Bodek gave him a few up-to-date graphs of this ratio, explaining that his smearing corrections usually changed the values by only a few percent—and never more than 20 percent. So the steep dropoff seemed real, not an artifact.

There was no group meeting to determine what to report at Cornell, or how it should be presented. Kendall flew back to MIT and consulted only briefly with Friedman. The unanticipated falloff was really quite startling. From a value of 1 at $x = 0$, *N/P* fell to about ¼ at $x = 0.8$. You could easily draw a straight line through these data that would fall *below* ¼ at higher values. To help guide the eye, Kendall took out a ruler and pen, drawing a straight line from the upper left to the lower right corner of the graph. Our measured values seemed to cluster around the line, or to fall slightly above it.

At the Cornell Conference, Kendall saved his big surprise for last, finally displaying two graphs of *N/P* with straight lines drawn from corner to corner. "It is seen that a fit consistent with the data is the straight line," he observed. "This fit raises

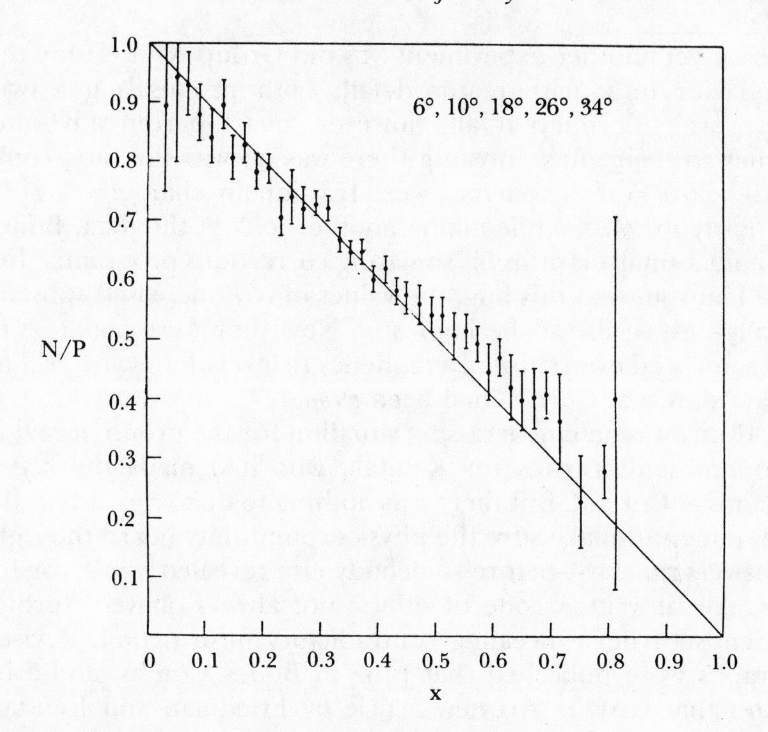

The graph of N/P *presented by Henry Kendall at Cornell. This ratio seems to plummet, as if the neutron were becoming transparent at large* x.

the unexpected possibility that the neutron's scattering may vanish in the limit." The audience had no way of knowing how crude a method had been used to make this "fit." For all they knew, *N/P* was headed for zero; the neutron might indeed be *transparent* in this limit.

Sitting in the audience, Feynman quickly understood the significance. "If the partons carry non-integral quantum numbers like quarks," he noted, "the ratio could nowhere fall below ¼. If it really falls toward zero at $x = 1$ then we would have to conclude that partons carry integral charges." Fractional charges would be ruled out if *N/P* continued its downward trend. Quarks seemed ready to slip from our grasp again.

The precipitous falloff of *N/P* was indeed the great surprise of the conference. That autumn, the preprint shelf at the SLAC library overflowed with papers from ambulance-chasing theorists trying to show how this ratio *had* to approach zero in this or that model of electron-nucleon scattering. We quickly proposed yet another experiment—as did Group A—to study this behavior in much greater detail; both proposals got swift approval. Through it all, however, quark-parton advocates stuck to their guns, insisting there was no way the ratio could fall below ¼ if the partons were fractionally charged.

Early in 1972, while taking another look at the data, Bodek found a small error in his smearing corrections program. After he had removed this bug, the values of N/P increased substantially—especially at the highest x. Now they were about ⅓ or greater, and even showed a tendency to level off toward ¼. The data shown at Cornell had been *wrong*!

If this was an embarrassing situation for the group, it had to be almost disastrous for Kendall, who had made the actual claims at Cornell. But there was nothing to do except admit the blunder and make sure the physics community heard the right answers quickly—before somebody else revealed our error for us. An unwritten code of ethics, not always obeyed, forbids scientists from concealing contradictory information. Revised graphs were published that June in Bodek's thesis, and a bit later that year in a review article by Friedman and Kendall. That fall, our follow-up experiment got the same results, with far better accuracy.

The whole episode was another stunning victory for the quark-parton model. Bjorken and Feynman had insisted that ¼ was the absolute limit on N/P if partons were to be identified as quarks; they had refused to follow the pack and modify the theory to fit our data. Anything less and the theory itself had to go. When the revised values came out in 1972, quark-partons emerged triumphant, while many other theories stood revealed as naked opportunism. More than anything else, this was the event that convinced me of the "reality" of quarks—that what we had been hitting deep inside protons and neutrons were indeed the fractionally charged entities of Gell-Mann and Zweig.

Hindsight is a remarkable human faculty. It allows us to recall past events unfettered by the diverse confusions actually attending those experiences. It creeps inevitably into attempts to reconstruct the process of scientific discovery. Those few intuitions later proved correct are remembered in sharpest

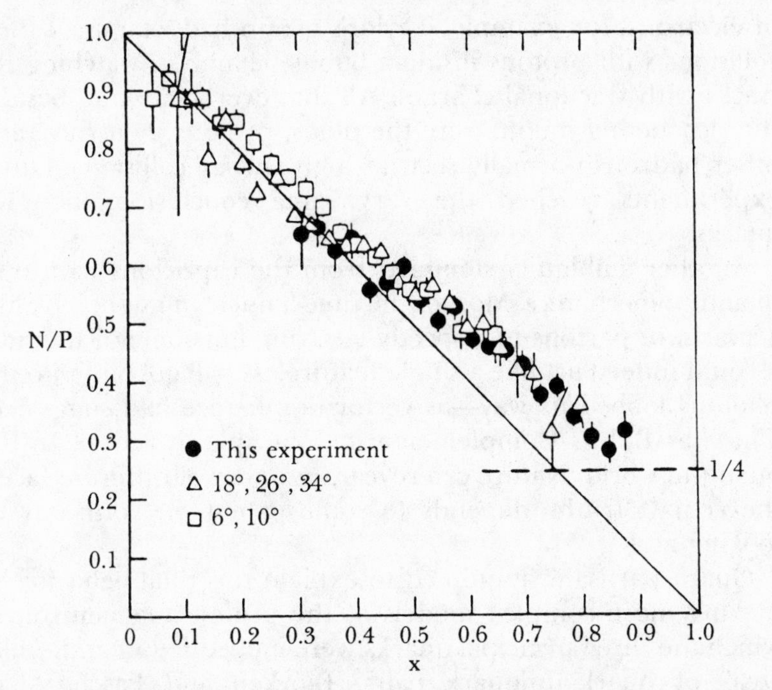

The revised graph of N/P, *after the computer bug had been corrected and a later experiment included. Now the ratio approaches ¼ at high* x*—suggesting fractionally charged partons, or quarks.*

relief; others evaporate from memory—or disappear into its mists. In a process historian Andrew Pickering has dubbed "retrospective realism," what had been merely a likely hypothesis at the time is later imbued with the solidity of an unshakable Law of Nature. Science does not happen like this.

Although the quark-parton hypothesis was emerging in the early 1970s as a strong contender for the true theory of nucleon structure, it was by no means unfettered by ambiguities and contradictions. Nucleon structure was a fascinating but baffling riddle whose solution seemed close but maddeningly difficult to pin down completely. Among the most serious problems was the well-documented absence of free quarks in Nature. We seemed to be hitting things inside nucleons that indeed behaved like quarks. Why, then, didn't we see them come flying out?

By 1972, other experimenters had begun to surround nucleon targets with detectors, hoping to witness a lone quark recoiling from such an encounter. Firing muon probes instead of electrons, for example, Taylor's group had examined their collisions with protons inside a bubble chamber, watching for tracks with fractional charges. All that ever came out, besides the ricocheting muon, were the pions, protons, neutrons, and other hadrons normally seen in high-energy collisions. Other experiments reached the very same conclusion: no free quarks.

Another ambiguity stemmed from the capricious nature of quantum mechanics. Most of the time a nucleon might look like a swarm of partons to a speedy electron, but some of the time it could indeed act like a single featureless ball gobbling up the photons lobbed its way—as vector dominance had suggested. This was Bohr's Complementarity Principle in action. In the quantum world, Nature can reveal a number of different faces; the actual "truth" depends to some extent on your way of looking at it.

Quark partisans attempted to explain this dual behavior by making more complex models of the proton and neutron in which the three principal quarks were embedded in an infinite "sea" of quark-antiquark pairs. Bjorken and Paschos had suggested such an idea in their 1969 paper; together with Julius Kuti, a Hungarian theorist visiting MIT, Victor Weisskopf published a detailed model late in 1971. Like what the outer, valence electrons do for atoms, three "valence quarks" in a nucleon were what caused its distinctive physical properties. The "sea quarks" were supposedly the same from one nucleon to the next—a vast reservoir of numberless potential pairs ready to spring out of the vacuum when enough energy was present. Valence quarks made the proton and neutron different; sea quarks made them alike.

For good measure, Kuti and Weisskopf also threw in a number of uncharged "gluons." These were the imaginary particles carrying whatever force it was holding the partons together in nucleons. Partons were the "parts" and gluons the "glue" that somehow kept them from flying apart—like the pions binding the protons and neutrons together inside atomic nuclei. Lacking any charge whatsoever, these gluons would be oblivious to high-energy electrons and so remained essentially invisible to the MIT-SLAC experiments.

By 1972, most of the startling surprises had already been

harvested from the MIT-SLAC experiments, and our work now became that of gleaning whatever results might have been overlooked in the early excitement. My own Ph.D. thesis was work of this nature—picking through the data of 1970 to see if I could come up with more accurate values of *L/T*, and hence a better idea of the parton spin. My measurements of this ratio turned out to be completely consistent with expectations if the partons had spin-½; it was yet another proof we had been hitting quarks inside protons and neutrons.

MIT and SLAC physicists had worked these fields essentially alone while the rest of particle physics had waited expectantly for our results. The great lore of parton ideas had sprung up largely in response to these experiments. But the time had now come to look elsewhere. The final test of these ideas had to come in their usefulness to *all* of particle physics, not just the small corner where they had first taken root.

Experiments underway in Geneva, almost halfway around the world from Stanford, soon took center stage. Among other things, neutrino scattering at CERN would be crucial in resolving the ambiguities confronting the quark-parton model. Where electrons are sensitive to the charges of any partons, neutrinos respond to the *numbers* of up and down quarks inside a nucleon. Thus, neutrino experiments could provide key tests of quark-parton ideas.

For years the CERN group had doggedly pursued neutrino research, making one improvement after another in their beams and targets. Like electron scattering, neutrino experiments had been considered somewhat of a backwater during the 1960s. While Director of CERN, however, Weisskopf had taken the neutrino group under his wing, making sure it had solid funding, good equipment, and excellent people. Their affairs took a big stride forward in 1971, when the huge bubble chamber named Gargamelle finally came into service. Its name was borrowed from the novel *Gargantua,* written in 1534 by Rabelais. Gargamelle was the mother of the giant Gargantua, having given birth to him through her ear.

Where previous large bubble chambers had been 1 or 2 meters long and held about 1000 liters of liquid, Gargamelle was almost 5 meters long and contained 12,000 liters—enough to hold 5 tons of propane or 15 tons of freon. The iron core of its magnet alone weighed 800 tons, about the same as SLAC's entire 8 GeV spectrometer. Truly a monster. It had been built in Orsay near Paris at a cost of some 25 million Swiss francs, or about $10 million. The arduous process of moving it to CERN, piece by ponderous piece, had taken many months.

Gargamelle's tremendous size was the key to its use as a target for CERN's neutrino beams. The expected number of neutrino collisions—then a scarce commodity at anybody's accelerator—is roughly proportional to the mass or volume of the target. The bigger the better. To get a good idea of how infrequently these spooks interact, consider that a 1 GeV neutrino can easily pass through 100 million miles (about the distance from the earth to the sun) of normal matter without ever stopping. So you had to make a beam with as many neutrinos in it as possible, put a huge target in its path, and then wait. With a billion neutrinos hitting Gargamelle every second, the CERN experimenters saw about one collision per minute. By the end of 1971 they had taken more than half a million photographs.

More than anybody else in the CERN group, Donald Perkins of Oxford had championed the use of neutrinos to probe nucleon structure. In the game of particle physics, he had wagered early and heavily on quarks. This was a man who played his hunches and stuck by his favorites.

In September 1972, Perkins flew to Chicago to deliver a paper by the CERN neutrino group at the Rochester Conference, which was held that year at the National Accelerator Laboratory, the site of America's brand-new 200 GeV proton synchrotron. These first Gargamelle results included a wide variety of neutrino data. Of special interest to the quark-parton camp were the results on *inelastic* neutrino collisions. To preface his talk, Perkins claimed that this new data would "provide an astonishing verification of the Gell-Mann/Zweig quark model of hadrons."

Where earlier CERN experiments had only used neutrinos as probes, the Gargamelle results also included antineutrinos. There were about a thousand inelastic collisions in each category, all at energies between 1 and 10 GeV—roughly comparable to SLAC electrons. As earlier, the chances of a neutrino collision seemed to grow in direct proportion to the energy.

Antineutrinos behaved the same way, as expected if they were hitting pointlike entities inside a nucleon. And the ratio of antineutrino/neutrino collisions came in close to ⅓, exactly what had to happen if almost all these partons had spin-½.

The most striking new information, however, was his comparison of the Gargamelle results with our own electron scattering data. With some imagination, which he possessed in ample supply, Perkins could extract a new structure function from the Gargamelle data. Dividing his function by our own, he got a ratio fairly close to 18/5—the very same value expected in parton models with *fractional* parton charges. The likeliest conclusion? These neutrinos were hitting quarks.

People at MIT and SLAC were impressed but hardly overwhelmed by the Gargamelle data. It was comforting to have an independent confirmation of the parton model in a completely new arena. But it was a bit unnerving to see how much information Perkins could extract from a mere 2000 collisions, the great majority of them at comparatively low energy, when we had based our own cautious claims on many *millions.* "Considering the quite different techniques in the two sets of experiments," he admitted in his talk, "the measure of agreement is somewhat miraculous." Most of us had to agree.

But by mid-1973 the Gargamelle data were in better shape. The group had recorded another thousand events and done a more careful analysis that Perkins and others explained at the various conferences that summer. The neutrino/electron ratio now came in almost exactly at the value 18/5 specified by the quark-parton model. These data could also determine the number of valence quarks—which came out, as expected, to be 3. And they suggested that half of a nucleon's energy and momentum were tied up in the gluons that supposedly bound the partons together. However one looked at nucleons, with electrons or neutrinos, the answers all seemed to be coming up quarks.

What had sprung up ten years earlier as a mere quirk of imagination was close to becoming objective reality by 1973. Quarks had been "seen" in two independent experiments using different probes that interacted with nucleons through two different forces. Compare this situation to the use of two separate senses to perceive a single common object. Our eyes alone can easily be deceived by a mirage or hallucination, but if we can also reach out and *touch* this vision, our recognition of its "objectivity" is greatly enhanced. Without such tactile reinforcement, we might conclude that our candidate for reality did not "exist" in a normal sense.

The existence of quarks was on fairly solid footing in 1973, after they had been "seen" with electrons at Stanford and "touched" by neutrinos at Geneva. Many other experiments were reporting in with still other, independent evidence for quarks—additional perceptions that helped reinforce this emerging picture of hadron structure. But nobody had yet witnessed even a single, solitary quark drifting along in space all by itself—at least not very convincingly. If these curious objects "existed," it was only *inside* the pions, protons, and other hadrons that we did actually see.

MESOLOGUE

Paradox is an important element in physics. When a paradox occurs, it usually means we are getting somewhere, about to learn some truly new lessons about what we call reality. The shopworn conceptual framework that we had been trying to extend by forced analogy into new realms of experience breaks down in the wider context, and no amount of patchwork can save it.

The latest paradox began raising its head in 1970, when partons started looking a lot like Gell-Mann's fictive quarks. We seemed to be hitting *something* tiny with fractional charge and spin-½. And the quark idea also explained the huge variety of visible mesons and baryons found in Nature. But then why didn't quarks ever come out and show themselves, like other well-behaved corpuscles?

Richard Feynman was one of the first to recognize that we were brushing up against a paradox here. In a free-wheeling panel discussion held at Irvine, California, in December 1971, he returned to this theme again and again:

> All these quark rules are a new discovery since that time, and I believe that what we have really discovered since 1961 is much simpler than we thought it ought to be. There was no reason to think there should be so many rules that work so closely underneath the thing. We still don't understand why, but the fact that there's this simplicity, to me, means that we're getting close to a paradox.

His words capture the excitement—and confusion—of those times. Asked about the "reality" of quarks, Feynman repeated his theme:

> Oh yes, yes, if you suppose all partons are quarks, and you cannot produce real quarks, you're in the face of a possible paradox that I mentioned. It would be very pleasant! I think we would make real progress if we would discover ideas such as "All charged partons are quarks," and "You cannot produce real quarks," are both correct. . . . Then we would have a paradox.
>
> We are not very clever. We know that in the past history of physics we made our greatest progress when we were presented with two things that were mutually impossible, but really mutually, and one of the difficulties of this subject has been that we have vague ideas. We have not gotten up to anything which is just impossible, absolutely impossible; and I'm waiting for something impossible.
>
> I suspect it would be a thing like this, that every experiment indicates that the quark view is right and that partons are quarks. And let's say that the quark mass is zero, which is impossible, or that the partons are quarks, but you can't make individual quarks, which is impossible. Something like that is going to come out, and then we'll get somewhere.

These words came a few months after the Cornell Conference, when everybody was still thoroughly confused by the spurious *N/P* ratios Kendall had reported there. So nobody could then claim a paradox. But by the following summer, that wrong turn had been admitted and corrected. With a great amount of data from a great many experiments, all coming up quarks, Feynman had his paradox at last.

To Murray Gell-Mann, however, there was no paradox, only a lot of unnecessary confusion. From the outset, he had insisted that his quarks were not "real" entities at all, that they would *never* appear in detectors, no matter how hard we bashed our particles together. The confusion arose because physicists were unwittingly trying to apply old, everyday concepts in a strange new realm where they had limited meaning, if any. It was like using the classical idea of an orbit to describe the quantum motion of an electron within an atom. You inevitably encountered a paradox.

No, quarks were mathematical objects, Gell-Mann claimed, not Feynman's "put-ons," as he liked to ridicule them. The best way to work with quarks was by using not the crude, billiard-ball analogies of the parton model but the more rigorous mathematical equations that arose within the framework of current algebra, the original source of Bjorken's scaling hypothesis. The ultimate locus of reality was to be found in *form,* not substance.

Better than any other living scientists, Feynman and Gell-Mann personify what I identify as the two major schools of physical thought: the materialists and the idealists. Materialists find their ultimate reality in "stuff"—in atoms, nucleons, or partons. Ever since Leucippus and Democritus, and perhaps before them, complexity at one level of experience reduces to simplicity at a deeper level of substance. The idealists, by contrast, find their reality not in things but in *relationships*—in geometry, symmetry, and group theory. Since Pythagoras and Plato, they have been struggling to reduce Nature to pure number.

Neither of these schools make statements about "the way Nature *is*"; instead, they speak about the appropriate way to perceive Nature and talk about it. These prejudices are what science historian Gerald Holton has dubbed the "thematic content" of science—the psychological predispositions that arise more from aesthetic than scientific concerns. They lie at the root process of actually *doing* science, are its heart and soul, but rarely surface explicitly in the published literature.

These two polar opposites—Feynman and Gell-Mann—work just a few doors apart along the same corridor in Caltech's Lauritsen Laboratory, with only an ever-bustling, overprotective secretary parked between their two spacious suites. Talking with both of them in March 1984, I was delighted to hear each poking fun at the other's opinions. "You know that Gell-Mann character down the hall?" Feynman quipped, "He never really caught on to my parton ideas." Gell-Mann countered that "Feynman's 'put-ons' were just a crude, fake scheme to help people remember what is really going on inside nucleons."

It is difficult, some would say impossible, to find two scientists more different, stylistically and temperamentally. The wisecracking Feynman is the quintessence of informality, dressed in loose-fitting gray slacks and a white shirt open at the neck, talking like a Brooklyn cab driver. He sits precariously in a metal chair propped up against the wall and gesticulates as he talks, drawing pictures with his hands. Gell-Mann, by contrast, wears a well-tailored woolen sports coat over a V-necked sweater and tie. With thick, black-rimmed glasses and short-cropped white hair framing his eager face, he seems the paragon of orderliness. He sits calmly behind his desk in a plush blue swivel chair, hands folded, never once lifting one of them to make a gesture during the entire conversation. Information is exchanged by words and numbers, not hands or pictures.

For lunch, Gell-Mann took me over to Caltech's Atheneum, its faculty club, where an exquisitely appointed table is reserved for his use, visible from all parts of the room. Feynman was nowhere to be seen; my Caltech friends say he prefers to eat lunch at "The Greasies"—the college cafeteria, where he can rub shoulders and swap yarns with grad students and postdocs. While Gell-Mann is a man of lofty ideals, Feynman is a man of the earth.

Their personal styles spill over into their theoretical work, too. Gell-Mann insists on mathematical rigor in all his work, often at the expense of comprehensibility. The current algebra he fostered in the 1960s was nothing if not rigorous, but only a small group of physicists really understood what was being said. Where Gell-Mann disdains vague, heuristic models that might only point the way toward a true solution, Feynman revels in them. He believes that a certain amount of imprecision and ambiguity is *essential* to communication; it is the very soul of metaphor. He also insists on simplicity in his work. "I have a principle in strong interaction theories," he once observed. "If the theory is complicated, it's wrong."

Both approaches are essential to the progress of science. Nature is multifaceted, and different people will perceive the same "reality" in very different ways. In the ongoing search for solutions to the knotty problems confronting them, scientists repeatedly have to fall back on their own individual styles—their favorite ways of perceiving and confronting experience. Holton calls this process "going off the contingent plane," returning again to those deep personal reservoirs of unscientific prejudice we all harbor within us. Here is where art and culture have their greatest impact on science.

Unfortunately, the stylistic content always seems to get winnowed out of the final product, the scientific publications reporting the work and the textbooks that explain accepted dogma to students. This content is the principal difference between "personal science" and "public science," the best-foot-forward version of events that usually appears in print. Reading such desiccated accounts, you get the impression that science is done by robots, cut off from all contact with the larger culture. It's just not so.

My own sympathies, you may have guessed, reside with the "materialist" school. I really began to believe in quarks as "things" only in 1972—right after the *N/P* crisis had been resolved and our experiments taken as solid evidence for fractional parton charges. Other physicists sympathetic to the quark-parton interpretation were similarly convinced. By granting quarks a measure of objective reality, however, we had finally come face to face with a full-blown paradox.

Why didn't the little beggars come out?

The resolution of this paradox came swiftly, from an unanticipated theoretical quarter, bringing with it a whole new picture of subatomic reality. Particle physics shifted into high gear during the mid-1970s. But, to recount those exhilarating times, I have to retrace my steps first and pick up a few loose threads—important strands of theory that had been largely ignored for years.

CHAPTER 9

THROUGH THE LOOKING GLASS

"It seems very pretty," she said when she had finished it, "but it's rather *hard to understand! Somehow it seems to fill my head with ideas—only I don't exactly know what they are!"*

—Lewis Carroll, THROUGH THE LOOKING GLASS

THE weak nuclear force was a profound mystery to many physicists of the 1950s and 1960s. The cause of radioactivity and the slow decay of many subatomic particles, it had been a puzzle ever since it was distinguished as a separate force, a *fourth* force of Nature, distinct from the strong force that held nuclei together. By far the most disturbing feature of the weak force was its ability to distinguish left from right. It provoked certain particle decays, but not their mirror images.

When you look in a mirror, the objects you see there have traded left for right, or vice versa. Your right hand looks like a left hand, and printed words viewed in the mirror read like gibberish. If you could peer a lot closer, you'd notice that the spiral DNA molecules in your mirror hand curl up in the opposite sense from those in normal earthlings. This looking-glass world may seem a bit odd, but nothing in the laws of physics ever said it was not a *possible* world. Or so scientists had thought until 1956.

That summer, T. D. Lee and Chen Ning Yang, working as visiting scientists at Brookhaven, predicted that the weak force violated "parity," a cherished symmetry between left and right which physicists had previously never questioned. Their bold prediction was borne out later that year and in early 1957. By the next December, Lee and Yang had received the Nobel prize.

This overthrow of parity, of left-right symmetry, was difficult for many physicists to swallow. They had come to expect that the subatomic world would be a utopian refuge from everyday chaos, a special place where symmetry reigned supreme. Though the macroscopic world of our five senses might be subject to turmoil and irrationality, surely this microscopic realm would be different. But by a year after Lee and Yang's prediction, the experimental evidence for parity violation was indisputable.

A bit of normalcy returned, however, after the meaning of subatomic "reflection" was redefined in 1958. "If one performs a mirror reflection *and* converts all matter into antimatter," Yang observed, "then all physical laws remain unchanged." For example, the disintegration of a negative pion into a muon and an antineutrino, viewed in the looking-glass world, follows the exact same rules as that of a positive pion in our normal world. Similarly, neutrino spins always point opposite their direction of motion, while antineutrinos point along this direction. We say neutrinos are "left-handed" and antineutrinos are "right-handed." Reflection now meant the two *combined* operations of spatial reflection and the interchanging of matter with antimatter. Symmetry still reigned. Sort of.

Unfortunately, this fledgling new symmetry was struck down almost immediately by a group of Princeton experimenters working at Brookhaven. Led by James Cronin and Val Fitch, they discovered in 1963 that a certain feeble decay of neutral kaons violated this symmetry. Such oddities happened only rarely, less than one percent of the time, but the damage was done. When the weak force got into the act, irrationality always seemed to creep in too.

During the 1960s, there were three effectively separate disciplines in particle physics, each with its own distinct retinue

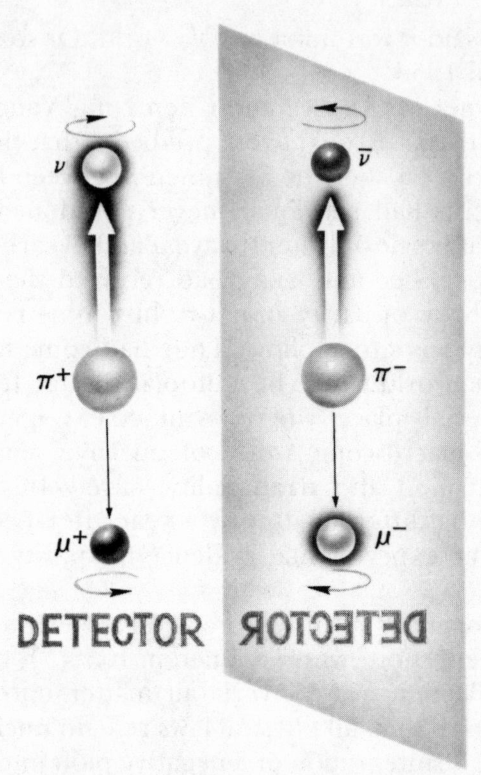

Symmetry in the looking-glass world. If you replace particles by antiparticles, the laws of physics remain unchanged—most *of the time.*

of physicists, studying the strong, weak, and electromagnetic forces. Each group went about its fragmentary ways, paying polite attention to one another's research, with hardly an inkling that all these seemingly disparate phenomena might somehow be just the different manifestations of a single fundamental force. Only a far-sighted few had begun to grasp the connections.

Chen Ning Yang was born in 1922 near Nanking, China, and grew up in a nation torn by almost constant revolution and war. Still he managed to obtain a college education and a degree in physics before he was twenty, emigrating to the United States in 1945. As a graduate student at the University of Chicago, he became very interested in quantum field theory, a popular subject after the war. The great success of quantum electrodynamics (QED), a field theory that yielded answers accurate to many decimal places, encouraged physicists to apply the same methods to the weak and strong nuclear forces too.

Field theory traces its roots back over a hundred years to Michael Faraday, the British experimenter who had established several important laws of electricity and magnetism. He believed that electric charges and magnets exerted their forces by means of field lines emanating outward from these sources. To Faraday, these field lines were real, physical entities—not just some mathematical construct. Their footprints could be seen by sprinkling iron filings in the vicinity of a magnet and watching them line up in an orderly pattern extending from one pole to the other.

In *quantum* field theory, the lines of force become particles like photons, which carried electromagnetic influences from point to point in space and time. Action at a distance has no place in field theory. Some *particle* has to bear the message, carrying on its back all the energy, momentum, spin, and whatever else might be transferred in the process.

Now quantum electrodynamics is a very special kind of field theory called a "gauge theory," in which some kind of "charge" is absolutely conserved. Electric charge, for example, never appears or disappears in *any* interaction—whether strong, weak, or electromagnetic. The total charge always remains the same after a collision as before.

Whenever physicists witness something being conserved like this, we say, "Aha! There is a symmetry of Nature at work here." Conservation of energy, for example, means that a system is symmetric in time: one moment is as good as the next. A similar observation holds true for the conservation of angular momentum, or spin, which tells us the system has no preferred direction in space, either.

Conservation of charge has been likewise interpreted as evidence for an *internal* symmetry called "gauge symmetry," related in quantum mechanics to the internal parameters of the wave function ψ, primarily its *phase.* Loosely speaking, gauge symmetry means a system's intrinsic properties do not depend on the calibrations of our measuring sticks—their "gauges." Well before this century, classical electromagnetism had been known to possess such a gauge symmetry.

At the Institute for Advanced Study in Princeton, where he worked in the early 1950s, Yang began trying to derive a gauge field theory of the strong nuclear force between protons and neutrons. To carry the force he introduced a triplet of spin-1 particles which he denoted by B^+, B°, and B^-—analogues of the spin-1 photon. While at Brookhaven in 1954, he worked out the details of this approach with his officemate Robert Mills, and they published their work that fall.

But the analogy with QED and the requirements of gauge symmetry meant the B's had to have zero mass, like the photon, which posed a severe problem for the Yang-Mills theory. The strong force was known to be a very short-range force extending outward no more than a quadrillionth of a centimeter. According to Werner Heisenberg and Hideki Yukawa, short-range forces had to be carried by heavy, massive particles. On the court of subatomic physics, the players had to swap cannonballs, not tennis balls.

Yang acknowledged this shortcoming and hoped it might be solved once these gauge theories were studied more thoroughly. But such a dodge did not satisfy stern critics like Wolfgang Pauli, who confronted Yang that year at a Princeton seminar.

"What is the mass of this B field?" asked Pauli.

"I don't know," came the feeble reply.

"What is the *mass* of this B field?" he demanded again.

"We have investigated that question," Yang now offered. "It is a very complex question, and we cannot answer it now."

"That is not a sufficient excuse!" snorted the aging but still redoubtable theorist.

The problem of mass plagued Yang-Mills gauge theories continually for over a decade before imaginative solutions began to appear in the 1960s.

Sheldon Glashow and Steven Weinberg had been boyhood chums at the Bronx High School of Science during the late 1940s. With an eager group of bright young friends they organized the science fiction club there, and taught each other scraps of calculus and quantum mechanics from College Outline books on math and physics. Perhaps more than the regular classes themselves, where such topics hardly ever came up, these informal discussions were the principal learning grounds for the budding scientists.

Following four unrewarding years at Cornell, the two friends parted company—Glashow going to Harvard and Weinberg to Princeton. "We were forced to take all kinds of revolting classes at Cornell," Glashow recalls. "I spent a good deal of time in the poolroom."

In graduate school, Glashow and Weinberg were naturally exposed to, if not immersed in, the midfifties enthusiasm for quantum field theory. A question that intrigued them both was the problem of infinities, which always seemed to arise in quantum field theory. In the collision process, not just one, but two, three, and more particles might well be swapped. And because of the Uncertainty Principle, any one of them could momentarily become a virtual *pair* of other, very different particles. But when theorists tried to calculate the extra contributions from these more complex exchanges, they often got an infinite result—in obvious contradiction with experiment. How could this be possibly be true?

Richard Feynman, Julian Schwinger, and Sin'itiro Tomonoga had rescued quantum electrodynamics from such nightmares by the sleight-of-hand known as "renormalization." Infinities in electrodynamic calculations were purged by burying them in the electron's mass and charge—the *m* and *e* in the equations—which were then redefined to accord with the observed, finite values. Few physicists at the time were completely comfortable with renormalization, but the agreement with experiment was absolutely breathtaking.

As part of his thesis research, Glashow worked on renormalization of the weak force, which had resisted previous attempts. The theory then emerging from the ferment over parity violation was still plagued by infinities that could not be conjured away by the same easy magic that had worked so well for QED. He submitted his thesis in 1958 without solving this problem.

Glashow was heavily influenced by his thesis adviser, Schwinger, who at the time was deploying gauge theory in an attempt to unify the weak and electromagnetic forces. It was an audacious, far-sighted project, because the weak force is thousands of times feebler—and it violated parity as well. Borrowing

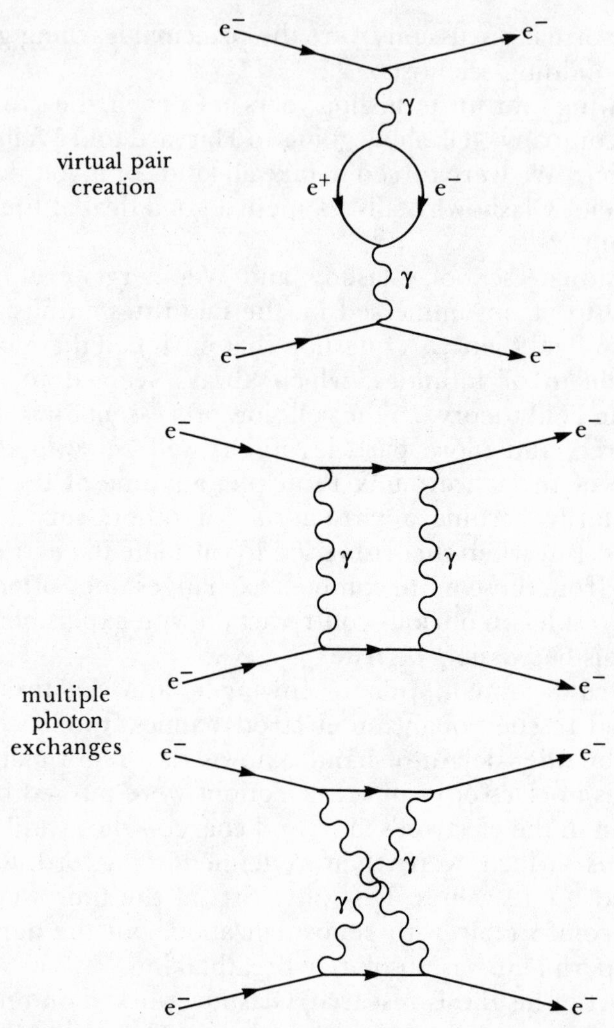

Feynman diagrams depicting some of the more complex exchanges possible when two electrons collide.

the triplet of Yang and Mills, Schwinger suggested the weak force might be carried by the two charged members, which he called the W^+ and W^-, while their third, uncharged partner was the familiar, massless photon. By assigning the charged pair tremendous masses, many times that of a proton, he arranged that the force they transmitted could act only over extremely short distances. Hence, it only *appeared* to be very weak. Others had already proposed a W^+ and a W^- as weak force carriers, but Schwinger was the first to suggest they might both be siblings of the photon in the same happy family of spin-1 bosons, a common name for force-carrying particles.

In 1960, working as a postdoc in Copenhagen, Glashow thought he had a better solution, and wrote it up for the European journal *Nuclear Physics.* "At first sight there may be little or no similarity between electromagnetic effects and the phenomena associated with weak interactions," he began. "Yet certain remarkable parallels emerge with the supposition that the weak interactions are mediated by unstable bosons." To the W^+ and W^-, Glashow added a third massive, spin-1 particle, the neutral Z°. Together with the photon, these three particles formed a family of *four* spin-1 "intermediate vector bosons" that transmitted the electromagnetic and weak nuclear forces. By judicious choice of masses and other parameters, he evoked both a short-range weak force that violated parity and a long-range electromagnetic force that conserved it. Unity seemed at hand.

But Glashow's theory had the same obnoxious infinities that had long been the bane of quantum field theories. And they could not be conjured away unless the masses of the weak force carriers were all zero, like the photon. That, however, destroyed Glashow's delicate scheme by making the weak force a long-range force, which it most definitely was not. His proposal soon joined the growing obscurity then enveloping most of quantum field theory.

At Princeton, Weinberg had also worked on a theory of the weak force, trying unsuccessfully to purge it of the accursed infinities. He submitted his dissertation on the subject in 1957, a year before Glashow's.

In late 1960, while an assistant professor at Berkeley, Weinberg learned about the new ideas of "broken symmetry." An idea borrowed from solid state physics, broken symmetry meant that the underlying equations of motion might well be symmetric even though their physical manifestations (the subatomic forces and particles) were anything but. The members of a meson octet, for example, have very different masses even though they were all supposedly close relatives. The symmetry was right in front of our eyes all the time—just hidden from view, like proverbial diamonds in a pile of coal. "Hidden symmetry" was a better, but less commonly used, phrase for the same phenomenon.

An everyday example of broken or hidden symmetry is a bar magnet. The equations governing electricity and magnetism favor no particular spatial direction; they are completely symmetric in this regard, like a spherical rubber ball. But the bar magnet itself *has* a special direction, namely the arrow pointing from south pole to north. It arises because the iron atoms in the bar are themselves tiny magnets whose preferred state occurs when they are all aligned along a single, common axis. Though the underlying equations may be symmetric, the physical state is definitely not.

At first glance, broken symmetry seemed a grand idea. The apparent chaos creeping into particle physics from all sides might have a ready explanation that spared a small but important reserve where symmetry still held absolute sway. "As theorists sometimes do, I fell in love with this idea," Weinberg recalled. "But as often happens with love affairs, at first I was rather confused about its implications."

In 1961, Jeffrey Goldstone at Cambridge University showed that broken symmetries required the appearance of massless spin-0 particles later known as "Goldstone bosons." But no such mote had ever turned up in an experiment. During the next year, Goldstone, Weinberg, and Abdus Salam, a Pakistani theorist, worked together on this problem at Imperial College in London, but to no avail. The zero-mass ghosts simply refused to go away. To the paper summarizing their work, Weinberg added an epigram borrowed from *King Lear:* "Nothing will come of nothing: speak again." The editor dropped it.

Not until 1967 did Weinberg make a virtue of necessity. In the meantime a number of theorists, most notably Peter Higgs at the University of Edinburgh, had been working on ways to cure the disease of massless Goldstone bosons. In field theories with broken *gauge* symmetry, they discovered, these troublesome bosons might well disappear while the spin-1, force-carrying particles acquired unexpected mass, as if they had somehow "gobbled up" the unwanted ghosts and gotten fat and bloated from their dinner.

Weinberg was the first to realize that such a "Higgs mechanism" might provide the long-sought key to unifying the weak and electromagnetic forces. It could generate the needed masses of Glashow's two W's and the Z° while eliminating the bothersome Goldstone bosons all in one fell swoop, with a minimum of artifice. Instead of having to insert these masses by hand, as Glashow had done six years before, they simply fell out of the equations when gauge symmetry was broken via the Higgs mechanism. Using just a single arbitrary parameter, Weinberg could predict the mass of the two W's to be over 40 GeV while the Z° mass exceeded 80 GeV.

Weinberg's great flash came to him in the fall of 1967 while driving to MIT, where he was a visiting professor. In mid-October he wrote a brief paper with the innocuous title, "A Model of Leptons." It was duly published the next month by the *Physical Review Letters* and soon forgotten by everybody, including its author.

Over the next four years, this article was cited in the scientific literature a grand total of only *five* times—including Weinberg's own citations. "Nobody paid any attention to it," recalled Harvard theorist Sidney Coleman in an excellent review of these breakthroughs many years later. "Rarely has so great an accomplishment been so widely ignored." Even Samuel Treiman, Weinberg's one-time Princeton thesis adviser, passed it over at the 1968 Vienna Conference, where he was reporting the significant theoretical work of the previous year. Field theory was in deep decline at the time, and most physicists were seeking salvation elsewhere.

A different approach to unification was taken by Salam and his British coworkers. Where Glashow and Weinberg followed the path of renormalization by trying to purge infinities, this group of theorists used the high road of gauge field theories.

A short, compact, bearded man of enormous humility and graciousness, Salam combines the best qualities of his Oriental heritage with a formal Western education. Born in 1922 in the Pakistani town of Jhang, he attended Cambridge after World War II and became immersed in field theory, calculating many of the finer details of quantum electrodynamics. There field theory became a leading passion in his life. Under his guidance,

Ronald Shaw developed the same gauge field theory as Yang and Mills but did not publish it after learning about their 1954 paper. In 1957, Salam became professor of theoretical physics at the Imperial College of Science and Technology in London, where he did his pivotal work.

Like Weinberg, Salam was enamored of broken symmetry in the early 1960s. He likens it to his own favorite analogy of guests sitting down to eat at a circular dining table, one which has been set in a symmetrical manner by the hosts, with salad dishes placed halfway between adjacent settings. The underlying symmetry remains perfect, even though the hungry diners will inevitably "break" the symmetry the moment one of them chooses to eat salad from the left or right. Then all others have to follow suit, and an asymmetric pattern emerges. At the dinner table of existence, Nature has to make many such choices.

In collaboration with theorist John Ward, Salam published a series of papers on gauge theories of weak and electromagnetic forces, trying to incorporate the ideas of broken symmetry. They culminated in a 1964 contribution to *Physics Letters* that used many of the same ideas of Glashow's then obscure 1961 paper. It contained the same three massive bosons, two charged and one neutral, transmitting the weak force plus a massless photon carrying electromagnetism—all members of one happy family. As Glashow had done before them, Salam and Ward gave the bosons their masses by hand, artificially. Unfortunately, this led to similar infinities they could not conjure away.

When Salam finally learned about the Higgs mechanism in 1967, he realized this might be the way to generate the W and Z masses *without* introducing the incurable infinities. He began lecturing on the subject at Imperial College that fall and delivered a paper about it at a symposium in Sweden early the following year. But his comments garnered scant notice. They were published in an obscure collection and soon forgotten by all but a few.

Both Weinberg and Salam conjectured that their unified theories might indeed be amenable to renormalization, to purging of their inevitable infinities, a crucial requirement before other theorists would pay any real attention. But saying so and proving it are two different things altogether. Renormalization requires some of the most horrendously complex calculations in all of particle theory, involving many delicate cancellations of the various infinities that crop up, any of which can be easily destroyed by the slightest error.

The graduate students to whom Weinberg suggested this problem could not handle such awesome calculations, and he had little luck himself, either. Likewise, Salam made scant progress with his own version of the theory. By 1970, they had abandoned the effort and turned to other ventures.

Unknown to both of them, a Dutch theorist named Martin Veltmann had been developing a new and idiosyncratic approach to these very same kinds of calculations. After working at CERN during the early 1960s on the theory of neutrino scattering, Veltmann returned to the Netherlands as professor of physics at the University of Utrecht. There he reached much the same conclusion as had everybody else at the time: whenever he gave the force-carrying particles a mass in Yang-Mills theories, he destroyed their symmetry and got the same incurable infinities.

But Veltmann had a clever graduate student named Gerard 't Hooft who used his odd methods to solve this dilemma in 1971 and, in Coleman's words, "revealed Weinberg and Salam's frog to be an enchanted prince." What 't Hooft did, in effect, was use the Higgs mechanism to generate the masses instead of inserting them by hand, artificially. That way the symmetry necessary to ensure all the delicate cancellations of infinities remained intact, albeit hidden. Renormalization was, in fact, possible for Yang-Mills theories.

Veltmann and 't Hooft began to show these results that summer at the 1971 European Physical Society Conference in Amsterdam. The American theorist Benjamin Lee learned about their success there and brought back word to the United States that fall, translating the obscure methods of the Dutch theorists into more conventional terms familiar to other theorists. With his convincing proof of renormalization, the interest in gauge theory swelled overnight. The trickle of publications on the subject swelled to a torrent.

During the early 1970s, Weinberg was professor of physics at MIT, where he had come after leaving Berkeley in 1969. Then a ruddy, middle-aged man of medium height and build, well dressed and clean shaven, he gave forceful and entertaining lectures, if not always the best prepared. He always seemed in a great rush to get it over with and on to his next project. It was a time of intense productivity for Weinberg, as he spearheaded the new burst of interest in gauge theory that began right after 't Hooft's pivotal work.

Once its infinities had been purged, the theorists had a theory they could finally *calculate,* and calculate they did. The citation history of Weinberg's 1967 paper helps to underscore the sudden surge of activity. Whereas his article had been mentioned only once in 1970 and four times in 1971, it received 64 citations in 1972 and 162 in 1973. Weinberg himself did plenty to foster this rejuvenation, writing three important articles in the last three months of 1971 alone.

That fall, SLAC and the entire quark-parton camp was in an uproar over the spurious *N/P* ratios Kendall had shown at Cornell. Our group was too busy preparing follow-up experiments to pay much attention to the resurgence of quantum field theory going on behind our backs. These were very abstract calculations, after all, while we had (or thought we had) a real, live, throbbing anomaly on our hands. Experimenters in general were still largely unaware of gauge theories.

But Weinberg and his comrades were about to change all that. By building *specific* theoretical models that incorporated gauge fields, broken symmetry, and all the Higgs paraphernalia, these theorists could make detailed predictions for experimenters to test. If the predictions were borne out, unity might be near at hand.

All unified gauge theories predicted the existence of a very heavy, neutral boson—Glashow's Z° ("zee-zero") or something like it—as an additional carrier of the weak force. The traditional weak theory, as enunciated by Fermi in 1934 and elaborated further in 1958 by Feynman, Gell-Mann, and others, needed massive charged W's, to be sure, but had little use for such a neutral partner. If found, the Z° would be the real clincher, and take us a long way down the road to unity. But with a tremendous mass of 80 GeV or more, it was awfully unlikely to turn up at any atom smasher then built—or even planned, for that matter.

So Weinberg and the growing legions of gauge theorists

turned next to the *indirect* effects that a Z° might have upon experimenters. Though we didn't yet have the energy to create a real, live Z° in our detectors, they reasoned, the Uncertainty Principle offered a loophole. We could make fleeting, *virtual* copies of the Z° that influenced the way our lighter, visible particles collided with one another. And the best place to look for such effects was in neutrino scattering.

When a neutrino collided with anything, according to the existing theory of the weak force, it coughed up a charged W and changed into a lepton of the opposite charge—an electron, positron, or muon. Like the hypothetical Z°, these W's would be far too heavy and live far too briefly to be seen, serving only to carry the weak force between the protagonists. But the charged leptons emerging from such a changeover left very distinctive tracks in particle detectors, footprints experimenters used to discover that a neutrino collision had indeed occurred.

At Brookhaven in 1962, Melvin Schwartz and his Columbia coworkers had used the long penetrating tracks left by muons in spark chambers to prove that there were two different kinds of neutrinos. Bubble-chamber physicists later used the same long muon tracks as evidence that a muon-type neutrino had collided with a nucleus in their chambers. The neutrino itself, remember, is invisible—being neutral and interacting only weakly. We know it has collided only because a long, bright track suddenly emerges out of nowhere in our detector, along with a spray of other debris from whatever object it strikes.

Now if these new gauge theories were correct, Weinberg reasoned, then the neutrino should sometimes disgorge a Z° during a collision, too. But since the Z° is neutral, and the total electric charge must not change in any collision, the neutrino *cannot* change into a charged lepton before its getaway. It must flee the scene as its same old neutral, invisible self. All an experimenter would ever see is a spray of debris emanating from the hapless victim.

Here was a key signature that could put the new unified gauge theories on a sound experimental footing: neutrino-nucleon collisions *without* any electrons, positrons, or muons emerging from the point of contact. If a Z° were involved in the exchange, all one should see would be a spray of hadrons.

* * *

Experimenters had already looked for such tendencies before, in neutrino scattering and elsewhere, but with little enthusiasm. Weak "neutral currents," an unimaginative name for weak interactions transmitted by neutral intermediaries—where no electric charge was swapped in the exchange—had appeared on the "wish lists" of theorists since the early 1960s. But they had received little emphasis, and so were not a high priority for experimenters. Nobody had any pressing need for neutral currents until the 1970s.

In his classic two-neutrino experiment of 1962, Schwartz had witnessed a number of odd events in his spark chambers that had no obvious muon or electron emerging. Such events he dismissed as a background contamination produced not by neutrinos but by neutrons liberated just upstream of these detectors, in the iron shielding used to filter out other spurious particles. Such neutrons supposedly crept into the equipment unseen, until they collided with a nucleus and left a visible spray of debris.

Donald Perkins and other CERN physicists had also noticed similar "muonless" events when they passed a beam of muon-

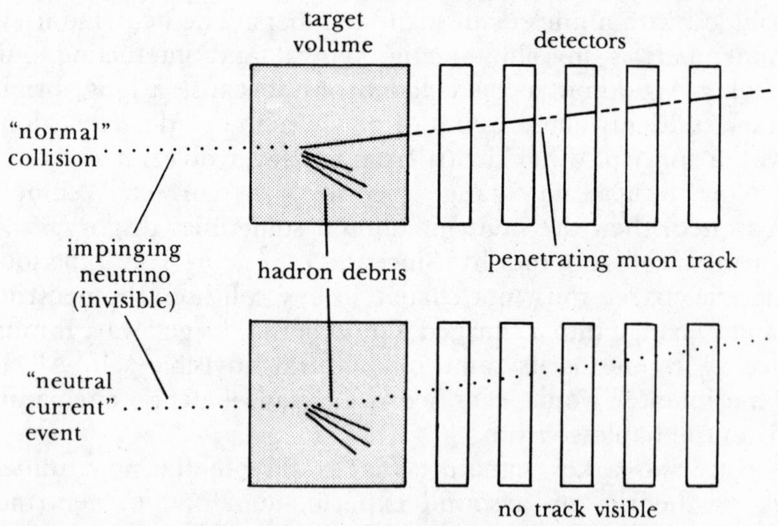

A normal neutrino-nucleon collision and a "neutral current" event. In the former case, the neutrino becomes a muon, which leaves an obvious penetrating track in addition to the expected shower of hadrons from the struck nucleon. In the latter, only the hadron shower is visible because the departing neutrino exits without leaving a trace.

type neutrinos through a freon-filled bubble chamber in 1963. But once again, these oddball events were dismissed as neutrons sneaking in. The CERN group concluded that genuine neutral intermediaries could be responsible for no more than 3 percent of the neutrino collisions, a figure later revised upward to about 10 percent.

In 1970, Schwartz had shipped his spark chambers to SLAC and installed them in a hole carved into the hillside behind End Station A. It was an imaginative, speculative experiment in which he hoped to trap any neutral particles created in our beam dump (the tank of water absorbing our electron beam) that somehow managed to tunnel all the way through the hill to his equipment. Most of the particles making it through were neutrinos that, if they interacted in the chambers at all, sported a long, penetrating muon track with a shorter spray of hadrons.

"So we did that experiment, ran a couple of weeks," Schwartz recalled, "and lo and behold, we see these two events that don't have any muons on them. Good, energetic, solid events, with *no* muons. And of course lots of events with muons on them—just the neutrino background. And the two events with no muons, God knows what to do with them."

Schwartz went back to the SLAC program committee and implored them for another month on the machine, this time as prime user in a full-fledged experiment. He wanted to dig a deep hole right smack behind the beam dump, where the event rate had to be much higher, and put in a lot of iron shielding to filter out backgrounds. Schwartz figured he needed about a hundred such muonless events to determine what in fact they might be.

But Panofsky granted him just a few more days in the existing hole, tacked onto the end of our own experiment. We had to wait several days before taking down our own equipment while Schwartz and collaborators literally *poured* an intense, high-energy beam of electrons through End Station A and into the beam dump. The radiation levels around SLAC rose alarmingly, and the experiment ended abruptly when a vacuum pipe burned out suddenly with an earsplitting bang.

All told, Schwartz had four or five "oddball events," which he showed the next year at Amsterdam, the same meeting where 't Hooft and Veltmann excited field theorists with their news of renormalization. But nobody there drew any connection. "I was puzzled by those events, everybody was puzzled by those events," Schwartz recalled. "In retrospect, they were probably neutral currents." Playfully dubbed "melons" by snickerers, the oddball events became one more unexplained curiosity among many.

The lack of enthusiasm for neutral currents was partly the result of their almost complete absence from other kinds of experiments. Any such intermediaries should have influenced the slow disintegration of neutral kaons, for example, which would have decayed rather frequently into a pair of oppositely charged muons. But such a route was taken thousands of times less often than expected.

The mysterious absence of this decay posed serious problems for the quark model. Two quarks supposedly jitterbugging around inside a kaon should occasionally annihilate one another, with their total energy emerging as a pair of muons. Why then did it happen so rarely, if at all?

In 1970, a possible solution occurred to Sheldon Glashow, who had returned to Harvard as professor of physics and was working there with two visiting European theorists, John Iliopoulos and Luciano Maiani. A tall, robust man with a strong penchant for pungent cigars and playful banter, Glashow was already well known and liked in the Cambridge physics community. Iliopoulos and Maiani joined him there in 1969 to finish up work begun at CERN the previous year.

While a visitor himself at the Niels Bohr Institute in 1964, Glashow had written an article with Bjorken suggesting a possible *fourth* quark, to complement the up, down, and strange quarks just proposed by Gell-Mann. "We called our construct the 'charmed quark,'" recalled Glashow, "for we were fascinated and pleased by the symmetry it brought to the subnuclear world." There were four known leptons at the time—the electron, the muon, and their two different neutrinos (plus antiparticles). Symmetry between leptons and quarks meant a fourth quark was needed, too.

Not much had been made of the charmed quark until 1970. Early one blustery February day, Glashow, Iliopoulos, and Maiani suddenly realized that a fourth, heavy quark magically suppresses neutral kaon decay by opening up a second pathway that cancels out the first. Quantum mechanics allows, and even *encourages,* such sleight of hand. Once they had this flash, Iliopoulos recalled, "The solution was trivial." They grabbed their coats, hopped in a car, and sped across Cambridge to MIT, where theorist Francis Low had been working on the same problem.

An impromptu seminar soon gathered in Low's office, with Weinberg listening intently and asking some probing questions. "We argued a lot," Iliopoulos remembers, "and spent all morning discussing these things"—gauge theories, symmetry breaking, the whole lot—but nobody made a peep about Weinberg's 1967 paper, "not even Steve himself." Glashow himself was "totally unaware" of that particular paper at the time. Here was a marvelous explanation for the apparent lack of neutral currents in kaon decays, the very same currents *required* by Weinberg's theory, and nobody in the room made the connection.

Years later, Iliopoulos quizzed Weinberg about the oversight. "Steve, why didn't you put us on the right track?" To which Weinberg confessed that he'd had a "psychological barrier" against his 1967 paper. He had encountered enormous frustration in trying to purge the theory's infinities, and somehow just did not want to think about it in 1970.

After the summer of 1971, however, it was a new ballgame: 't Hooft's renormalization of Yang-Mills gauge theory loosed a flood of theoretical activity, with Weinberg taking the lead. He was among the first to realize that the mechanism of Glashow, Iliopoulos, and Maiani had solved the puzzling absence of neutral currents in kaon decays. But their effects should still turn up, he predicted, in neutrino scattering.

The best place to search, Weinberg figured, was in bubble chambers, where a muon-type neutrino might occasionally knock out an atomic electron. This can occur if the neutrino swaps a Z° with the electron, but it will never happen through the exchange of a charged W. Here was a clear-cut test for neutral intermediaries. If possible at all, such a collision would appear as a single tenuous electron track emerging out of nowhere.

When Gargamelle experimenters were approached with this idea in November of 1971, however, they gave it a cool reception. Though a "clean" test with very little background, neutrino-electron scattering was an incredibly feeble process. Andre Lagarrigue, Paul Musset, Donald Perkins, and the other Gargamelle leaders understood the theorists' enthusiasm for the Weinberg-Salam theories. But a decade of unsuccessful searching had made them leery of neutral currents.

A much bigger signal was expected in neutrino-nucleon scattering, but the noise was a lot higher there too. Here were the infamous neutron backgrounds that had troubled neutrino experimenters during the 1960s. Instead of interacting inside the chamber, a neutrino could also ricochet off a nucleus in the shielding, the floor, the magnet, or the walls of the detector. Neutrons emerging from these encounters then crept unseen into the chamber liquid, where they would collide with another nucleus, giving rise to a visible spray of particles with no obvious muon track. Such neutron events mimicked the muonless events that were the definitive test of neutral currents. How could one tell the difference?

But by April of 1972 the Gargamelle experimenters were more hopeful that these bothersome neutron backgrounds might be manageable, and the search for neutral currents began in earnest. The group had had plenty of experience dealing with such problems in earlier experiments, and Gargamelle's sheer size gave them an added advantage. By then, too, several theoretical calculations suggested that neutral intermediaries would crop up fairly often—as much as 20 percent of the time. When you encounter a lot of noise, it helps to have a large signal.

Ever since Gargamelle had begun operating in 1971, in fact, the CERN experimenters had seen neutrino collisions without any muon shooting away. But these curious events had been casually dismissed as the neutron background—until theorists had a pressing need for such oddities. By July of 1972, one small team under Musset was carefully going back through all the muonless events, trying to determine their true origin. Another group led by Charles Baltay and Helmut Faissner looked for the rare neutrino-electron encounters. The search was on.

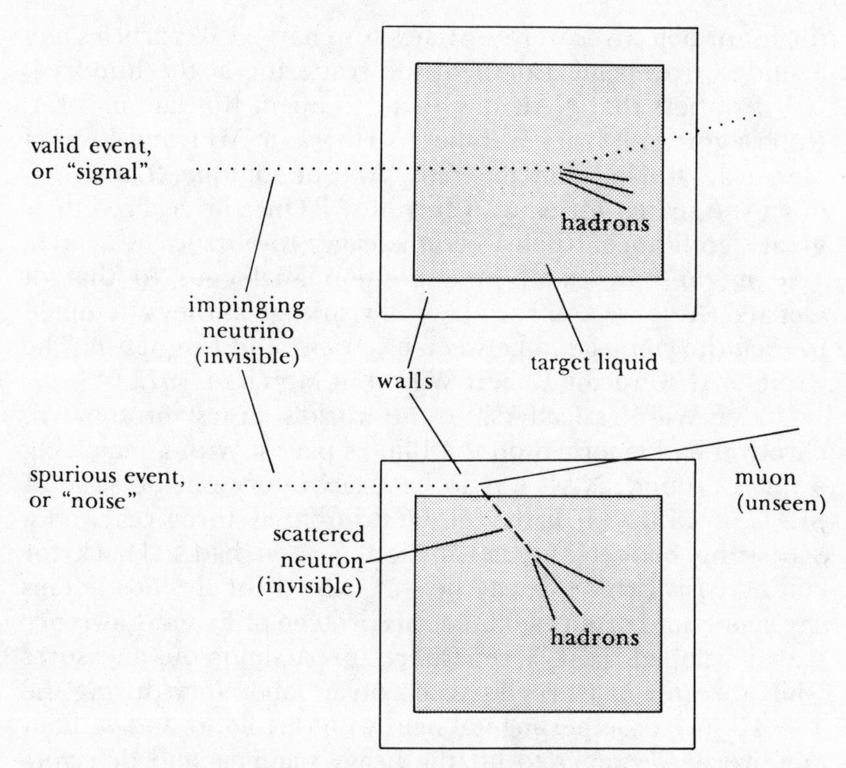

How a stray neutron can mimic a muonless neutrino event in bubble-chamber experiments.

Gargamelle physicists were not the only experimenters thinking about neutral currents in 1972. Another collaboration from Harvard, Pennsylvania, and Wisconsin was busy readying the first neutrino scattering experiment that year at the National Accelerator Laboratory (NAL) west of Chicago. The idea for their large detector had germinated three years earlier in the minds of Alan Mann of Pennsylvania and David Cline of Wisconsin. To add leverage to their proposal and help them build the device, they had joined forces with Carlo Rubbia, a Harvard experimenter who spent much of his time at CERN. In a fashion becoming typical of high-energy physics, their alliance was forged in a 1969 meeting held at New York's Kennedy Airport.

When they submitted their proposal in June 1970, it made no mention whatsoever of neutral intermediaries. Uppermost on the list of objectives were searches for charged W particles and a study of deep inelastic neutrino scattering at the hundred-GeV frontiers that NAL was about to open. But late in 1971, Rubbia got a call from Weinberg. "He was at MIT and I was at Harvard," Rubbia remembered. "And he told me, 'Look, why don't you search for neutral currents.' " Once he realized their great significance, Rubbia became eager to pursue the search. "We might now stand in a position analogous to that of Oersted, Ampere and Faraday 150 years ago as they attempted to elicit the connection between electricity and magnetism," he wrote NAL Director Robert Wilson in March of 1972.

Under Wilson's leadership, the world's largest proton synchrotron had emerged on the Illinois plains. With a main ring 4 miles around, NAL was to be a laboratory on the scale of SLAC or CERN. It had been built in barely three years on a shoestring budget of $250 million. Wilson had a knack for cutting costs (which is why he was named for the post). This shy, aloof, and often cantankerous protégé of Ernest Lawrence had developed quite a repertoire of parsimonious measures while director of Cornell's synchrotron laboratory during the 1960s—such as experimental halls with dirt floors and without any overhead cranes to lift the heavy magnets and detectors into place.

Shaking their heads, physicists often recall his cost-cutting measure on the magnets used to guide protons around the NAL ring. Wilson had ordered his staff to cut back sharply on the amount of epoxy they applied to the magnet coils as a binder and insulator. When the magnets were powered up, however, the coils shifted under the stresses, and tiny cracks developed inside the epoxy. Moisture began to work its way into these cracks during the humid summer months of 1971, and the magnets started to short out and explode. All the hapless technicians could do was wait for another magnet to blow up, then go rip it out of the ring and repair or replace it. Hundreds blew up that first year; eventually all 1000 magnets had to come out.

The frustrated experimenters, many of whom had worked around the clock getting ready to begin taking data that year, ahead of schedule, had to wait and watch helplessly as the huge machine came apart literally at its seams. Finally things got drier during the winter, and the blowouts became less frequent.

By late January, they were getting steady proton beams of 100 GeV in the ring, the first time that milestone had ever been passed. On March 1, 1972, it reached the design energy of 200 GeV. Wilson and his staff could celebrate at last.

NAL made a mixed beam of neutrinos and antineutrinos by slamming the high-energy protons into metal targets, thus creating a plethora of pions and kaons. These decayed into, among other things, muons and muon-type neutrinos or antineutrinos. After the muons had been filtered out by passage through a kilometer-long earthen mound, all that remained was a pure beam of the ghostly neutral particles.

It was Thanksgiving before the Harvard-Penn-Wisconsin (HPW) experimenters began their first few days of data-taking. By then, they were a year behind their CERN competitors. Perkins had already presented the earliest Gargamelle neutrino data in the Rochester Conference held at NAL that September,

NAL Director Robert Wilson leads his staff in a champagne toast to their successful acceleration of 200 GeV protons.

but the CERN studies of neutral currents were so far inconclusive. Rubbia and company still had an outside shot.

What had long been considered noise by CERN scientists was gradually becoming a signal in late 1972 and early 1973. Detailed analyses of their many muonless events showed that some, but probably not all, could be explained away as neutrons infiltrating Gargamelle and colliding inside. But then there were other backgrounds to consider—like neutral kaons creeping in and colliding, or muons misidentified as pions. So when Paul Musset presented some preliminary data at a conference in early 1973, he made no overt claim for the discovery of neutral currents. It was more a progress report, in which he explained the group's efforts to cope with their capricious backgrounds.

Musset did not mention, however, the discovery that had excited him the most. In January, Franz Hasert, a graduate student from Aachen working with the group, had been reexamining some of the older Gargamelle photos and in one photograph had spotted a single, spindly track appearing out of nowhere, which the scanners had mistaken as a muon. Hasert recognized it instead as an *electron*—and the event as a possible occurrence of neutrino-electron scattering, which suggested the existence of neutral currents.

The excitement intensified the next week as copies of the photo were passed around the group for comments. Here at last was the gold-plated event they had sought for so long. "The event has excited us a great deal," wrote Aachen leader Helmut Faissner to Lagarrigue after conferring with Perkins. "It is in effect a lovely candidate for an example of the neutral current."

But the cautious Gargamelle experimenters did not publish anything about this event right away, preferring instead to make some additional measurements with their bubble chamber and to continue studying background processes. There was a small fraction of electron-type neutrinos passing through the chamber, for example, which could generate an electron track *without* the help of any neutral intermediary. Another five months passed before the physicists finally published.

Musset and the others continued analyzing the neutron background in excruciating detail that spring, finally conclud-

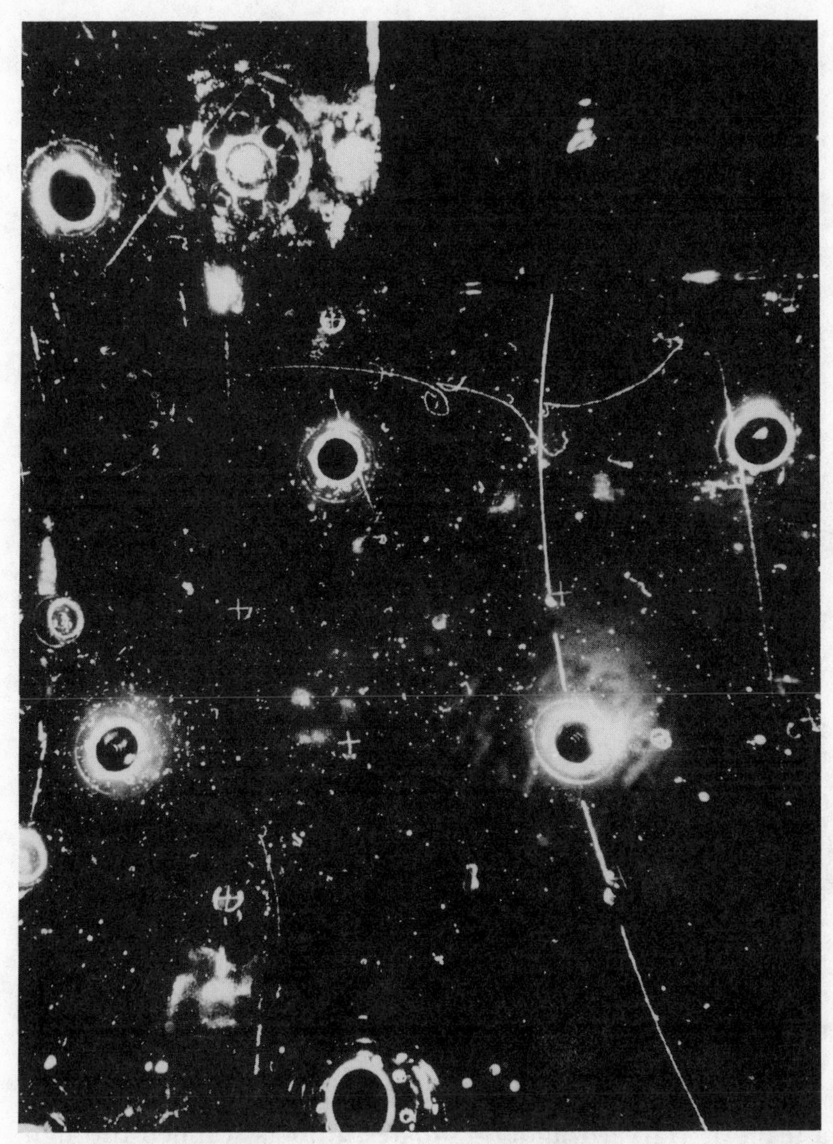

The first Gargamelle photograph of a neutrino-electron collision. The horizontal track emerging at left is that of an electron kicked out of an atom by a neutrino.

ing that it contributed no more than 20 percent of all the muonless events observed in Gargamelle. But arguments still erupted almost daily over other possibilities. The CERN physicists might have continued their deliberations the rest of the year had they not learned in July that NAL was hot on their heels.

In February 1973, on one of his frequent CERN visits, Rubbia also heard the rumors of Gargamelle's gold-plated event. He returned to Harvard determined to discover neutral currents himself. A thick-set, double-chinned slab of a man who earned his doctorate at Columbia before heading for CERN, Rubbia maneuvers his way through particle physics with all the delicacy of a Mack truck in a flowerbed. Dubbed "the Alitalia Professor" by Harvard students because he spends so much time on jumbo jets, this indefatigable Italian native lectures at break-neck speed, flipping from one graph to the next before his listeners have had much chance to comprehend anything.

Living his life at the frontiers and in the fast lanes of particle physics, Rubbia hates to be scooped. Some physicists grumble that he would rather be *wrong* than come in second, but that is a little unfair. When you swing for a home run as often as he does, you expect to strike out a lot.

The HPW detector built by Rubbia and company had more than twice the target mass of Gargamelle, and it received neutrinos at ten times the energy, too, where they collided ten times as often. So these physicists had an advantage of almost *twenty* in counting rate. Thus the few weeks of running time they had eked out in late 1972 still gave them a comparable number of neutrino events.

Together with Larry Sulak, a young assistant professor, and a team of Harvard undergraduates, Rubbia scanned through the pictures of their events that spring, searching for those without any muons flying off. They indeed found muonless events—"lots of them," one student recalled. The only question was how to interpret them. Were they neutron background? Or events where a muon had simply escaped detection? Or solid evidence for neutral currents?

At NAL's extremely high energies, neutron backgrounds

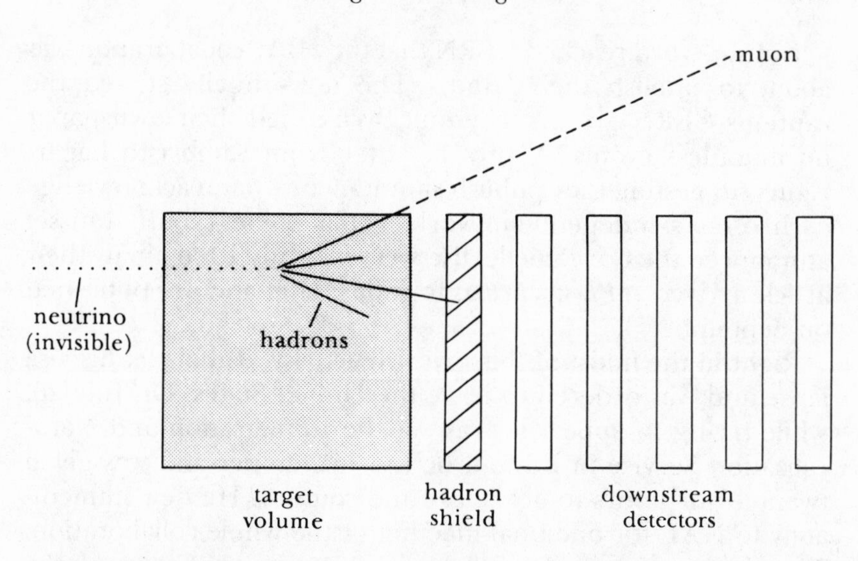

A large-angle muon emerging from a neutrino collision in the HPW detector. Such an event could be mistaken as a muonless event because there was no long, penetrating track in the rear portion of the detector.

were not the problem they had been for Gargamelle. But the HPW detector had its own nemesis—the possibility that some muons emerging at large angles might exit the detector or be misidentified as hadrons. Whenever this happened, a routine collision could easily be mistaken as a muonless event and taken as evidence for neutral currents. So Rubbia and Sulak hastily prepared a "Monte Carlo" computer program to simulate this false contribution, and convinced themselves it could only explain half the muonless events they were seeing. The other half, they concluded, had to be a true signal.

But they still had to convince the rest of the group, particularly Cline and Mann, who were far more skeptical. Before the NAL experiment, both of them had searched fruitlessly for neutral currents in kaon decays. When evidence for neutral currents began to appear in neutrino scattering, their natural inclination was to doubt the Monte Carlo program, to suspect the backgrounds had not been understood properly, to ask for more studies. "They had been vaccinated, so to speak," quipped Rubbia. He was getting impatient, knowing full well that Gargamelle had solid proof they might release any day. "If you've got results," he insisted again and again, "you publish them!"

In July, word reached CERN that the HPW collaboration was about to publish their "find." The news finally stirred the cautious CERN group; they quickly finished their own paper on muonless events. A July 17 letter from Rubbia to Lagarrigue, suggesting they publish simultaneously and acknowledge each other's independent work, got a polite rebuff. Musset announced the Gargamelle discovery in a July 19 seminar; their article arrived at *Physics Letters* four days later and was published on September 3.

Right in the midst of this feverish activity, Rubbia let his visa lapse and was ordered to leave the United States. On July 26, while trying to appeal his case at the Immigration and Naturalization Service in Boston, he lost his temper and was given twenty-four hours to get out of the country. He flew immediately to NAL for one final meeting of the whole collaboration. Then he sped away, with Sulak at the wheel, racing through the Chicago suburbs to O'Hare Airport, arriving there just in time to catch a flight to Europe.

With Rubbia deported, Sulak took charge of getting the Harvard studies published. He finished the paper and carried it himself to the Brookhaven offices of *Physical Review Letters* on August 3. Meanwhile, preprint copies had been circulating since late July, encountering widespread criticism—particularly of the Monte Carlo program used to simulate muons escaping the detector. Sulak did his best to answer these qualms, reworking the analysis and submitting a revised paper. But Rubbia, the driving force behind the discovery, had been cut off from the action.

Aix-en-Provence, an ancient town near Marseille, was the site of the September 1973 International Conference on Elementary Particles, another off-year alternative to the Rochester Conference. Abdus Salam and a colleague arrived by train and were lugging their heavy baggage on foot through the cobblestone streets when a car lurched up from behind and stopped abruptly. The shaggy-haired, bespectacled driver leaned out the window and inquired, "Are you Salam?"

"Yes," he replied, a little perplexed at such a greeting and weary from the arduous walk.

"Get in the car," ordered Paul Musset, the driver; "I have

news for you. We have found neutral currents." (Recalling this moment, Salam could not remember whether he was more pleased by the exciting news or by the offer of relief from his heavy luggage.)

The Gargamelle results took center stage both in Aix and at the Electron-Photon Symposium held in Bonn, West Germany, that summer. This was the first time neutrino physicists had even been allowed to *speak* at the Electron-Photon Symposium, and they almost stole the entire show. Oxford's George Myatt, one of the Gargamelle experimenters, recounted their twin discoveries—the single neutrino-electron event and the hundred-odd examples of muonless neutrino-nucleon scattering. Roughly between 20 and 30 percent of the time, he noted, a neutrino caromed off a nucleon without any muon zooming away. Handed the HPW paper just before his talk, Myatt included a brief mention of their results, too, but only as a *confirmation* of the Gargamelle discovery.

At Aix it was much the same story. Musset gave a summary of the year's work in neutrino physics, with the Gargamelle discovery and HPW's confirmation of neutral currents the highlight of his talk. Following him immediately, however, Steven Weinberg was cautious. "It is perhaps premature to conclude from all this that neutral currents have really at last been observed," he warned. "There may be some mysterious source of background contaminating all these experiments." And there were other ways to explain these results without having to invoke the Z° required by the Weinberg-Salam theory. But Weinberg took heart that, based on these new findings, his proposed unification of weak and electromagnetic forces "may not be so far from the truth."

With Rubbia still in Europe during September and October, however, the conservatism of Cline and Mann began to take over at NAL. All along they had been very uneasy about the muonless events and the neutral current paper, preferring instead to study the more "normal" neutrino collisions where a muon came shooting out of the debris. Both Cline and Mann never really trusted the slick Monte Carlo program Rubbia and Sulak had used to simulate muons escaping at large angles. So early that fall they rebuilt the HPW detector for one more experiment intended to settle this question once and for all, by making a *measurement.*

There had already been troubling indications, from runs at higher energy made earlier that year, that the muonless events were not quite as large a fraction of the total as it had first seemed. Because of these data the first paper, which reported a 42 percent ratio of events without muons to those with muons, had been revised downward to 26 percent by the Bonn Conference. The mere fact that this ratio had shifted so much in a single month was cause for worry.

When preliminary results began coming in from the rebuilt device in early October, this ratio seemed to plunge still lower—less than 21 percent and possibly even *zero.* The muonless events were now beginning to disappear, and along with them the evidence for neutral currents at NAL.

To make matters worse yet, the referees' comments on their paper arrived from *Physical Review Letters* in mid-October. The Monte Carlo simulations were harshly criticized, with both referees advising against publication until their objections had been met. They echoed the general sentiment of the particle physics community, which at the time felt that the muon identification problems had been swept under the rug by the HPW collaboration. Rather than contest the referees' judgments, the experimenters simply shelved their paper for a few months.

By early November, Cline and Mann were convinced that the muonless events had disappeared for good. Just then returning from Europe, Rubbia went along with their conclusion. They began preparing a different paper that contradicted the earlier one, as well as the published Gargamelle results.

When word of this null result reached Geneva later that month, it caused widespread consternation. Worried that CERN might be embarrassed by the latest HPW findings, Director Willibald Jentschke called a meeting of the Gargamelle team to challenge their earlier conclusions. Though thoroughly shaken, they refused to back down.

Meanwhile, all the word-of-mouth reports and rumors were causing great confusion in the physics community. What was the true answer? Did neutral currents exist or not? Some began to chuckle aloud that Cline, Mann, and Rubbia had discovered "alternating neutral currents."

By mid-December, the HPW group was changing its mind again. The rebuilt detector, it seemed, had allowed pions emanating from neutrino collisions to "punch through" into the rear portion of the detector, where they were misidentified as muons. When this happened, a genuine muonless event could acquire a spurious "muon" track and fall into another category. All that fall, the HPW group had underestimated such a possibility, with the result that muonless events had "disappeared" from the rebuilt device.

When the "punchthrough" was finally understood properly, muonless neutrino events reappeared at NAL. By February 1974, Cline, Mann, and Rubbia were confident of their positive findings and agreed to resubmit the original paper, along with suitable modifications and the comment that further studies had confirmed these results. In March, they sent in another paper reporting the results with the rebuilt detector, in which the all-important ratio of muonless events to those with muons had settled out at 20 percent. After a lot of second-guessing, neutral currents seemed here to stay.

In retrospect, all the confusion over neutral currents at NAL had occurred because of a rush to judgment. Prodded hard by Rubbia, the HPW collaboration had released their first results well *before* they completely understood their equipment. So they had to endure the embarrassment of seeing their numbers shift all over the map while they struggled to come to grips with its idiosyncrasies. The process of winnowing out fact from fallacy, which normally occurs in the private discussions of a scientific collaboration, went on essentially in public.

When pressed on this question, Rubbia preferred to place the blame on the Immigration and Naturalization Service. Right at the most critical juncture, he had been forced to leave in a hurry, depriving the collaboration of its most forceful neutral current advocate. Asked whether events might have gone differently if he had remained in the U.S., Rubbia replied:

> Had we [had] the possibility to stay there and work all together and hammer it through in a closed room, rather than with the world listening? Well, that probably should have happened, right? That's what we really *wanted* to have happen. That would have been the natural course, but external circumstances didn't let it work that way.

One might add, however, that the accelerated publication schedule was Rubbia's own choosing, and it became a major reason for the subsequent embarrassment.

The Gargamelle collaboration, by contrast, published only after a year of soul-searching analysis—and then only because they feared being scooped. Preceding that had been a decade of bubble-chamber experiments in which these experimenters gradually came to grips with the many difficult backgrounds encountered in neutrino scattering. In the final analysis, neutral currents were discovered at CERN and confirmed at NAL.

By the Rochester Conference that summer in London, there could be little doubt about weak neutral currents. A Caltech group also reported positive findings at NAL, as did another team at Argonne. Neutral currents were here for good, and with them the unified quantum field theories of the weak and electromagnetic forces.

The seeming irrationality of the weak nuclear force was finally beginning to dissolve. Its manifest asymmetries could be interpreted as a freak of Nature, the result of inevitable choices made at the very birth of the universe. In unified gauge theories, the electromagnetic and the weak forces are just two separate manifestations of a single fundamental force. They appear so different because seemingly very different but closely related spin-1 intermediaries are swapped in the two cases: a massless, parity-conserving photon that works exactly the same in the looking-glass world, and three massive, parity-violating bosons that do not. Beneath it all rests a perfect but *hidden* symmetry of the underlying equations, a serene refuge safe from everyday chaos. Nature was beginning to make sense.

CHAPTER 10

THE DYE THAT BINDS

By faith we understand that the world was created by the word of God, so that what is seen was made out of things which do not appear.

—Saint Paul, HEBREWS

NOON at CERN is a time to relax over lunch. In clement weather, physicists from many nations cluster about the tables on a broad patio outside the cafeteria, enjoying light sandwiches or full meals, sipping wine or coffee, and engaging in lively debates about the latest discoveries, theories, or rumors. Beyond the patio lies a grassy field, bordered by a row of tall poplars riffling in the breeze, where two or three players are often kicking a football back and forth. On a clear day, snow-encrusted Mont Blanc, Europe's tallest peak, can be seen peering through a cleft in the nearby hills.

The revelations at the recent Bonn Conference were the favorite topics of luncheon conversation when I arrived at CERN in September 1973. Strolling out on the patio in search of a table, I happened upon Robert Jaffe, a young MIT theorist and friend of mine, who was visiting CERN after the Conference. The previous spring he had helped me interpret the results of my thesis experiment, which was reported at Bonn and taken as convincing evidence that the partons were spin-½ entities—another important stone in the quark-parton edifice. Apparently my thesis had been a big hit at the affair.

Jaffe then recounted all the excitement at Bonn over neutral currents and gauge theories—the unification of weak and electromagnetic forces had now become a distinct possibility. What's more, he added, these new theories and an even more recent idea called "asymptotic freedom" might hold a key to solving the baffling paradox posed by the MIT-SLAC experiments.

In some gauge theories, Jaffe noted, forces become quite *feeble* at short distances—the very opposite of commonsense expectations based on the behavior of gravity or the electric force, which get *stronger* when two objects are closer together. If true of the forces holding a nucleon together, asymptotic freedom might explain why quark-partons behaved like free particles when struck by an electron, yet quarks never shot out after a collision. At large distances the force could simply get stronger and stronger, yanking them back into the fold.

The most appealing feature of the new idea was how this seemingly paradoxical behavior arose in a very natural way, with a minimum of artifice. And, if such theories were true, it meant that the strong force was a close cousin of the electromagnetic and weak forces, all of them arising in gauge theories of the type originally written down by Yang and Mills. It could answer so many stubborn questions all at once, and do so in the context of a grand and strikingly beautiful synthesis. Here, at last, was a theory worthy of extra attention.

Ever since quarks had first been proposed, the force tethering them inside hadrons had been a major stumbling block. To believe that quarks really existed, we needed some kind of force that *never* let go. Otherwise, they should escape the confines of their parent hadron, and their fractional charges would leave obvious tracks in particle detectors. Some theorists had developed ersatz models in which the quarks were attached by elastic bands or were trapped at the bottoms of deep wells or in steep valleys—kind of like billiard balls confined to a pool table, but free to move upon it. But these crude models were hard to take seriously. They were only answering a question with another question. There had to be a more fundamental, intellectually satisfying solution.

Gell-Mann had skirted this thorny issue in his current-algebra approach to hadron behavior. Although it might not be clear exactly what it was tethering quarks together, there were certain abstract features that were *independent* of the exact nature of this force. Such a "veal and pheasant" approach allowed field theorists to concentrate on the apparent symmetries of the hadrons without getting too bogged down in the messy details of their internal dynamics. It also helped to foster the notion, prevalent in the late 1960s, that quarks were mathematical entities—not real, material particles.

As interpreted by Bjorken, the flat, scaling behavior observed in the MIT-SLAC experiments provided an important hint: whatever force held nucleons together, it was surprisingly weak at short distances. If there were any fundamental constituents "inside," they were pointlike and stuck rather loosely together. Impinging electrons could easily pick them off, one at a time.

But when theorists used the available tools of ordinary field theory to *calculate* such a force, difficulties arose. Either they had to impose hobbling restrictions on their theory, or they could not obtain such a pointlike internal structure. Sidney Drell and a small army of SLAC coworkers, Jaffe included, had worked on these problems in 1969 and 1970, without much success. Other theorists attacked the same question elsewhere, with the same general result: scaling did *not* arise in field theories—at least not without a lot of gimmickry. All the while, the MIT-SLAC group kept reporting better and better data, all of which bore out Bjorken's original prediction. So much for ordinary field theory.

In the current-algebra picture, scaling was interpreted as "the behavior of two nucleon currents separated by light-like distances," evidence of a "singularity on the light cone." It was all very abstract and confusing. Experimenters could work through all the detailed mathematics and convince ourselves it was reasonable, but we preferred to use the good old familiar parton model in doing our own calculations. Later, Jaffe demonstrated that these two approaches were essentially equivalent, as Heisenberg's matrix methods and Schrödinger's Wave Equation are two equivalent formulations of quantum mechanics. Two facets of the same gem.

In the quark-parton approach, the forces between quarks required another kind of parton—the spin-1 force carriers dubbed "gluons." Julius Kuti and Victor Weisskopf had developed the idea extensively in their quark-parton model of 1971—portraying a nucleon as a swarm of quarks and antiquarks, with gluons flitting betwixt and between, keeping it all together. Though a convenient mental image, however, this model gave us hardly an inkling of how to *calculate* gluon properties—that is, the actual nature of the forces binding the innards of a nucleon.

And in a parton picture, perfect scaling meant that the quarks flying around inside a nucleon had to be *absolutely* free. But if free, these fractionally charged grains should then have emerged when struck, leaving obvious tracks in particle detectors. What was going wrong here?

The solution to this conundrum emerged from a number of different directions, with contributions from many theorists. "It was kind of a resonance in the physics community," Jaffe recalled. At the focus of this activity was Murray Gell-Mann, the man who wove the different threads into one coherent pattern. Having recently received the Nobel prize for his development of the Eightfold Way, he had been sitting contentedly on his laurels, watching the theoretical explosion coming in the wake of the MIT-SLAC experiments. But in 1971 he leapt back into the action.

His collaborator in this work was Harald Fritszch, a headstrong young theorist who had defected from East Germany and come to Caltech to work as a postdoc. With the secret police on his tail, he had kayaked off into the Black Sea from Bulgaria, eventually reaching freedom on the shores of Turkey. Gell-Mann remembers their first encounter fondly, coming as it did a scant three hours after the great San Fernando earthquake of February 1971.

In the autumn of that year, while Gell-Mann was at CERN on sabbatical, they began trying to understand (within the context of current algebra) how neutral pions decayed. These fragments seemed to decay about nine times too quickly if one allowed only three fractionally charged quarks—the minimal set. Fritszch and Gell-Mann realized that such a problem did not arise if each up, down, or strange quark itself came in *three* different kinds, increasing the decay rate ninefold.

In this picture, each quark had to have some new physical

property, something that made them in fact *different,* which Gell-Mann with characteristic aplomb dubbed merely "color." It was "a natural choice," he recalled. With Feynman in the late 1950s he had advocated "red" and "blue" neutrinos to distinguish two different kinds of these ghosts, names that never caught on. Now he resurrected the old terminology and applied it to quarks, which henceforth came in red, white, and blue varieties. He did not mean that quarks actually had *visible* colors; that was absurd. "Color" was just another name among many, plucked from everyday discourse to help physicists communicate. Like strangeness, it later acquired a much deeper meaning through repeated usage and further elaboration.

One immediate reward of the color idea was that it solved the problems with Pauli's Exclusion Principle, which the simple three-quark model had always encountered (see Chapter 4, p. 109). Recall that this principle would not allow two *identical* quarks to occupy the exact same quantum state, as apparently had to happen if there were two up quarks in a proton. With each quark in a proton sporting a different color—red, white, or blue—no two of them were ever identical. Thus, they could occupy the same turf simultaneously. So Pauli's sacrosanct principle was not violated as long as the quarks were colored.

The idea of color helped explain why, out of all the many imaginable combinations of quarks, Nature seemed to favor only two: the pairing of a quark and an antiquark to form a meson, and the grouping of three quarks (or three antiquarks) to make a baryon (or an antibaryon). These combinations arose quite naturally if the observable mesons and baryons had to be "colorless"—that is, have *zero net color.* In mesons, a quark of one color combined with an antiquark of the anticolor (red with antired, for example) to yield zero net color. In baryons, the combination of three different quark colors somehow canceled to give a colorless result.

Physicists later adopted a new terminology of red, green, and blue quarks over Gell-Mann's original, more patriotic choice because red, green, and blue light can add up neatly to give white, colorless light. In a television set red, green, and blue combine in many different proportions to yield the complex spectrum of colors you actually see, but in the subatomic world Nature is far simpler. She always uses *equal* proportions of each color, or one color with its anticolor, to paint the colorless screen of the observable mesons and baryons.

The essential idea of color—but not the name—had occurred to other theorists before Fritszch and Gell-Mann, in different guises. "Paraquarks," advocated by O. W. Greenberg in 1964 as a way to sidestep the thorny problems with the Pauli principle, had been an equivalent formulation. Gell-Mann remembers toying with this idea late in 1963, before writing his first paper on quarks, but he was not too enamored of it and omitted any mention of paraquarks.

In 1965, M. Y. Han and Yoichiro Nambu had suggested the possibility of *three* fundamental triplets with whole number charges as one way to evade the then unpleasant requirement of fractional charge. To accomplish this, they endowed their motes with a new physical property Nambu later dubbed "charm"—appropriating a word from Sheldon Glashow. Han and Nambu even suggested that this new property might give rise to interquark binding forces.

Neither of these two proposals caused much of a stir before 1972, however. Both had introduced an extra degree of complexity to evade knotty problems of the quark model, without making any new and different predictions that could be tested by experiment. So they had languished largely unnoticed until Fritszch and Gell-Mann revived the idea of three quark triplets in late 1971. At two conferences early the next year in Italy and Austria, these theorists began to preach the virtues of color. Not only did the new property now have experimental consequences; not only did it have an appealing formulation and a catchy name; this time it also had the backing of the world's premier theorist.

Quark colors arrived at an auspicious time in high-energy physics. Our measurements of the *N/P* ratio in the second generation of MIT-SLAC experiments were showing that any partons most probably had fractional charge. Gerard 't Hooft's proof that gauge field theories could be purged of their infinities had been announced the previous summer, and Steven Weinberg was busily deriving the consequences of his own unified theory of the weak and electromagnetic forces. And just then Italian physicists were puzzling over the surprising results of collisions of electrons with positrons observed at Frascati, near Rome.

Electron-positron "storage rings" were becoming a major growth industry as the seventies began. In these machines, separate bunches of electrons and positrons zoomed repeatedly around a single racetrack in opposite directions—one clockwise, one counterclockwise—at essentially the speed of light. As they shared a common path, the bunches encountered one another at selected locations called "interaction regions." Most of the time the bunches passed right through each other with only a few minor deflections occurring, but occasionally there was a devastating head-on collision between an electron and a positron, which spewed debris out at large angles. Like the crowd of eager onlookers at a demolition derby, experimenters set up their detectors at these regions to watch the fun.

When an electron and a positron collide head-on, they often annihilate each other to form a momentary state of pure light, a short-lived virtual photon whose energy equals the sum of the energies of the luckless pair. This fireball quickly materializes into massive particles—another electron-positron pair, two

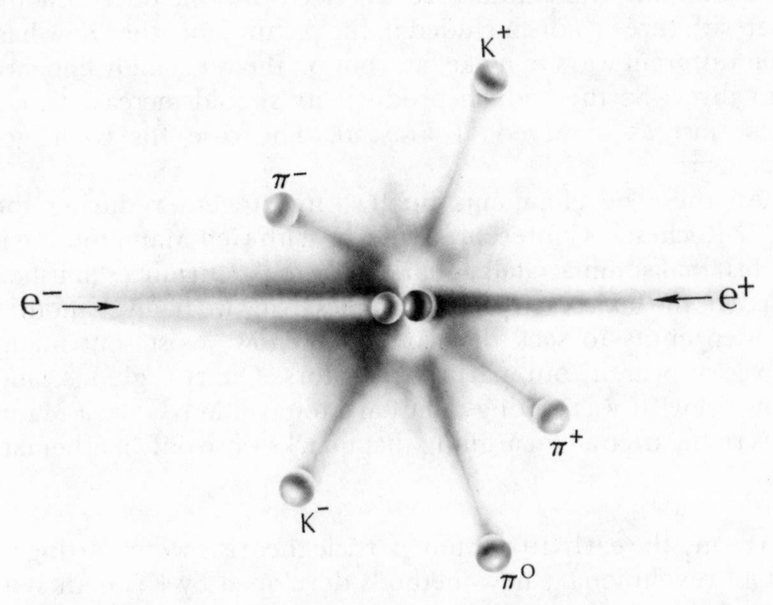

Annihilation of an electron and positron, creating a spray of hadrons from their combined energy.

muons, or a spray of hadrons. As long as there is enough energy around and no conservation laws are violated in the process, almost anything is possible.

In the late sixties, Italian physicists had built such a colliding-beam storage ring, called ADONE, in Frascati. Rotating bunches of electrons and positrons, each with energies up to 1.5 GeV (for a total of at most 3.0 GeV), met at four different points around the doughnut-shaped ring—where particle detectors sat watching for the inevitable debris. The first preliminary results, presented at Kiev in 1970, had shown that smash-ups yielding hadrons were occurring far more often than had ever been anticipated—"a garden where a desert was expected," to borrow the words of one Italian physicist. It was yet another nail in the coffin of J. J. Sakurai's vector dominance model, which had predicted exceedingly few such collisions, and yet another bonus for the parton camp.

With much better data by 1971, however, the Italians were putting even the simple quark-parton model in jeopardy. Using only the minimal set of just the up, down, and strange quarks, they still fell short by about a factor of *three* in predicting the rate of hadron manufacture by electron-positron collisions. But with three colors included in the picture, too, they now had *nine* different ways to make hadrons on their assembly line, not just three. So the hadron productivity should increase threefold, just as observed at Frascati. The case for color got stronger.

All these developments finally came together during the 1972 Rochester Conference at NAL. With Gell-Mann touting it in his final summary talk, quark color took a firmly established place in the lexicon of particle physics. Now we had yet another hidden entity to seek that could somehow "exist" but might never appear in our particle detectors. Quarks, gluons, and now color! It was getting a bit hard to take. Maybe Gell-Mann was right, after all, in arguing that quarks were only mathematical.

During the early 1970s, too, particle theorists were starting to adopt revolutionary new methods developed by Kenneth Wilson, then a professor of physics at Cornell. Ever since the late 1950s, as one of Gell-Mann's graduate students, Wilson had been struggling to attain an understanding of the strong force within the context of quantum field theory. Success had eluded him for a decade, a time when most of his contemporaries were deserting field theory in favor of the S-matrix approach of Geoffrey Chew and the Berkeley school. But in 1969 his labors began to bear fruit.

The thorniest problem about the strong force was the simple fact that it was so terribly strong. To pack a throng of unruly nucleons, about half of which sported a positive electric charge driving them astray, into a space less than a trillionth of a centimeter across obviously took quite some doing. Field theorists had a few tried-and-true methods called "perturbation techniques" for calculating the behavior of fairly *feeble* forces—like the electromagnetic and the weak forces. Such methods treated the effects of a force upon a system as small perturbations upon its normal states. They worked well as long as the force was relatively weak and the perturbations were indeed small. But they failed miserably for the strong force.

In perturbation theory the chances of two subatomic particles colliding are calculated by taking a sum over the various possibilities for their interaction. This is easier to understand in Feynman's diagram approach, where the complex exchange between the two particles breaks down into the separate events where one, two, three, and more intermediate particles are swapped. To resurrect the basketball analogy, this is like trying to find the chances of scoring on a fast break by making separate calculations for those plays where two teammates pass the ball one, two, three, and more times. In both cases we add up the individual results to get the overall chances.

In such a "perturbation expansion," we get an infinite sum of numbers corresponding to an infinite number of possible exchanges. Cases where the force is feeble, or where the teammates don't pass the ball very often, give a sum like $1 + 1/10 + 1/100 + \ldots$, where successive contributions become smaller and smaller because the chances of multiple swaps drop off rapidly. Fortunately, such sums converge to a finite number, in this case $1.111\ldots$, and all is okay. This is essentially what happens for the electromagnetic and weak nuclear forces, once renormalization has purged the infinities from individual terms. When the force is *strong,* however, the sum is more like $1 + 1 + 1 + \ldots$—clearly an infinite result. So these perturbation techniques failed miserably for the strong force.

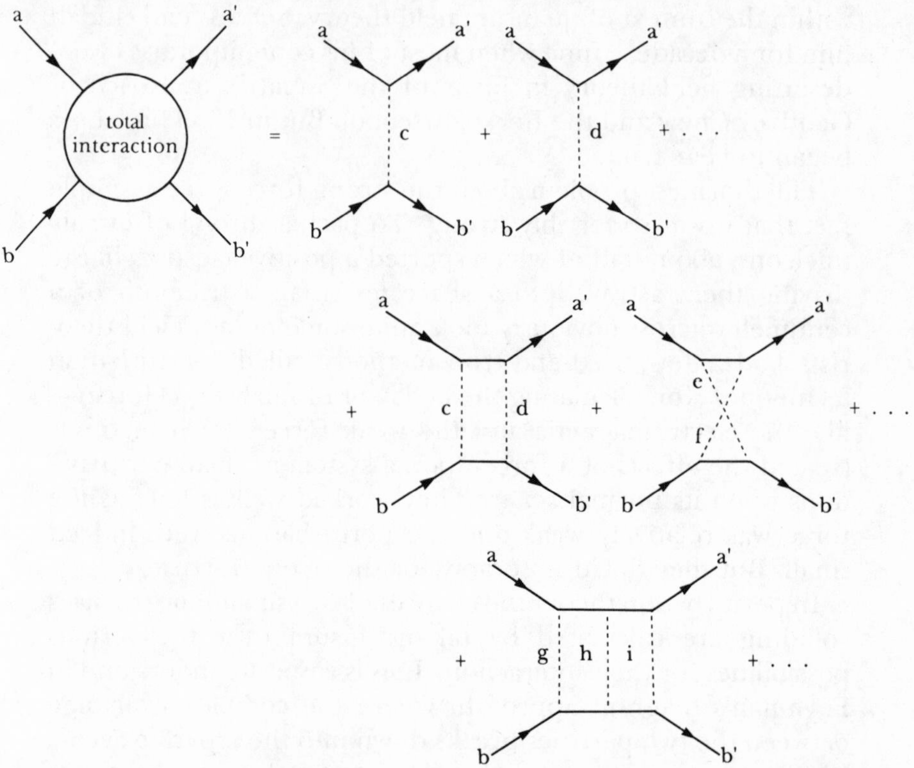

The "perturbation" approach to calculating the chances of particle collision, expressed in Feynman diagrams. You take a sum over all possible particle exchanges.

For years, Wilson sought a different way to calculate the effects of the strong force, finally returning to a method Gell-Mann and Francis Low had considered briefly in the 1950s. The "renormalization group" is a way of getting results at one energy scale if they are known at another, adjacent scale. If you can do your calculations on the basketball court, in other words, you can apply them to what happens on a tennis court, where the ball is ten times lighter. When combined with other work that Wilson was doing at the time, this approach had immediate application to the scaling behavior witnessed in the MIT-SLAC experiments. In essence, we were then learning that nucleon structure does not change very much from one energy scale to the next. The partons seemed pointlike, no matter how close we looked.

For quantum field theorists conditioned to working with

perturbation theory, Wilson's idiosyncratic methods had little impact until they were combined with more traditional approaches. In 1970, Curtis Callan at Princeton and Kurt Symanzik at DESY independently developed an equation based on Wilson's work that told how the strength of a force *changed* in going from one scale to another. In field theory the strength of a force is specified by its "coupling constant"—a pure number that is typically 1 for the strong, 0.01 for the electromagnetic, and 0.00001 for the weak force. The magnitude of this coupling constant determines whether or not perturbation techniques can work at all for a particular force—whether the sum of terms converges to a finite result or explodes into infinity.

According to the Callan-Symanzik equation, different kinds of field theories give different behaviors of the coupling constant as you move from one scale to the next. In 1972, Gerard 't Hooft speculated that, in Yang-Mills gauge theories, this all-important number might actually *decrease* as the energy went higher and higher. In effect, the corresponding force might get weaker at small distances. But 't Hooft did not pursue the idea further.

Two American graduate students finally performed the detailed calculations in the spring of 1973—Frank Wilczek at Princeton and David Politzer at Harvard. Entering theoretical physics after beginning his graduate studies in mathematics, Wilczek wanted an important problem he could solve quickly, and his adviser David Gross suggested he determine the behavior of the strong force in gauge theory. The same problem occurred to Politzer at about the same time, and by April he had a result: the force indeed became weaker at small distances.

Showing this result to Sidney Coleman, his adviser, Politzer was dismayed to learn that Wilczek had already done the calculation and gotten the opposite behavior. He checked his work and found no error. By then, Wilczek had found an error in his own calculations which when removed yielded the same results as found at Harvard. Two consecutive papers in the June *Physical Review Letters* announced this startling behavior, dubbed "asymptotic freedom," to the physics community. After decades of frustrating attempts, the strong force was finally beginning to succumb to quantum field theory.

* * *

It is difficult to overstate the importance of this one small step in the history of particle physics. The fact that a force could actually become *weaker* at short distances meant that an entire galaxy of computational techniques developed over the past few decades might now be brought to bear upon it. If asymptotic freedom were indeed true for the strong force, then broad new vistas were about to open up for particle theorists. And if it could be described by a gauge field theory, that meant the strong force was a close cousin of the weak and electromagnetic forces, too. What more tantalizing unity could one ask?

As Jaffe was quick to point out in October 1973, our MIT group might hold the key right there in our very own hands. By that autumn we had the world's best data by far on nucleon structure functions—as yet unpublished. If asymptotic freedom held true for the strong force, he said, then these functions should decrease ever so slowly with higher energies—just like the coupling constant itself. Instead of Bjorken's perfect scaling behavior, we should see small violations. It was a key test.

Fortunately, we were already searching for features like this. With characteristic insight, Friedman had suggested I start looking for them over a year earlier. We were motivated not by gauge theory but by people like Drell and Wilson, working in ordinary field theories—and by our own gut feelings that this was the best way to test for parton substructure. Preliminary results were reported in a short section of my thesis, and had been favorably received at Bonn.

That same fall Gross, Wilczek, and Politzer were writing and publishing longer, more detailed papers on the experimental consequences of asymptotic freedom, making public what Jaffe had already told us in private: look for small violations of scaling in the structure functions. In particular, they should drop off slowly in one region and rise slowly in another.

A quick glance at our ongoing studies indicated just this kind of qualitative behavior—a slow fall-off in one region and a slow rise in the other. It was a tremendous eye-opener. Gross, Wilczek, and Politzer had no possible way of knowing what our data looked like, yet here they were predicting essentially what we were just then noticing. Other theories usually had left enough loose ends hanging so that they could later be modified to fit the data eventually measured. Asymptotic freedom, by contrast, had made a hard-and-fast prediction that could well have been used to shoot it down. Trial by fire. Its authors had run the gauntlet and come through our first crude test unscathed.

The Aspen Center for Physics is a favorite summer rendezvous for physicists, especially those with a weakness for hiking or climbing. Set in a world-class resort village near Colorado's highest peaks, it is a place where they can continue plying their esoteric trade while taking an occasional relaxing day off amid the visual splendors of the Rockies. In June of 1973, Gell-Mann arrived there for two months of stimulating encounters. He departed in August with a new theory of the strong force.

That summer the work of Gross, Wilczek, and Politzer was a favorite topic among theorists trying to crack the code of mesons and baryons. If true, asymptotic freedom promised a revolution in the way one treated the force tethering these heavy particles together. Of paramount importance was the question of what might actually be carrying this force. In Gell-Mann's abstract methodology, this was a matter of the true "group structure of the underlying field equations." To theorists who spoke in partons, it was more a question of, "What are the gluons holding everything together?"

If there indeed were three colors of quarks, then there were two major options. Either the gluons were drab, colorless motes that did not carry any color with them from one quark to the next, or they formed an octet of eight multicolored droplets that in fact *did* carry color. Colored gluons sported *mixtures* of the three primary quark colors—red, green, and blue—and their anticolors, generating a variety of eight possible hues. The interior of a proton might be a dazzling place indeed, with a riot of colorful characters dancing wildly about.

Gell-Mann had been aware of the second option for at least a year, and Yoichiro Nambu had suggested it as early as 1966 in connection with his own quark model. But nobody had made very much of this possibility until the summer of 1973, a pivotal time in the history of particle physics. At Aspen, in discussions with several other theorists, Gell-Mann finally resolved a troubling anomaly in his theory. He also recognized that colored gluons were the only choice if the asymptotic freedom of gauge field theories were ever to hold true. Among other things, colored gluons gave the small violations of scaling required.

Quarks, then, could attract one another by swapping their colors, which were carried on the backs of spin-1 gluons flitting betwixt and between. Color might be the *source* of the strong force, just as electric charge was the reason for electrostatic attraction and repulsion. This new property of matter, which had only been a gleam in Gell-Mann's eye hardly a year before, was now being drafted to explain how the mercurial quarks were bound together.

The emerging theory of the strong force, incorporating colored quarks and the asymptotic freedom of Yang-Mills gauge theories, needed a name. Never at a loss for words, Gell-Mann conjured up a splendid name—quantum *chromo*-dynamics—in a conversation with Rockefeller theorist Heinz Pagels. A play on the phrase quantum electrodynamics, or QED, the all-encompassing theory of the electromagnetic force, the name quantum chromodynamics, or QCD, was a truly excellent choice reflecting the hypothesis that color was the cause of the strong force.

Gell-Mann published these ideas that autumn, together with Harald Fritszch and Hans Leutwyler, in a paper titled "Advantages of the Color Octet Gluon Picture." As always, their arguments were couched in abstract language, making little use of the parton viewpoint. "The simplest and most obvious advantage" of an octet of colored gluons, they noted, "is that the gluons are now just as fictitious as the quarks." Just possibly, color might provide a way to keep quarks and gluons from ever appearing outside mesons and baryons—a means to imprison them forever inside these polychromatic dungeons.

Hindsight again makes crystal-clear what was so murky and confusing at the time it transpired. In truth, Nature was sending out thoroughly mixed signals in 1973 and 1974, at least as far as the strong force was concerned. "Quantum chromodynamics" was a phrase that languished almost completely ignored until 1975. And only a small contingent of particle physicists showed any real enthusiasm for asymptotic freedom. These ideas had to be fought for. They did not capture everybody's attention overnight.

By the end of 1973, scaling violations had emerged as the premier test of asymptotic freedom. Arie Bodek and I were now MIT postdocs, putting the finishing touches on the analysis of our two SLAC experiments and preparing a few publications to describe the results. Here was an unexpected bonus dropped into our laps. With first crack at some of the world's most accurate measurements of nucleon structure, we had the inside track on testing asymptotic freedom.

But there was a troubling ambiguity that first had to be resolved. In Bjorken's original formulation, the flat, scaling behavior of the structure functions was supposed to occur only in the so-called "Bjorken limit," where the energies became very "large." How large? Bjorken couldn't really say. He originally thought SLAC energies to be fairly moderate in this regard and was pretty surprised himself when scaling showed up there.

At these moderate energies, it turned out, whether the structure functions "scaled" or not depended a lot on the exact path one took to reach higher energies. One path, trodden most often by Group A physicists, seemed to yield a perfectly flat behavior almost immediately; another path, the one originally proposed by Bjorken, led to small deviations. This difference could be ignored during the late 1960s when the data were rougher and the theories less demanding. But during the 1970s, as scaling violations became a cause celèbre, such fine distinctions assumed central importance. Which was the correct path to take? Not even Bjorken could say for sure.

So we took *both* likely paths, each in turn, to see if such a choice made any difference. By early 1974 it had begun to appear, from our much more accurate data, that scaling was indeed being violated at SLAC. No matter which path we took, we found small deviations. They were much less striking with the Group A path, but they were still there. And, to my delight, these small violations were roughly consistent with asymptotic freedom. So I hurriedly wrote a short paper and sent it around the group for comments. By June, "Observed Deviations from Scaling of the Proton Structure Functions" was off to the *Physical Review Letters*.

To our great surprise and dismay, the paper came back about a month later, rejected. Both its reviewers had advised *against* publishing, on the grounds that we had not really resolved the ambiguity over the appropriate path to take. Rather than contest the decision, however, we sent the paper instead to *Physics Letters*, where it was quickly published in September.

And just as quickly forgotten. At the Rochester Conference in London that July, Friedman cautiously pulled in his sails and reported only our least speculative results on scaling—making no mention whatsoever of our tests of asymptotic freedom.

Few physicists really wanted to hear much about scaling violations that year. Perhaps it just had to be that way. SLAC, after all, had unearthed scaling in the first place, a discovery for which it had achieved world renown. How could it now go back on its word and say that scaling was violated? Only another accelerator operating at higher energies could do that.

The ambiguities plaguing our analysis were not the same problem at the hundred-GeV energies then becoming available at NAL. Alumni of CEA, HEPL, and SLAC had flocked there early in the 1970s to form two large collaborations that were measuring deep inelastic *muon* scattering from protons and neutrons. Point leptons like the electron, muons can probe the internal features of nucleons equally as well—perhaps better. The virtual monopoly MIT and SLAC had enjoyed for six years on this kind of work was finally about to end.

But the first muon scattering results from NAL, which Cornell's Louis Hand presented at London, were unconvincing, too. Though their energies reached as high as 150 GeV, almost ten times that available at SLAC, these physicists suffered from an abject scarcity of muon projectiles. During three whole months of running in 1973 and early 1974, they had managed to shoot only 2.5 billion muons through their targets. While this may seem like a huge quantity to the casual observer, it was only about 1 percent of the number of electrons found in *one* typical SLAC pulse. Hand and company had tried to compensate for this scarcity by using iron targets many feet thick instead of a few inches of liquid hydrogen. But it still would have taken them many *years* of running to match the accuracy of the MIT-SLAC data.

About all Hand could conclude at London was that NAL, too, was seeing small violations of scaling. And given their limited accuracy, these data still suffered from the same ambiguities that had been confounding our own studies. Later that year, these NAL experimenters also encountered severe criticism of their experimental methods. So the whole question of scaling violations emerged from the 1974 summer conferences largely unresolved. It would be a difficult, uphill battle.

The easiest way to interpret scaling violations, at least as they seemed to exist in 1973 and 1974, was in terms of partons themselves having structure. If partons were truly dimensionless points, so these arguments went, then we should observe perfect scaling: the structure functions should remain absolutely flat with increasing energy. But if the partons were spread out in any way—if they were "fuzzy" blobs, that is—then we should see small scaling violations when the energy of a probe was high enough to penetrate *inside* these unbelievably tiny snowballs.

Such an occurrence would only be the latest in a hallowed tradition that stretches back beyond Rutherford all the way to the Greeks. A new and apparently "deepest" layer of matter seems to be populated by "pointlike" entities until the resolving power of some probe becomes fine enough to distinguish a deeper layer beneath it—until a "knife" becomes fine enough to cut the "uncuttable." Was this cycle about to happen yet again?

During the early 1970s a small group of theorists began to think the answer to this question was *yes*. Perhaps the most prolific among them was Drell, who with several postdocs and graduate students worked out the detailed consequences of "parton structure." If their ideas were valid, then the apparent scaling violations we were beginning to witness at SLAC and NAL indicated a parton "size" of about 10^{-15} centimeter—about a hundred times smaller than a proton.

Another way to approach the same question is to determine the mass of any gluons responsible for parton structure. The "size" of a parton, so went this argument, came from the cloud of virtual gluons buzzing around it, just as, decades before, the size of a proton had been thought to arise from the virtual cloud of mesons associated with it. By the Uncertainty Principle, a parton size of 10^{-15} cm meant a gluon mass of roughly 10 GeV, in equivalent energy units, about ten times the proton mass.

This is what deep inelastic experiments seemed to be saying by 1974, if you ignored all the gritty ambiguities and simply took the data at face value. A number of physicists, myself among them, believed we might be seeing the first faint hints of structure at the next layer of the cosmic onion.

* * *

Despite their success in offering a natural explanation of Bjorken's scaling behavior, gauge theories of the strong force still could not answer the principal question puzzling physicists in the early 1970s: If quarks really "exist," why don't they ever come out? It was *the* crucial question that any complete theory of the strong force could not dodge for long.

Initially, there were fond hopes that once we calculated the interquark force at short distances, using asymptotic freedom and perturbation methods, we could then work backward to larger distances like the diameter of a proton, where the force became stronger. Perhaps somebody could *prove* from first principles that the force between two quarks of different colors was strong enough here so that they could never be torn apart—effectively imprisoning them within their parent hadrons. Weinberg's words, taken from a major review of gauge theory written in late 1973, are typical of these sentiments:

> I find it difficult to express strongly enough my enthusiasm for the discovery of asymptotically free theories of the strong interactions. Ever since the disappointing failure of field theory to account for the new facts of meson physics in the early 1950's, progress in elementary particle physics has been impeded by the inadequacy of perturbation theory to deal with strong interactions. Now in asymptotically free theories we see the fog of the strong interactions lifting here and there, revealing the underlying spectrum of the elementary hadrons.

But by mid-1974 nobody had *proved,* explicitly and rigorously, that the quarks were trapped—nor even come close. Whenever a force got truly strong, unfortunately, perturbation methods were useless. When pressed about this question of "quark confinement," theorists only shrugged their shoulders.

Another problem was the mass of the gluons supposedly holding quarks together. Yang-Mills gauge theories gave massless force particles like the photon. But short-range forces were usually carried by *massive* particles, according to Yukawa. Studies of scaling violations seemed to suggest gluon masses of 10 GeV or even more. How to give the gluons such a mass without obliterating the critical property of asymptotic freedom? The Higgs mechanism for symmetry breaking, which had worked so well in generating the masses of the W and Z particles, failed to work the same magic for this color force.

As the seventies progressed, quark confinement became more an act of faith than a scientific judgment. Although gauge theory could not actually *prove* quarks to be trapped, field theorists were reluctant to abandon its property of asymptotic freedom. So they contented themselves with ad hoc arguments that "somehow" the force became strong enough at large distances to imprison quarks forever—even though nobody could yet do a convincing calculation.

Instead, they increasingly relied on clever analogies to "explain" quark confinement. In one of these, the color force between two quarks resembles an elastic string tethering them together, one quark at each end. When the quarks are close together, the string hangs limp and exerts no influence upon them, so they are free to move essentially as they please: hence asymptotic freedom.

But to pull the two quarks far apart, this string must be stretched by applying force. At some point the string suddenly breaks into two pieces, exposing two fresh new ends. The energy applied to stretch the string to its breaking point materializes as a quark and its antiquark, one attached to each new end. So now there are two hadrons instead of one, and all quarks still remain sequestered within them.

A three-dimensional variant of this string model was the so-called "bag model," which visualized the quarks as trapped within some kind of closed surface, a "bag." Originally conceived by MIT theorists Kenneth Johnson and Charles Thorn

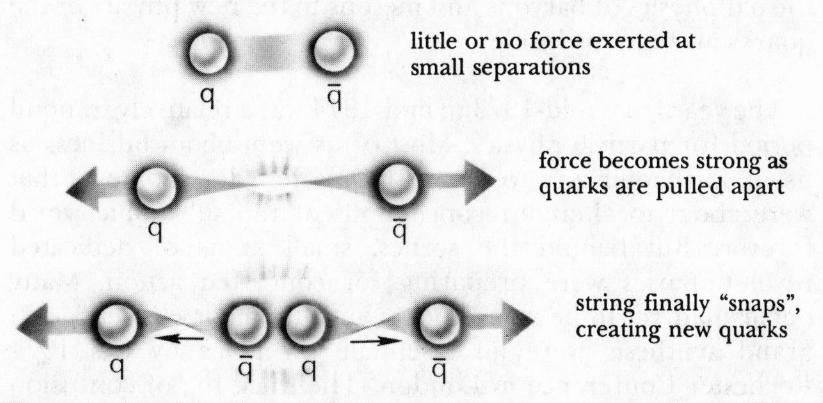

The "string" model of the interquark force, popular in the mid-1970s. Only at large quark separations can its influence be felt.

in the fall of 1973, the bag model was gradually elaborated over the next few years until it had considerable explanatory power. Jaffe and Weisskopf joined forces with them in early 1974 to publish a key article titled "A New Extended Model of Hadrons." Soon Drell and other SLAC theorists had produced their own version of the same idea, so we had an "MIT bag" and a "SLAC bag" to work with.

Bag models likened empty space to the interior of a liquid, say the hydrogen in a bubble chamber. Hadrons are like the bubbles that form inside this liquid whenever enough energy is injected to make it boil. Quarks then behave like the vapor molecules trapped inside these bubbles, free to roam about as they please—at least until they approach the outer surface again. Like the vapor molecules, quarks have no independent existence outside of a bag.

Strings and bags provided convenient metaphors to help us visualize the inner workings of hadrons during the mid-1970s, before quantum chromodynamics had become the widely established theory it is today. But there was always something lacking about these models. Instead of being derived from first principles, the puzzling property of quark confinement was *inserted* right at the outset. So they could never truly "explain" why it ever happened. Ever the idealist, Gell-Mann was particularly impatient with such "phenomenological" approaches, often deriding the bag model as "the shopping-bag model." But however crude, strings and bags were nonetheless important mental crutches that helped us commoners grope our way from the old physics of baryons and mesons to the new physics of the quarks and gluons inside.

The year from mid-1973 to mid-1974 was a relatively tranquil period for particle physics. Most of us went about business as usual, fairly oblivious to some striking new developments that were about to alter our conceptions of the subatomic world forever. But behind the scenes, small knots of dedicated revolutionaries were preparing for concerted action. Many details had yet to be worked out, but the major elements of a grand synthesis were all essentially in place by the 1974 Rochester Conference in London. The thick fog of confusion over neutral currents had lifted that spring. Now only one last major hurdle remained in the way.

CHAPTER 11

THE MOMENT OF CREATION

Like artists, creative scientists must occasionally be able to live in a world out of joint.

—Thomas Kuhn,
THE STRUCTURE OF SCIENTIFIC REVOLUTIONS

NINETEEN SEVENTY-THREE had been a banner year for Burton Richter. The squat, cigarette-smoking head of SLAC Group C, a man fond of ill-fitting sports coats, baggy gray pants, and scruffy tennis shoes, was finally taking data on SPEAR—the electron-positron collider he had for more than a decade been planning, proposing, redesigning, and at long last building and testing. When the earliest results were ready for dissemination that fall, they bore out his longstanding hunch that electron-positron scattering would provide extremely fertile territory for research.

Richter had been an early pioneer of colliding-beam machines. After getting a Ph.D. from MIT in 1956, he began work on the Mark III linear accelerator at Stanford's High-Energy Physics Laboratory (HEPL). He had come there keenly interested in checking quantum electrodynamics. "This was a marvelous theory, but it had some puzzling features," he recalled. "In some calculations infinity had to be subtracted from infinity to get the answer. I thought that this couldn't be quite right and that the place to look for disagreement would be in experiments with high-energy electrons."

At first Richter thought of bouncing electrons off the atomic electrons of a stationary target. The main limitation of this approach is that the target's electrons will refuse to sit still for such a pummeling—and can be dealt only a glancing blow. It's like smashing a speeding automobile into a bicycle. Only a tiny fraction of its available energy of motion can be imparted to the bike.

In 1957, Gerard O'Neill arrived at Stanford on leave from Princeton with a brilliant new idea. Why not instead make *two* circulating beams of electrons and smash them into one another? That way, two electrons, one in each beam, would stop each other almost dead in their tracks, like two automobiles colliding head-on. Virtually *all* the available energy could be tapped for the collision process.

Then the Director of HEPL, Wolfgang Panofsky liked this idea enormously. After O'Neill, Richter, and others had drawn up a plan, he flew to Washington to secure funds from the Office of Naval Research, eventually obtaining a healthy commitment of $800,000. The new science of colliding-beam, or storage-ring, machines was born at Stanford in 1958.

The Princeton-Stanford collider was a rather ungainly beast, quite unlike anything high-energy physics had ever seen before. Two separate rings about ten feet in diameter sprawled out in a figure-eight pattern at the end of the Mark III, which fed them both with electrons at energies as high as 500 MeV. Collisions occurred at the point where these rings touched, allowing the electron beams to clash at essentially the speed of light.

Making the new collider work properly was not easy. To yield reasonable results, the electron bunches had to continue circulating millions of times per second for several hours at a stretch. Many instabilities—some anticipated and others a total surprise—had to be conquered before this goal could be realized. But by 1962 the HEPL physicists had cured these ills and begun taking data. In 1965, they published one of the most definitive proofs of quantum electrodynamics, showing that electrons behaved exactly like point charges, at least down to the level of 10^{-14} cm—a tenth the size of a proton—that could be studied with these storage rings.

By then, several other groups were planning or building their own storage rings. In 1961, Italian physicists led by Bruno Touschek had shown that separate bunches of electrons and positrons could be stored in *one* ring, rotating in opposite directions. Their first prototype, hardly a meter across and storing beams of only 250 MeV each, was called "AdA," for Annello di Accumulazione (literally, "storage ring"). Based on its success, the Italians were hard at work in the midsixties building ADONE ("big AdA"), a 1.5 GeV, scaled-up version of the same essential device.

In 1961, Richter began planning, with Stanford's David Ritson, a large electron-positron collider to be built at SLAC after the linear accelerator was finished. They sent their first formal proposal to the Atomic Energy Commission in 1964, a one-ring design with an estimated cost of $20 million. To be fed by the main accelerator, this collider was intended to store beams with energies up to 3 GeV.

Unfortunately, their request reached the AEC just as another, bargain-basement proposal arrived from the Cambridge Electron Accelerator (CEA). A panel of physicists reviewing the two requests thought the SLAC idea sound but recommended waiting a year before funding it. "But we had missed the boat," Richter lamented later. "Funding for particle physics had begun to get tight." With the Vietnam War heating up and the Apollo moon shot getting into high gear, money for pure science was becoming scarce. Despite favorable committee recommendations in 1965 and 1966, he left Washington empty-handed both times. "In 1967 we got tired of writing new books and resubmitted the same book," Richter added. "Still no money."

Meanwhile, storage rings had begun to proliferate around the globe. The French now had a small electron-positron ring at Orsay, near Paris, and the Russians were building a continuing series of colliders at Novosibirsk in Central Siberia. CEA was starting its collider and DESY eventually began one, too. At CERN a proton-proton collider called the Intersecting Storage

Rings, or the ISR, was underway by 1966. Storage rings were popping up everywhere but at Stanford, the very place where this vigorous new branch of accelerator physics had originated.

Down but not out, Richter continued building his own experimental group, collaborating with Ritson and Richard Taylor on the design and construction of the huge spectrometers for End Station A. When the first SLAC beams became available in 1967, he began a series of high-energy photon-nucleon scattering experiments using the 20 GeV spectrometer. It was solid but not earth-shaking work, concentrating mainly on tests of the then fashionable theory of vector dominance.

Richter had not given up yet on his collider; he was just marking time, waiting for his big opportunity. In 1969 and 1970, he redesigned the collider twice to slash its initial construction cost. The first attempt was a two-ring collider he dubbed the Stanford Positron-Electron Asymmetric Rings, or SPEAR, projected to cost only $9 million. The next year he returned to the single-ring approach but kept the name, cutting the cost still further to a paltry $5 million—less than what had been spent to equip End Station A. But despite strong recommendations from the AEC's advisory panel, both proposals were rejected. All new construction funds were then being absorbed by Robert Wilson's mammoth NAL project.

The last time around, however, Richter found an angel inside the AEC bureaucracy. Comptroller John Abbadessa pointed out a loophole in the AEC regulations that allowed SLAC to build SPEAR out of its normal operating budget—instead of seeking separate construction funds. The key was to consider SPEAR as a *detector,* not a new particle accelerator. Always a strong supporter of the collider idea, Panofsky decided to build SPEAR on the sly. He tightened the purse strings everywhere possible, and money began flowing in August 1970. "I've had a soft spot in my heart for Mr. Abbadessa ever since," Richter admits.

In May 1971 the outlines of a 600-foot oval began to take shape on a vacant expanse of asphalt just north of End Station A. A year later, SPEAR was complete and Richter's group was hard at work tuning it up. In April 1973, they were taking data at last.

* * *

By the early seventies, results were already pouring in from European storage rings. Besides ADONE, where the Italians were seeing a surprise abundance of electron-positron collisions, Europe also had the Intersecting Storage Rings at CERN. Here, protons with energies of 8 to 28 GeV circulated in opposite directions within two interlaced beam pipes, both housed inside a large circular hall about a thousand feet in diameter. Protons smashed into protons in eight interaction regions, where the beams crossed paths at an angle of about 15 degrees. These encounters began in January 1971, well ahead of schedule; by late that same year, the first survey experiments were actively logging data.

This process is comparable to firing two shotgun blasts at one another, slightly off axis. The vast majority of these particles, whether protons or buckshot, continue speeding forward without any deflection whatsoever. Only the scantest few protons ever collided in any single pass, but when they did, virtually *all* of their combined energy was available to create new particles. To get the same equivalent energy by smashing protons into stationary targets would have required energies of 2000 GeV—a factor of ten beyond the proton beams then coming on line at NAL.

With all this energy lying about, the ISR was a natural place to set up and look for suspected heavy particles, like the elusive quarks or the ephemeral W bosons that supposedly carried the weak force. A number of the early experiments began to try just that.

Leon Lederman arrived at the ISR from Columbia in 1971, part of a collaboration that included Rockefeller University and CERN itself. These scientists began searching for W's by detecting electrons or positrons shooting out at large angles from the intersecting beam pipes. Such a geometry was necessary, they thought, to avoid the intense background of pions that would be thrown forward in the same general direction as the clashing proton beams, close to the pipes. To their great surprise, they found their equipment being swamped, even at large angles, by a blizzard of pions—over a *million* times beyond expectations. So intense was this background that clear glass blocks used in their detector began to turn yellow from radiation damage.

Other ISR experiments were also being swamped by similar backgrounds that first year, as physicists struggled to chart this new territory. By late 1972, it had become increasingly clear that CERN had an entirely new phenomenon on its hands. These scientists, who had been inured to thinking of protons as soft, mushy snowballs, slowly recognized that there were tiny objects inside that could occasionally have a hard, violent collision—spewing out debris at large angles. The same partons that had turned up in electron and neutrino scattering, where only the electromagnetic and weak forces came into play, were now showing their faces in proton collisions, where the strong force held sway.

Though this large-angle scattering was cause for great excitement at CERN, it elicited little more than a few mild yawns and knowing nods at SLAC. By 1972, partons were pretty much taken for granted there as the reason for almost everything that went on inside nucleons. Based on the parton model, in fact, Bjorken, Drell, and other SLAC theorists had already predicted plenty of debris at large angles. The only true surprise for them was the stunning *size* of this effect—more than a hundred times beyond even what they had imagined. It could only mean that the strong force, too, acted as if it originated from a point source. *That* was news.

What was really grabbing SLAC's attention right then were the ADONE results on electron-positron collisions. These leptons seemed to be annihilating one another much more often than many theorists had expected, giving rise to sprays of pions and other hadrons about as often as they created a pair of muons. It was a crucial piece of evidence supporting the colored quark idea of Harald Fritszch and Murray Gell-Mann. It also meant that SPEAR would have interesting territory to explore when finally ready to log data.

Characteristically, Bjorken had foreseen it all in 1966. In a *Physical Review* article, he conjectured that hadrons might be produced about as often as muons in these smash-ups. There is a curious footnote there that seems to indicate Richter was aware of this possibility, too. But when I asked Bjorken about it, he set me straight. After finishing the calculation, he recalled, he had strolled over to Richter's office and asked, leading the witness, "Burt, is it folklore that hadrons may be created about as often as muons in electron-positron scattering?"

"Yeah, BJ," he answered, unaware of the tug on his leg. "That's folklore."

During the late 1960s the only ongoing U.S. work with storage rings was CEA's Project Bypass, the proposal that had been approved instead of Richter's. To this circular, 6 GeV electron synchrotron, physicists added a special loop about 130 feet long. Electrons and positrons were first accelerated in the main ring, where they circulated in opposite directions within two distinct orbits. When they reached the desired intensity, "kicker" magnets booted them into this alternate path. Here the electron and positron bunches finally clashed with one another, registering occasional collisions inside a surrounding detector.

The CEA Bypass seemed a conceptually simple idea, but it proved to be a technological nightmare in actual operation. The complicated process of boosting first the positrons and next the electrons up to the required energy, and then diverting them both into the Bypass, was fraught with instabilities that constantly threatened to destroy the delicate beams. Although electrons circulated through the Bypass as early as 1968, serious data-taking had to wait until spring 1972.

Colliding two particle beams is a little like shining a pair of flashlights at each other. Scattering rarely happens. Experimenters try to better their chances by maximizing a key parameter of the storage ring called its "luminosity"—roughly the compound intensity, in terms of the number of particles per second passing through a single unit area, of the two colliding beams. If the colliding bunches can be squeezed down to a very small size at the moment they cross, this luminosity gets a lot better. Provided this does not make the beams unstable, that is.

Luminosity was always the major problem with the CEA Bypass. Despite repeated attempts to improve it, the luminosity still fell about a factor of a thousand below its intended value. This meant that experiments had to be run a thousand times longer than planned to get the same accuracy of data, or experimenters had to settle for far worse data than they'd planned.

CEA itself had had a checkered career in high-energy physics. Not only had a technician died there in a 1965 hydrogen explosion which shut the whole place down for over a year, but an experimental group under Harvard's Francis

Pipkin had reported a famous quantum electrodynamics violation that later proved erroneous. Results from CEA always had to be taken with a grain of salt.

When the budget for this ill-fated atom smasher was slashed by 30 percent in 1970, its directors decided to scrap the rest of the experimental program so they could keep the Bypass project going. The last scheduled experiment before they flicked off the main switch for good was a study of electron-positron scattering.

When the first CEA results came out in summer 1972, few physicists paid much heed. The data had been hastily analyzed to be ready for the Rochester Conference at NAL. Better to wait for more definitive results. But the electron-positron data presented at Bonn the next year by Harvard's Karl Strauch, leader of the group, made everyone sit up and take notice. Where ADONE had witnessed hadrons emerging roughly twice as often as muons, at the higher CEA energies this seemed to be happening even more frequently—perhaps four to six times the rate of muon creation. Colored quarks could explain a factor of two, but it was difficult to justify anything higher.

To understand this better, consider what happens when an electron and a positron annihilate. They give birth to a photon whose total energy equals their two energies combined. Literally seething with energy it cannot long contain, this flash of light materializes almost instantaneously as a particle-antiparticle pair—another electron and positron, two muons, or (in the quark-parton model) a quark and its antiquark. In a very real sense, these colliders simulate the conditions that must have occurred during the earliest moments of the Big Bang, when matter first coalesced from pure energy.

Whether actual pairs of point particles *can* materialize depends only upon their masses and spins. If enough energy is present to create a given pair, and the two spins can add up to the unit spin of the photon that gives them birth, then this pair will indeed be formed in electron-positron collisions. How *often* it is formed depends upon the square of the charge on either point particle. Scanning the diagrams here, then, you might guess that a *u*-quark would be created four-ninths as often, and a *d*-quark or an *s*-quark one-ninth as often, as a muon. If quarks

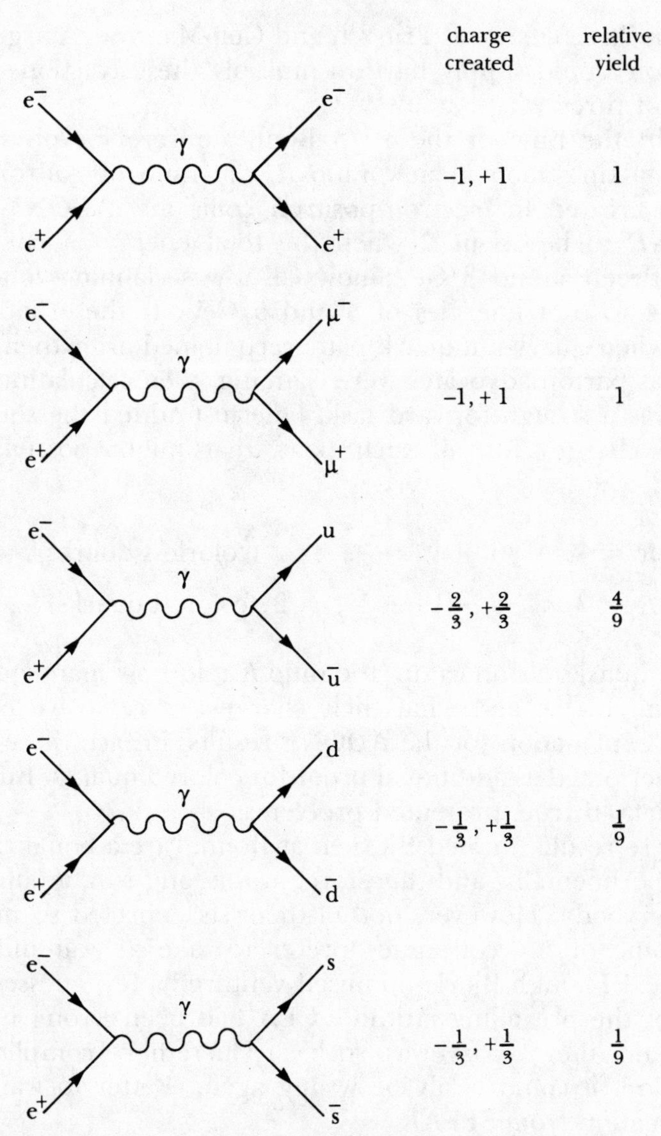

Feynman diagrams for some possible outcomes of electron-positron annihilation. Under normal circumstances the chances of creating a particle-antiparticle pair are proportional to the square of their electric charge.

came in three colors, as Fritszch and Gell-Mann were arguing, then you would simply have to multiply these fractions by a factor of three.

About the time of the 1973 Bonn Conference, physicists began talking about a new ratio *R*, the ratio of hadrons to muons created in electron-positron collisions. ADONE had shown *R* to be about 2 when the total energy of collision fell between 1 and 3 GeV; now CEA was claiming *R* to be about 4 to 6 at energies of 4 and 5 GeV. If these hadrons arose when quark-antiquark pairs recombined after their creation, as parton advocates were claiming, then calculating this ratio was a straightforward task. One just added the squares of the charges for all such pairs that might possibly be created:

$$R = 4/9 + 1/9 + 1/9 = 2/3 \qquad \text{(colorless quarks)}$$

$$R = 3 \times (4/9 + 1/9 + 1/9) = 2 \quad \text{(colored quarks)}$$

To the quark-parton camp, the ratio *R* told how many quarks there might be, and what their charges were. So we had a simple explanation for the ADONE results; in fact, these data were welcomed as additional proof for colored quarks. But the CEA data, if true, presented problems.

"These results," noted Bjorken at Bonn, "are a bonanza for the experimentalist and therefore, in the long run, a bonanza for everybody. However, not all theorists expected so much, and some of my colleagues prefer to take a 'wait-and-see' attitude." Though Bjorken himself ventured a few guesses, his was not the prevailing attitude. CEA had been wrong in the past, and the Bypass was such an incredibly complicated operation, it could easily be wrong again. Better to wait for confirmation from SPEAR.

Over ten years later, SLAC physicist John Rees was recounting the history of electron-positron colliders. The man who had been instrumental in bringing SPEAR to life could only shake his head regretfully when he came to speak of the Bypass:

> The saga of CEA is the Book of Job of the accelerator builders. They were afflicted by every handicap that could have been visited upon them. Yet they persevered, and in the end the Lord

loved them, and they got the right value of *R*. But of course nobody believed it, the machine was so hard to operate.

By August 1973, experimenters had been logging data at SPEAR for four months. One group in particular, of which Richter was the guiding force, had also seen hadrons appearing in profusion. So they had a solid hunch that the CEA results for *R* were not that far-fetched, after all. But their data was only a few months old, much too preliminary to be revealed at Bonn.

The bunches of electrons and positrons speeding around SPEAR met repeatedly at two interaction regions, where separate experimental collaborations set up particle detectors. The "East Pit" closest to End Station A was occupied by Pennsylvania and Stanford physicists including Robert Hofstadter, working mainly on tests of quantum electrodynamics. In the West Pit sat the Mark I detector, a mammoth cylinder teeming with wires, magnet coils, iron, phototubes, and cables that surrounded the beam pipe on all sides. Nothing quite like it had ever been built before.

Richter had wanted a detector surrounding as much of the interaction region as possible and having a magnetic field inside to help distinguish the various types of particles spewing forth in all directions. That way, he could capture most of them and also be able to tell immediately what they were. Earlier devices at ADONE and CEA had been lacking on both counts, with consequent ambiguities in the interpretation of their results. But Richter was far too busy building SPEAR to pay sufficient attention to the many detailed problems of detector design.

So he tapped the old-boy network of high-energy physics to find his champion. In 1970, Roy Schwitters had been an MIT graduate student doing his Ph.D. research under Louis Osborne, Richter's thesis adviser. The two of them joined forces with Group C to do a photon scattering experiment in End Station A just ahead of our own 1970 experiment. When Schwitters finished his thesis the next year, he had a job offer to come and work as a SLAC postdoc.

With a team that soon included Martin Breidenbach and a few others, Schwitters took charge of designing the new detector. He figured out the tricky geometry necessary to combine all its many diverse components without leaving any seams or

The Mark I Detector, with Roy Schwitters standing near its axis. Electrons and positrons collided within this device, which recorded the tracks and properties of the debris.

blind spots—except at the two ends, where the beams had to pass. The Mark I began to take shape in 1972, a colossal, million-dollar retina for staring into the heart of matter. A group of former bubble-chamber physicists from the Lawrence Berkeley Laboratory (LBL) joined the effort that year, forming what became known as the SLAC-LBL collaboration.

SPEAR had far higher luminosity than the CEA Bypass, so the data came pouring in during the latter part of 1973 with hadrons aplenty showing up in the Mark I. By late fall it was pretty clear that CEA had been right, after all: the ratio *R* seemed to be large and getting steadily larger at higher energies. Hadrons were being created four to six times as often as muons. Quark-partons were in trouble.

When Richter first revealed these results publicly, in a SLAC seminar and an informal workshop that December at Irvine, California, they threw the quark-parton camp into immediate disarray. We could handle large values of *R* simply by adding more quarks. By including Glashow's charmed quark, for example, we could push *R* up to 3.3; Han and Nambu's quarks with integer charges gave an *R* as high as 4.0. Stir in a few extra leptons and, presto, we could easily cook up values like 5 or 6. But this ratio still had to level off *somewhere* along the line. A smoothly rising *R*, such as witnessed at CEA and SPEAR, was simply not in the cards.

"It has the theorists running for cover," admitted a bewildered Sidney Drell. "You really don't know," quavered Bjorken, "how the quark fits into the dynamics of things."

Richter was loving every moment of it. Stumping the theorists is a favorite sport among experimenters. Rather than latch onto any particular theory, he preferred to sit back with an impish grin on his face, wave his hands in the air, and observe that these uppity electrons and positrons seemed to behave a lot like hadrons whenever they were smashed together hard enough. Couldn't *that* explain why the ratio kept rising?

He did have a point. The subatomic fragments spraying into the Mark I looked an awful lot like the fireworks seen in the ISR at CERN, where protons were bashing into other protons, flinging all kinds of frightful garbage outward. Maybe the electron had a tiny core deep inside where the strong force ruled the roost. Were electrons really hadrons at heart?

Whatever was going on, particle physicists agreed it had to be something *new*. But that was about all they agreed on. During the first half of 1974, theorists ransacked their minds for explanations, coming up with many, none terribly convincing.

The trouble was that almost every new idea proposed to

explain a rising R had implications elsewhere, especially in our own electron-nucleon scattering experiments. The two experiments were brother and sister to one another. Both were taking a detailed look at what happens when electrons and hadrons get very close together—in space or in time. To explain one usually involved making predictions for the other, like it or not. And, in the spring of 1974, those predictions too often proved embarrassingly wrong.

Like the scaling of the structure functions, the anticipated flat behavior of R was a consequence of an assumed pointlike nature of the quark-partons, the putative ingredients of hadrons. So one way to interpret a rising value of R was to say that these quarks were *not* mathematical points, after all; they had some internal structure. But that meant our structure functions had to plummet steeply with increasing energy. Though our group could by then make a solid case for a small drop-off, say 20 percent, we did not see the wholesale collapse required for compatibility with the electron-positron measurements.

True to form, two Stanford theorists found a way around this difficulty by making tricky cancellations that gave a rising R and preserved the structure function behavior observed thus far. But their predictions in other areas were quickly proved wrong by further measurements.

Those who tried to explain the growth of R by invoking new quarks or other particle species were also puzzled by the MIT-SLAC results. If new particles could be created one way, why couldn't they be created in the other?

And the gauge theories of the strong force, with their tantalizing prospect of asymptotic freedom, were completely at a loss to explain this striking behavior. In fact, they predicted that R would *fall* slowly at high energy, like the structure functions—the exact *opposite* of what had been measured at CEA and SPEAR.

Early 1974 was a time of turmoil in the economies of the Western world, too, as the full effects of the "energy crisis" began to be felt after the Arab oil embargo had driven oil prices up fourfold. It was an equally confusing time in particle physics, at least for those who espoused the quark-parton world view. The problem of a rising R refused to submit to any glib solutions. Soon dubbed the "R-crisis," this perplexing state of affairs carried right up to the London Conference that July.

"It is clear there is no consensus among theoreticians working on electron-positron annihilation," concluded John Ellis there, "not even about such basic questions as . . . whether or not to use parton ideas." The young, red-bearded theorist, owner of the best T-shirt collection in high-energy physics, one that he frequently wore to these august affairs, had spent a few years at SLAC and Caltech before moving on to a permanent job at CERN. One of the staunchest advocates of the quark-parton viewpoint in all of Europe, he had been asked to discuss the many ideas that had surfaced in the past six months to explain the baffling rise of *R*. At the close of his talk, Ellis displayed a laundry list of models, twenty-three in all, with predictions for *R* ranging from a low of 0.36 all the way to 70,383 and even *infinity*. Hardly a consensus.

Now it was Richter's turn. As the man with most of the new results, he was scheduled to review the entire subject of hadron creation in electron-positron collisions. "This subject is of particular interest at the present time," he began, "because the results of recent experiments flatly contradict all previous models of hadron production available up to about half a year ago." Richter then warned his audience, representing the best of high-energy physics, that the CEA and SPEAR data "are in violent disagreement with the predictions of the simple quark model," adding that various attempts to patch up this model had so far been unconvincing.

Turning to new ideas, Richter unveiled his favorite—that electrons and positrons were actually hadrons at heart. Abdus Salam, together with his countryman Jogesh Pati, had even written down such a theory in the past year. In it, electrons could change *directly* into quarks, and positrons into antiquarks, without taking the intermediate step of annihilating one another first and making a flash of light. That meant the creation of hadrons could occur much more rapidly once the total energy rose high enough to ignite this process. Hence the abrupt rise being observed in *R*.

Such a theory, Richter continued, made the "spectacular prediction" that *R* would begin to increase even *more* rapidly at still higher energies. With SPEAR that very moment being upgraded to run at higher energy in the fall, he was anxious to return to SLAC. "Theory is in a confusing state," he concluded, "and it is up to the experimenters to clear the air a bit."

One theorist, at least, was not confused by the R-crisis. Or if he was, he certainly didn't show it. John Iliopoulos had recently collected several bets on the existence of neutral currents, which no longer seemed in doubt that summer. (The most memorable was a bottle of wine served him by Jack Steinberger, a CERN experimenter who chose to believe the null results of his former Columbia colleagues over the findings of his new European associates; when he could not produce the specific wine they'd wagered, Steinberger graciously reached into his cellar for a fine bottle of Chateau Margaux "worth many cases" of the other.) Flushed with success and brimming with confidence, Iliopoulos strode to the lectern to review the prospects of gauge theories.

A smallish, slender, dark-haired man, Iliopoulos began his talk on solid ground, reviewing briefly the brilliant successes of gauge field theory in the ongoing unification of the weak and electromagnetic forces. But he soon skated out onto thin ice, arguing passionately that with gauge theory, physics might at long last have stumbled upon "the ultimate model of the world" that could also explain the strong force and maybe, just maybe, even gravity itself. In a speech that can only be described as clairvoyant, this Greek zealot enunciated one by one the desirable features of such a model, "for which the gauge theories have given at least the glimmer of a possibility." Colored quarks, charmed quarks, electroweak unity, asymptotic freedom, and some far more speculative things, all rolled into a single, all-encompassing package.

Even some of the true believers in the audience were aghast at the enormous risks Iliopoulos was taking. "It was all presented with absolute conviction," remembered Bjorken ten years later, "and sounded at the time just a little mad, at least to me."

The linchpin of the entire argument was charm. Now that neutral currents had been confirmed, we desperately needed a fourth, charmed quark to explain why nobody had ever witnessed neutral kaons decaying into a pair of muons. Iliopoulos had an unshakable confidence it would soon be found—and would even solve the bewildering R-crisis in electron-positron collisions. "Asymptotic freedom has a real meaning there," he argued, "and the hadron production cross-section, which absolutely refuses to fall, creates a serious problem. The best explanation may be that we are observing the opening of the charmed thresholds, in which case everything fits together nicely."

In an attempt to keep the conference running on schedule, the organizers had offered each speaker a bottle of wine if he finished on time. There the bottle sat, tauntingly, on the table before Iliopoulos, as he ran two, then five, then ten minutes over, still hardly halfway through his transparencies. Finally he relented. He grabbed the bottle and opened it, pouring a generous libation and lifting it, offered yet another wager. "I have won already several bottles of fine wine by betting for the neutral currents," he boasted, "and I am ready now to bet a whole *case* that the entire next Conference will be dominated by the discovery of the charmed particles."

The gauntlet was down. Experimenters thought Theory was pretty confused, and theorists—at least those who trafficked in gauge field theory—felt Experiment was the one befuddled. The stage was set. In this overheated, supercharged atmosphere, something had to crystallize, and crystallize fast. But none of us, not even John Iliopoulos, had any inkling then of what a gem it would prove to be.

CHAPTER 12

THE NOVEMBER REVOLUTION

The man was tired, for he had diligently worked the area for weeks. He stooped low over the pan at the creek and saw two small glittering yellow lumps. "Eureka!" he cried, and stood up to examine the pan's content more carefully. Others rushed to see, and in the confusion the pan and its contents fell into the creek. Were those lumps gold or pyrite? He began to sift through the silt once again.

—Harvey Lynch, SPEAR Logbook, November 9, 1974

By 1974 Samuel Chao Chung Ting, MIT Professor of Physics, had earned a reputation as a painstakingly careful scientist. His critics, of whom there are many, usually put more emphasis on the *pain* than on the care. Stories abound of how his graduate students and postdocs toiled seven days a week and sixteen hours a day, of how they performed menial tasks of little consequence, of how they were forbidden to eat, drink, smoke, or read while on shift.

But everybody agreed that this stern taskmaster produced results. After a brief stint at Columbia in 1965, doing experiments at Brookhaven, he had taken a leave of absence to head a DESY experiment testing predictions of quantum electrodynamics. That was the year Harvard's Pipkin had reported a disturbing violation of this sacrosanct theory. Ting made another, far more detailed study of the same process without finding any such violation. His experiment won the day largely through its sheer attention to detail. It put DESY on the map, and Ting's name into the annals of high-energy physics.

Following that experiment, these physicists modified their equipment to measure the production of neutral vector mesons—the rho, omega, and phi—in collisions of high-energy photons with nuclei. These short-lived mesons often disintegrate into electron-positron pairs; Ting's group detected each of the two offspring simultaneously in the two eyes of a double-eyed magnetic spectrometer.

The key to the measurements was an extremely fine energy resolution: the ability to distinguish very narrow peaks only 5 to 10 MeV wide—about one part in a hundred. The human eye, by analogy, has exceedingly fine *spatial* resolution; it can resolve a strand of hair at arm's length. With characteristic attention to detail, Ting's group gradually upgraded their detector until it could readily distinguish the 10 MeV-wide peak of the omega and the even narrower phi.

By 1970, the group had almost exhausted the potential of DESY to study these kinds of particles. Exhausted himself, Ting took a year off to review his accomplishments and to survey his options. This intense, workaholic son of Chinese intellectuals had come a long way since 1956, when he arrived in Ann Arbor from Taiwan with a hundred dollars in his pocket, knowing only a smattering of English. Six years later, he had a Ph.D. in physics from the University of Michigan, an architect wife, and a Ford Foundation fellowship to do research at CERN. In 1967, Victor Weisskopf noticed the rising young star and offered him a faculty position at MIT.

My own first encounter with Ting occurred about this time, in a seminar he gave early in 1969 about his work on vector mesons. He presented a very short description of these experiments, then in progress at DESY, devoting no more than two minutes to each, scribbling just a few diagrams on the blackboard. Then he turned and asked if we had any questions, his dark, piercing eyes flashing about the room.

Dumbstruck, we all sat there mute. How could we have any questions? He hadn't really told us anything yet. "Well, if there aren't any questions," Ting continued after an awkward moment of silence, "there's no point in my being here." And left the room. He must have had more important things on his mind.

What to do next? And where to go? Powerful new proton accelerators were then under construction at CERN and at NAL. How to secure a foothold there, when his experience was predominantly with photons and electrons? By the end of his sabbatical, Ting had decided that his group would employ its fine-resolution spectrometers to search for massive particles produced at the high-energy frontiers these huge machines were soon to be opening up. So he formulated an ambitious three-pronged plan of attack, with experiments at DESY, NAL, and CERN, to study the entire range of particle masses attainable—from 0.5 to 50 GeV.

Unfortunately, Robert Wilson refused to go along with this tidy scenario at NAL, where Ting wanted to construct a high-energy photon beam and use it to do experiments like those he had done at DESY. Wilson wanted Ting to work there, of course, but he didn't want him building an empire that might conflict with his own dynasty. The two locked horns through much of 1971, until Ting finally abandoned the struggle late that year.

His situation was now getting "a little desperate," as one physicist familiar with the group put it. He was stymied at NAL; his experiments at DESY were winding down; and any possible work at CERN was still years in the future. So he turned back to old friends at Brookhaven and in January of 1972 proposed to build a double-eyed spectrometer and study pairs of particles—especially electrons and positrons—created in collisions of protons with nuclei. It was a hastily written proposal firmly grounded in the "old physics" of the bootstrap model and vector meson dominance: the way to resolve all the recent discrepancies and rescue these doddering old theories was to find still more vector mesons at higher masses. Only the briefest mention is ever made of quarks or partons.

Here Ting met a more favorable reception. A Columbia team under Leon Lederman had noticed a curious enhançement (later known as the "Lederman shoulder") in a similar though cruder 1968 experiment. But instead of returning to check it out further, he had pushed on to the higher energies then becoming available at CERN and NAL, leaving a loose thread hanging that had always nagged the Brookhaven directors.

Ting's was an enormous and difficult experiment. To pry a scant few electron-positron pairs out of a tremendous background that included billions of pions, kaons, and other hadronic garbage required the very utmost in sophisticated detectors. Most of the staff physicists at the AGS—Brookhaven's 30 GeV proton synchrotron—shook their heads in disbelief and said it simply couldn't be done. Ting found it difficult to convince any physicists besides his own group to join the effort. Others criticized the extremely fine resolution planned for their detectors, claiming it added needlessly to the cost. But Ting ignored their qualms and barged ahead with his preparations.

Finally, in April 1974, the MIT physicists were ready to begin. They surrounded their beryllium target and detectors with 10,000 tons of concrete blocks and added another 5 tons of borax soap to absorb neutrons. But when the intense proton beam began striking the target, the radiation level in the trailer where the physicists sat logging data leapt to dangerous levels—high enough to give them all the maximum yearly dose in twenty-four hours.

The experiment went on standby until this problem could be solved. Ulrich Becker, a tall, outgoing German physicist who had worked with Ting since the two first met at DESY, finally found the leak by poking around the concrete shielding with a Geiger counter. They plugged the gap and got going again late that month.

May was devoted to routine tune-ups and check-outs, including measurements at low mass. Using the energies and angles of the electron and positron emanating from the target, the MIT scientists could reconstruct the mass of the parent particle from which the two offspring had emerged; the results they then plotted as a histogram versus mass. As expected, the phi meson showed up as a narrow peak at 1020 MeV. The equipment was indeed working correctly.

After a shutdown that lasted through mid-July, Ting's group set its equipment to detect pairs emerging from parents with masses between 3.5 and 5.5 GeV, but found disappointingly few. Then came a two-week layoff while the proton beam went to an adjacent experiment led by Melvin Schwartz. Getting the beam back on August 22, Ting continued scanning this mass range. On August 31, four days before another scheduled accelerator shutdown, he decided to take a look between 2.5 and 4.0 GeV, too.

The small PDP 11/45 minicomputer logging the data and monitoring the equipment lacked sufficient memory to do a thorough analysis of the events rolling in. To dig true electron-positron pairs out of the tremendous background, the physicists had to reconstruct the exact trajectory of each sibling and trace it back to the target, making sure they both emanated from the same point—a task that required the full power of a mainframe computer. So the experimenters lugged their data tapes over to Brookhaven's CDC 7600 computer and did a more thorough analysis there.

Ting distrusted data analysis. Too often computer bugs had given experimenters spurious results that later proved embarrassingly wrong. So he always had two separate teams doing two independent analyses of the same data—and tried to keep them from influencing each other's work. They were to report any findings only at the regular group meetings.

On September 2, Terence Rhoades, a hot-tempered postdoc doing one analysis, began to notice electron-positron pairs piling up at the same place in his histograms. This bump began sprouting at roughly the same mass as Lederman's "shoulder" had six years earlier—suggesting a parent particle with a mass of about 3 GeV. Unsure of what he was observing, however, Rhoades did not inform Ting, who left Kennedy Airport for Geneva that same day. After the experiment stopped two days later for the accelerator shutdown, most of the others retreated to Cambridge.

Unknown to the MIT group, Stanford and Berkeley physicists working on SPEAR's Mark I detector had noticed something odd about their data earlier that year. They were studying collisions that were essentially the reverse of Ting's: instead of smashing two hadrons together and seeking pairs of electrons and positrons in the debris, they smashed electrons and positrons together and detected emerging hadrons. During 1973, these scientists had scanned energies from 2.4 to 4.8 GeV in steps of 0.2 GeV—or 200 MeV. They found no obvious bumps or dips, but the counting rates at 3.2 and 4.2 GeV did seem a little high, perhaps worth a second glance.

At the urging of Martin Breidenbach and John Kadyk, the SLAC-LBL scientists took another, more careful look in late June of 1974. This time they took smaller steps of 100 MeV, making careful measurements at energies of 3.1, 3.2, and 3.3 GeV, and at 4.1, 4.2, and 4.3 GeV. If there were indeed a bump only 100 MeV wide, it should reveal itself in these more detailed studies.

The XDS Sigma V computer controlling the SPEAR collider and the Mark I detector was powerful enough to sample the data as it rolled in and do a cursory analysis on line. During these special June runs, Breidenbach recalled, nothing extraordinary was seen. The rate of hadron production remained disappointingly flat. In the last few hours before shutdown, he finally penned a cryptic note into the logbook: "The Whiz-Bang Analysis Team says there is *no bump* at 4.2 GeV."

So Burton Richter flew to London with no new surprises to divulge, and Roy Schwitters started a paper summarizing the past year's work. On July 5 he sent a draft around to all his collaborators. "The total cross-section [i.e., the rate of hadron production] is a rather smooth function of energy over the range measured in this experiment," began his summary paragraph; "There is no strong evidence for resonance peaks or production thresholds."

By the beginning of classes on Tuesday, September 10, Ting had returned to MIT. Both of his analysis teams were now seeing strange bumps near 3 GeV; but, suspecting computer bugs, they refrained from telling him. The following Monday he flew to Brookhaven to lobby for eight more weeks on the AGS at the earliest possible date. The reasons he gave were financial: he was about to run out of funds and needed to show some results before he could ask for more money.

Late that month Ronald Rau, the director of high-energy physics at Brookhaven, wrote back granting Ting most of his request. "I understand perfectly well," he began, "why you so desperately would like to have eight weeks of running time in this fiscal year." But there were only ten weeks left and they had to give Schwartz a fair share, too. So Ting got six weeks, with three of them coming in late October and early November. He also agreed to put his group on twenty-four-hour standby: should another experiment break down, they would be ready to use the available beam within a day.

Meanwhile the two teams continued their separate analyses. Confined to the Boston area with a serious leg ailment, Becker worked with Rhoades. Min Chen, a young assistant professor who had joined the group after getting his Ph.D. from Berkeley, did the other analysis with the help of Glen Everhart, a postdoc. Highly competitive with one another, Becker and Chen did not discuss their results at all. So they remained largely unaware of the funny bumps and blips appearing near 3 GeV in each other's histograms.

During the second week of October, Chen began combining all the good events—the electron-positron pairs—witnessed in the various experimental runs of early September. Now that the time-consuming task of reconstructing particle tracks was done, he could determine the energy of each pair with great precision. About a hundred events in all were plotted in a single histogram showing the number of events witnessed versus energy. In keeping with the fine resolution, he put the pairs into narrow bins typically 50 MeV wide.

At first, Chen could not believe his eyes. Almost *all* the events were falling into a single narrow bin. "It has to be rubbish," he thought, a computer bug or something like it. He immediately made a series of checks to see what had gone wrong. When the odd spike refused to budge, he finally began thinking that perhaps it might not be so spurious, after all. "Suddenly I realized it could be *real!*"

Unknown to Chen, Becker was having a similar epiphany. His first reaction was to suspect a computer error, too, but a thorough check of his programs revealed nothing amiss. Perhaps this was physics! His heart racing with the thrill of discovery, Becker now hobbled down to the computer center with Rhoades and locked the doors, taking it over for the weekend so they could run a whole series of further tests.

What was so startling about this peak was not just its size but the fact that it was so *narrow*—less than 50 MeV wide. A hundred events was nothing to get excited about, but such a slender peak *was.* Such a small energy spread meant, according

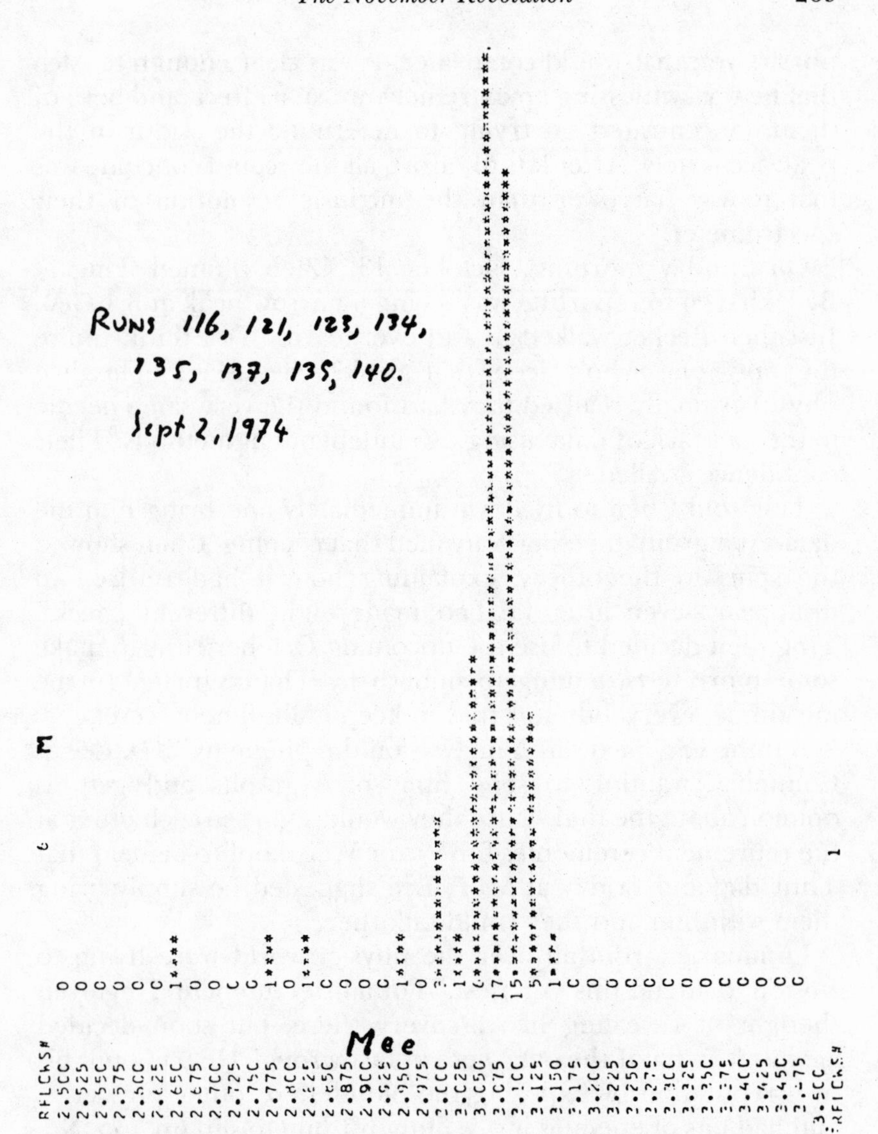

One of Ulrich Becker's early computer plots revealing a narrow peak at 3.1 GeV—about October 11, 1974.

to the Heisenberg Uncertainty Principle, that this resonance had a very long lifetime, far longer than almost every other resonance. For some unknown reason, this odd corpuscle was awfully reluctant to disintegrate. This was no ordinary particle.

But Becker and Chen weren't paying much attention to the

physics yet; that would come later. It was clear enough to each that he was witnessing an extremely unusual effect, and both of them concentrated on trying to determine the width of the peak accurately. After lots of effort, all they could conclude was that it was narrower than the intrinsic resolution of their spectrometer.

On Sunday morning, October 13, Chen phoned Ting at Brookhaven to report he was seeing a narrow peak at 3.1 GeV. Just then Becker walked in and overheard. "You think you're the only one who's seen that peak?" he asked. The two physicists finally realized they had found the very same needle in the haystack of data, using two independent methods. Their confidence swelled.

Ting told Chen to fly down immediately and bring him the data. At a group meeting convened that evening, Chen showed the spire to the others, explaining how it had refused to disappear—even after he had made eight different checks. Ting then decided to use the upcoming October runs to make some more tests, aiming to publish by Thanksgiving. In the meantime, everybody was told to keep it absolutely secret.

But the very next day Ting was on the phone to T. D. Lee at Columbia, wanting to send him some graphs and get his opinion about the find. Since they would soon see each other at the retirement ceremonies for Victor Weisskopf to be held that Thursday and Friday at MIT, Lee suggested he simply bring them with him and they could talk there.

Luminaries from all over the physics world were flying to Boston to attend this "Vikifest." For a brief moment Ting even thought of revealing his discovery there, but soon decided against it. What if the data were in fact wrong? He had built his own reputation by exposing the blunders of other physicists, and had lots of enemies just waiting for him to slip up, too. No, extreme caution was the watchword here.

Instead, Ting began giving private glimpses of the bump to some of his most trusted confidants there, including Willibald Jentschke, the Director of CERN who had backed his work since the early days at DESY. Lee finally got to see a graph during a cocktail party, but Ting must not have shown him the best data. "There was definitely a broad peak at about 3 GeV," Lee recalled, "much more pronounced than Lederman's shoulder." But it was nothing very earth-shattering, just another resonance. Unimpressed, he soon hurried away to chat with other physicists.

The Saturday after the Vikifest, October 20, Ting and a few others flew back to Brookhaven. A bubble-chamber experiment led by Nicholas Samios had failed to work properly, opening a gap in the AGS schedule. The MIT group might get beam early the next week if they could be ready. It was just the break they needed.

Roy Schwitters had made scant progress with the paper he had started back in July. With the death of his best friend and the birth of his second child, plus all the pressing responsibilities involved in upgrading SPEAR and the Mark I detector to run at higher energies, there was little time available for writing. The paper had been shelved until October.

By then the SLAC physicists had found a "glitch" in the on-line computer programs used to analyze the events as they rolled in during the June runs. Because of this error they had thrown away many good events they should have kept. When they corrected the glitch and recovered these events off-line, the extra June runs showed there might well be "structure" at 3.1 and 4.2 GeV after all. In both regions, the counting rate seemed to jump by about 30 percent above the surrounding data.

Schwitters realized he had to understand these details before the group could ever publish a paper. So he took it upon himself, in his few spare moments, to go back over the analysis. Sitting at a desk in his tiny Palo Alto home late one mid-October night, he began to examine the individual runs, one by one. Suddenly it "leapt out" at him! Of all the runs—numbers 1380 through 1389—taken at a total energy of 3.1 GeV, two were far out of line. Whereas an average run contained about 20 events, runs 1380 and 1383 had 61 and 104 events respectively. Otherwise, they seemed to be completely normal runs.

It was just these two runs, Schwitters discovered, that led to the 30 percent excess at 3.1 GeV. If he took them to be spurious runs and threw them away, then the excess disappeared and everything looked flat. Most experimenters confronted with such an abnormality would probably do just that. But Schwitters could find no convincing reason to reject them—and he refused to do so until he could. At a group meeting on Tuesday, October 22, he asked Gerson Goldhaber, a leader of the LBL contingent, to take a closer look and see if he could find "anything fishy" about runs 1380 and 1383.

That same Tuesday, Schwartz returned to Brookhaven after a month-long trip to China, one of the first American scientists to visit the country after its Cultural Revolution. The very first thing he heard upon entering his trailer was that Ting's group "had a very narrow peak at slightly more than 3 GeV." His assistant, an Indian woman named Jay Sharee Toraskar, had learned about it the day before from Glen Everhart, whom she had been dating off and on. Now things began to fit together for Schwartz. So *this* is why Ronald Rau had asked him to give up precious beam for Ting's benefit.

Schwartz is a pretty nosy scientist, one who likes to be in on a big discovery before it hits the preprint shelves. Any old resonance probably would not have pricked his interest, but a *narrow* resonance was a horse of a different color altogether. He immediately walked over to Ting's nearby trailer and found him standing inside.

"Sam, I hear you have a sharp peak at 3 GeV," Schwartz began, asking to see the graph.

"I have absolutely nothing," replied Ting, denying the request. "It's even less interesting than what Leon Lederman saw."

A keen judge of character, Schwartz knew when he was not getting the truth. "Sam," he continued, "I'll bet you ten dollars you've got a peak at 3 GeV." Still trying to stifle these nasty "rumors" before they got out of hand, Ting agreed to the wager, and they shook hands on it. A bit later, after Schwartz had left, he tacked a memo on his bulletin board declaring: "I owe Mel Schwartz $10."

Despite Ting's efforts to squelch them, rumors began circulating around Brookhaven that his group had made an important find. They were reinforced whenever the entire group marched as one to the cafeteria for a meal, with Ting sitting watchfully at the head of the table, making sure nobody blabbed.

Schwartz didn't exactly keep quiet, either. "Sam Ting has

found something," Rau recalls him saying, "and he won't tell us what it is." Later that week Schwartz cornered Y. Y. Lee, the only Brookhaven physicist to join the experiment, heading for his car in the parking lot. "I hear you've found a bump near 3 GeV," he pried.

"Who told you?" came Lee's unguarded reply.

Late that week Schwartz flew back to Stanford and stopped by his SLAC office to check on things there before flying off to Israel the next Monday. "I just won ten dollars from Sam Ting!" he told Jasper Kirkby, a British physicist then working with him as a postdoc. When asked how, Schwartz told him about the rumors of a narrow peak at 3 GeV, and how Ting had denied them. "And I know goddamn well," he concluded, "that Sam is lying."

That same Tuesday, October 22, Becker had to give a routine progress report on the Brookhaven experiment at a seminar for members of MIT's Laboratory for Nuclear Science. Unable to join the others because of his leg ailment, he was the logical choice for this thankless task. However much he wanted to, Becker could not duck the scheduled talk. But he tried his best to conceal the tall, narrow peak by displaying the data in such a way as to blur it out. It didn't take too much effort to make a lousy hundred events dissolve into the background. His ploy fooled almost all the physicists present. All but Martin Deutsch.

The goateed, pipe-chewing director of the Laboratory was "squirming in his seat," Becker recalls. Deutsch had backed the experiment sinçe its inception, and had gone out on a limb to get the needed funding. So he was far more familiar than the others with its fine details. He called everyone's attention to the odd feature near 3 GeV, but nobody else seemed very interested. Becker had put the graph at the end of a long, boring talk, so most of the others were nodding off by then, oblivious to what was right before their eyes.

Afterward, Becker visited Deutsch in his office and revealed the data plotted in a way that made the discovery obvious, cautioning him not to divulge anything yet. He did not share Ting's reservations and wanted to enlist the director's aid in getting the discovery published quickly. Still, Becker admitted the need to make a few more checks first.

That evening Ting and company got a feeble proton beam from the AGS and began using it to check out their equipment again. Where most other groups might take a day or two to make sure nothing had changed, the cautious MIT experimenters took almost a week—checking and rechecking the detector efficiencies, timing, and a myriad of other details.

On Sunday, October 27, they finally began logging data again. At first they set their spectrometer in the same configuration as used in September, but two days later they reversed its polarity. Now one eye detected positrons instead of electrons, and the other vice versa. This was the first of many tests that Ting required, in an attempt to see if the peak might disappear. With Chen and Everhart rushing the data tapes over to the Brookhaven computer for off-line analysis, the MIT group had rapid feedback on their efforts. By month's end another narrow peak could be seen sprouting at 3.1 GeV. The signal was still there!

The pressure to publish now increased tenfold: Becker and Chen urged Ting to write a short paper and take it to the nearby offices of the *Physical Review Letters.* They even began writing their own drafts. Urgent phone calls from Deutsch compounded the pressure. But Ting hesitated, wanting to make still more checks. Precious days slipped by while his group continued running tests and kept its nervous silence.

By late October, Schwitters was becoming convinced there was something awfully fishy going on in the SPEAR data around 3.1 GeV. But detailed studies of runs 1380 and 1383 failed to uncover anything amiss. They seemed completely normal, just like the other runs at the same energy, except for the fact that they contained three to five times too many events. "Here were a series of runs, all taken at the same energy, as far as we could ascertain, and yet they showed considerable inconsistencies," recalled Goldhaber. "This was an obvious clue that something was going on that we didn't understand."

Richter was away from SLAC at the time, giving the Loeb Lectures at Harvard. Subbing for him before a SLAC review panel on October 30, Schwitters told them there was "something funny" about the most recent measurements. The rate of hadron production was not quite as smooth as the group previously had been broadcasting. That evening he called Richter at Harvard and suggested he "cool it on the smooth cross-sections."

The following Monday, November 4, Schwitters went to the computer center with another physicist, carrying the computer tape on which the data from runs 1380 and 1383 had been written. Mounting it on a tape drive, they began to scan through these runs again, bringing up an image of each event on the scope—a picture of the tracks left by charged particles in the Mark I detector. Hadron events usually showed three or more distinct tracks, while muon pairs or electron-positron pairs were always two-pronged. Perhaps, Schwitters thought, the analysis programs had confused one kind of event with the other, yielding the apparent excess of hadron events. They scanned the two anomalous runs carefully, event by event, and verified that all the excess were indeed hadron events. Nothing amiss there.

The next day Schwitters rechecked all the physical parameters from the two runs. Then he asked Ewan Paterson, an accelerator physicist with the group, if the luminosity of the collider could have jumped abruptly by a factor of three to five during the odd runs—causing the observed excess. Paterson told him to forget it: the luminosity could not possibly be off by such a large factor.

That afternoon, November 5, Schwitters had to guide a small group of Stanford graduate students on a tour of the SPEAR facility. He was sitting in the warm sun on a bench outside the Administrative and Engineering Building, waiting for their bus to arrive, running all these recent observations through his mind, when suddenly it hit him. All the parameters were okay, the luminosity could not explain the excess, and the hadron events were indeed good hadron events. These two runs were perfectly acceptable. "There's something *real* going on," he thought, "that has to be *physics*."

By early November, there could be no doubt in Ting's group that they had discovered a new and unusual fragment. They had run with thicker targets, they had changed the magnet polarities and currents, they had altered the detector voltages, and the narrow peak still refused to budge. There it stood at 3.1

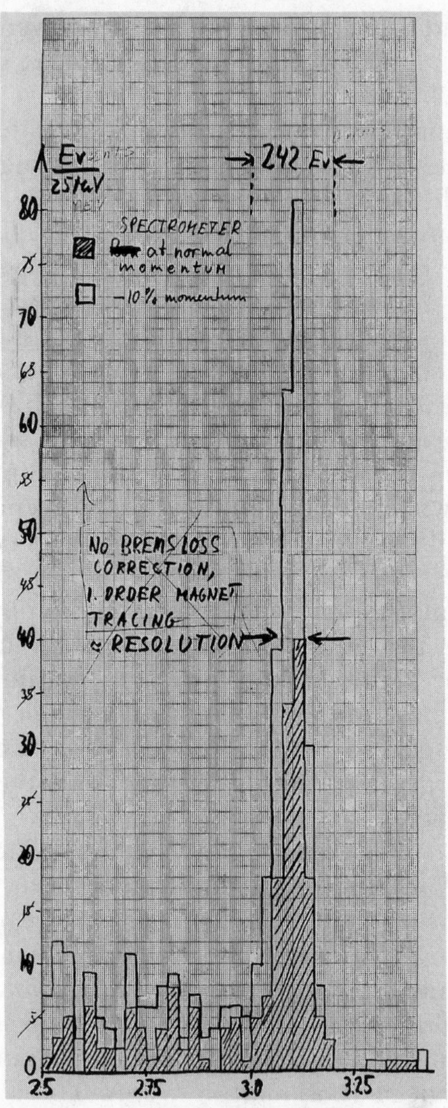

Histogram of the combined data plotted by Becker in early November. Despite a deliberate 10 percent shift in the spectrometer momentum, the narrow peak at 3.1 GeV refused to budge.

GeV, naked and unyielding, towering above the scant few background events to either side.

It was the runs taken with altered magnet currents that had finally convinced Ting. Just before midnight on Halloween, the MIT physicists lowered these currents by 10 percent and began run 185. If this peak were a real, physical phenomenon and not some fleeting "spurion," it had to remain fixed at a mass of 3.1 GeV. If it had instead been caused by particles glancing off the magnet poles, say, it would either disappear or shift to another value.

"I was *very* anxious to get that run back," Ting confided; "I just wanted to see whether there were a few events there." When the off-line analysis came in the very next day, showing a peak still sprouting near 3.1 GeV, Ting was finally convinced. This was no spurion. He remembers calling his daughter Jeanne in Boston to tell her about the discovery.

Now the pressure to publish really intensified. From the recent Schwartz encounter, they knew that word about their find had leaked out—especially at Brookhaven. Given the ease with which rumors travel in the particle physics community, it would not be long before others learned about the resonance and began looking for it themselves. Becker and Chen took every opportunity to urge Ting to begin a paper; Deutsch phoned them repeatedly from MIT. But Ting hesitated for a critical week.

Physicists close to Ting suggest a number of reasons why he held off publishing the discovery. An article he wrote two years later claims he was "puzzled" by the large fraction of electrons emanating from hadron collisions and also "wanted to know how many more [similar] particles existed." Becker and Chen echo this sentiment, recalling that the group had excellent reason to believe it had stumbled upon only the first in a whole *family* of new particles. They claim Ting wanted to unearth more of them so he could publish the discovery of an *entire family*—in what would be seen as a tremendous coup.

Once they published and other physicists knew where to begin looking, Ting realized, newer and far more powerful accelerators would beat them to all the follow-up discoveries. But this strategy was extremely short-sighted. Had Ting established sole claim to priority on the first family member, any other discoveries would have been seen as merely derivative of his own. He would have received the lion's share of credit for finding the entire family.

Other people close to Ting note that he had always been an extremely careful, cautious experimenter. Having built his reputation by disproving the results of other experimenters, he dreaded making a foolish blunder himself. In fact, Ting had never published a result in disagreement with accepted theory, notwithstanding his many protestations about ignoring what theorists had to say. But this time there was no theory available, at least not one Ting knew about, to accommodate such a strikingly tall and narrow peak. You also worry when you have *too* clean a signal, Y. Y. Lee noted. "What shall we do?" he recalls Ting asking the group as he paced to and fro about the trailer. "What shall we do now?"

Finally, there was the element of overconfidence. The peak at 3.1 GeV was *so* narrow, Ting felt, that nobody else would find it unless they somehow knew *exactly* where to look. "The storage ring boys will find it immediately," he told Rau on November 1, "if they know where to look." Indeed, Becker and Chen wanted to phone Richter while he was speaking at Harvard and suggest they together use SPEAR to examine this resonance in detail. The total SPEAR energy, they knew, could be set right at 3.1 GeV to produce a plethora of these corpuscles. But Ting and Richter had never gotten along well, they recalled, so this idea was scotched.

In retrospect, it was probably a combination of all these factors that led to Ting's hesitation. And though not a very long delay, it was a crucial delay.

On November 4 or 5, Ting showed his results to Weisskopf, who told him he was crazy not to publish immediately. Ting still maintained that there were "other effects" he wanted to check out first and to publish together with the narrow resonance. He was "absolutely sure" nobody else would find it because the peak was so incredibly narrow, a needle in a haystack.

But Weisskopf's admonitions must have had some effect, because Ting finally agreed to publish. On Wednesday, November 6, he visited Editor George Trigg at the *Physical Review* offices near Brookhaven to inquire about the rules for bypassing the referee process. Then, after conferring with Becker and Chen, he began a draft of the paper.

Richter returned to SLAC on Monday, November 4, to discover growing excitement over the curious effects at 3.1 GeV. On Tuesday, Schwitters was convinced these features were physics; with Breidenbach and Paterson, he began urging

Richter to go back and take another look before boosting SPEAR to higher energies. Harvey Lynch, Schwitters's co-spokesman for the upcoming experiment, was more skeptical.

The storage ring had been painstakingly upgraded that summer to run at energies above 5 GeV, so it was no small matter to crank it back down and run at 3.1. The natural inclination of any red-blooded experimenter was to forge ahead into the virgin, unexplored territories before backtracking for a second (here, actually a *third*) look at such oddities. And with DESY hot on their heels, about to commission its own collider (DORIS), able to explore the same region, it was also the wiser option. Richter held out for continuing as planned up to higher energies.

The debates got more heated as the week progressed. It was a very hectic time, what with all the last-minute preparations for the next running cycle. But Schwitters and Breidenbach kept pressing Richter to go back to 3.1 GeV, while Lynch and others kept urging him forward. Richter liked to manage like this, to sit at the focus and listen to the arguments of a group of bright young physicists before making the final decision himself. Until the week's end, however, he refused to budge.

Then on Friday morning, Goldhaber called Schwitters with some tantalizing news. He and Scott Whitaker, a Berkeley graduate student, had been identifying the individual particles produced at 3.1 GeV; there seemed to be a "remarkable increase" of kaons in the debris of run 1383. This was exactly what the advocates of charm had been suggesting that experimenters look for: a sudden increase in the production of strange particles. Perhaps the SPEAR energy had drifted slightly in those two runs, and they had grazed the side of a new resonance. "There simply had to be some new physics," Goldhaber recalled, "hanging on the ragged edge of 3.1 GeV."

Although Goldhaber's report later proved to be a "red herring," a spurious observation that eventually proved erroneous, it was the crucial piece of information that finally tipped the scales. Late that Friday afternoon, at the eleventh hour before the next cycle was due to begin, Richter convened a meeting of the collaboration in his office. Schwitters was prepared for battle this time, with his most recent graphs showing very suspicious bumps at 3.1 and 4.2 GeV. But it was Goldhaber's red herring, everybody recalled, that finally won

Richter over. They were given just the weekend to crank SPEAR back down and take another look.

The sun was setting as the meeting ended. Cirrus clouds were gathering above the Santa Cruz mountains to the west as another big storm approached the California coast. The sky was alight with a "weird, multicolored sunset," Schwitters recalled. Rudolf Larsen glanced out the window at the ominous sky and remarked, "It looks like the Messiah is coming."

Paterson, Schwitters, and Whitaker arrived at SPEAR before 8 A.M. the next day to steal an early look. They filled the ring with electrons and positrons at 3.12 GeV and began counting hadron events one by one as they burst out on the computer scope. They logged twenty-two events in just twenty-five minutes, more than a factor of three above normal. This was the same sort of behavior Schwitters had noticed in the "fishy" June runs. The signal was still there!

The rest of the day the scientists made sure that SPEAR and the Mark I detector were working properly. By evening the experimenters on shift were ready to begin logging serious data. After hearing about the morning runs, Goldhaber drove down from Berkeley to join in the hunt. Breidenbach was at the helm now, and a magnum of champagne was cooling down in the refrigerator.

At 11:15 P.M. that night, they once again began clashing the two beams at a total energy of 3.12 GeV and watched with glee as the counting rate more than doubled—clear proof that they were hitting something. With the excitement mounting, Breidenbach pulled out some graph paper and plotted all the data thus far measured. Sure enough, there seemed to be a bump growing. It had to be centered somewhere between 3.10 GeV, where the counting rate was normal, and 3.14 GeV—a gap of some 40 MeV.

The obvious thing to do next was to begin slicing into that interval in small steps, to determine the exact location of the peak's center. But before they could do that, the machine began acting up. In the early morning hours of Sunday, November 10, Breidenbach went home and Goldhaber drove to a nearby motel to catch some sleep. The champagne remained in the cooler, unopened.

* * *

At 8 A.M., the day crew came back on shift. Soon they were joined by Schwitters, Goldhaber, and others eager to be part of the final assault on the mysterious peak. At 9 o'clock they set up a base camp by making one last measurement at 3.10 GeV; the counting rate was normal, as before. The climb began an hour later, after Richter had arrived, when they bumped the energy up to 3.11 GeV—midway between two previous settings—and began run 1460.

Suddenly, riot broke out in the SPEAR control room. The scope began flashing repeatedly with events, many more than one per minute. Somehow Schwitters had the presence of mind to make a short entry in the logbook:

> This past fill has been incredible. While running at 3.10 GeV we saw essentially the baseline value. During the middle of the fill, we bumped the energy to 3.11 GeV, and the events started pouring in. This is a remarkable resonance indeed!

A quick calculation showed that the counting rate had leapt by an amazing factor of *seven* above baseline. No question about it, this was a very sharp and unusual peak.

Goldhaber and Richter agreed that they should start a paper immediately. So Goldhaber found himself a quiet table in a side room, away from the commotion, and began to write a draft on the back of some computer printout. But an hour later, William Chinowski burst excitedly into the room to tell him that the counting rate had jumped by *another factor of ten,* while measuring at an energy of 3.104 GeV.

Goldhaber dropped everything. Racing into the control room, he found everybody crowded around the scope gaping in disbelief at the event display. Hadron events were absolutely *flooding* in now, almost one every second! Awestruck, they watched the scope light up like the grand finale in a fireworks display. By now, Panofsky had arrived, too, and was pacing back and forth with his hands on his head, uttering over and over, "Oh my God! Oh my God! Oh my God!"

When they finally topped the peak that afternoon, at a total energy of 3.105 GeV, it stood an absolutely incredible factor of a *hundred* above baseline. And halfway up its steep sides it was only a few MeV wide. Compared to other resonance bumps, this amazing spire was a towering needle soaring far up into the stratosphere.

Early that afternoon Bjorken was sitting down with his family for Sunday dinner when the phone rang. It was Richter with the startling news. "I couldn't believe such a crazy thing was so low in mass, was so narrow, and had such a high peak cross-section," he recalled. "It was sensational." He returned to the table a few minutes later, seemingly in a daze. His wife and children then watched open-mouthed as he unthinkingly heaped a large tablespoon of horseradish onto his baked potato and quietly began munching away, staring absent-mindedly off into space. "BJ," his wife finally counseled, "I think you'd better go down to the lab now."

As word of the fantastic discovery leaked out into the SLAC community, a crowd of happy onlookers began to gather in the SPEAR control room. Somehow amid the euphoria, the physicists remembered the champagne, but now there was not very

Berkeley and Stanford physicists gather in the SPEAR control room as Martin Breidenbach plots the on-line data on Sunday evening, November 10, 1974. Just behind him are Gerson Goldhaber (left) and Ewan Paterson (second from left).

Burton Richter checks the SPEAR energy calibration late Sunday afternoon. Note the unfinished glass of champagne.

much to go around, plus a shortage of glasses. Recruits departed for more supplies.

Meanwhile, Goldhaber's draft had been made a shambles by the more recent data, so Richter began writing another version. The new particle needed a name, and he first tried calling it the "SP" after the first two letters of SPEAR. But nobody liked that very much, so Richter went over to Leo Resvanais, a Greek physicist with the University of Pennsylvania, and asked him what Greek letters were still unused. Resvanais went down the list alphabetically, until he came to *iota,* but Richter didn't like that very much because it also meant something small, which this discovery was clearly *not.* The next available letter was *psi,* or ψ,

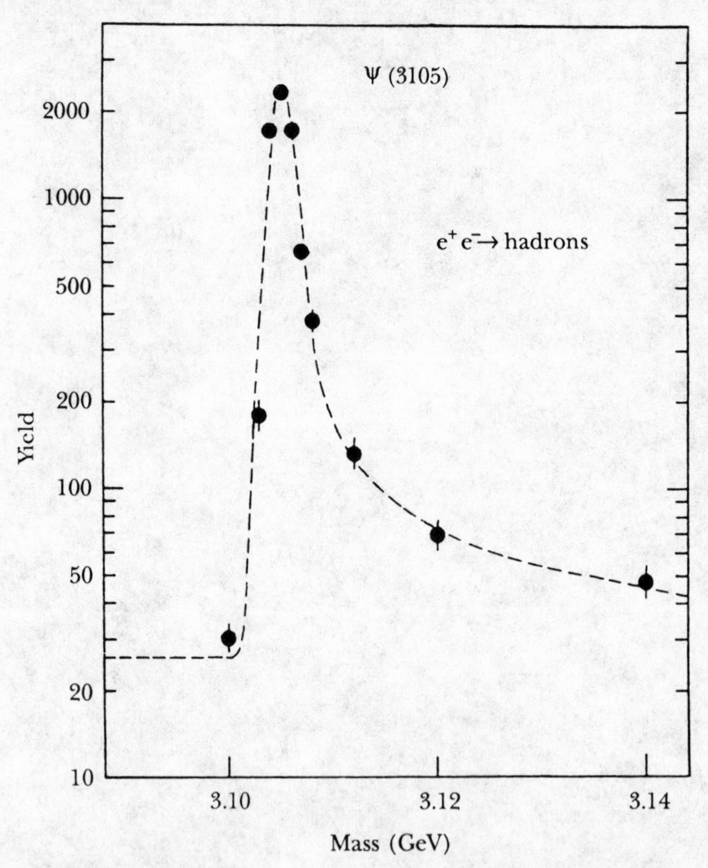

A finished graph of the narrow peak witnessed in the Mark I detector. The vertical scale is logarithmic, making the peak seem far shorter and broader than it actually is.

which Richter thought might be confused with the Greek symbol for the wave function, but Resvanais told him that made it a *good* choice because this discovery was certainly something *big*. Most of the others liked that name, especially after Goldhaber pointed out that it contained the letters *s* and *p* in reverse order.

Word of the find soon began to spread well beyond SLAC. Goldhaber phoned his Berkeley colleagues and they in turn called others. And so on, and so on. The phone in the control room began to ring and ring, again and again. "You could feel a broad wave surging across the country from west to east," Breidenbach recalled. "By that evening it had reached the East Coast; Sunday night it jumped the Atlantic."

Late that Sunday afternoon, Ting rode to Kennedy Airport with Ted Kycia, a particle physicist at Brookhaven. There they boarded a TWA flight bound for San Francisco. As members of SLAC's Program Advisory Committee, they were headed there for a meeting scheduled the next day. Unaware of the rumors circulating at Brookhaven, Kycia asked Ting if he had discovered anything yet. Ting replied no, he hadn't even seen the shoulder Leon Lederman had observed.

Little did Ting know then of the rude shock awaiting him in California. He continued reading his papers, serenely unaware of the great waves rolling back and forth across the country 30,000 feet below.

Sau Lan Wu, a postdoc with Ting's group, was working the swing shift that evening at Brookhaven when an unknown character poked his head into their trailer at about 9 P.M., apparently looking for Schwartz. "Have you heard the news?" he asked. "SPEAR has found a sharp resonance at 3.1 GeV! People there are celebrating with champagne!" He departed quickly, after noticing their crestfallen expressions.

"We all looked at each other, pale, and Min with tears in his eyes," recalled Wu. They immediately put in a call to TWA, leaving an urgent message at the San Francisco Airport for Ting to call them upon arrival.

By the time he called in at 1 A.M., these experimenters had convinced themselves this was just a lousy practical joke being played on them by Schwartz's troops; Ting agreed to check it out anyway when he got to SLAC the next day.

But immediately after he arrived at the Flamingo Motel in Palo Alto an hour later, Ting received a call from Deutsch. There was great excitement at SLAC, he'd just heard, about a resonance discovered there. Could it be the very same resonance? Worried now, Ting called Stanley Brodsky, a SLAC theorist he knew, and learned that yes, indeed, there had been a big discovery that would be announced tomorrow.

For Ting, it was the moment of truth. He was about to be scooped on the greatest discovery of the year, if not the decade. His weeks of cautious silence now had to be undone overnight.

First he rallied the troops at Brookhaven. Wu remembers waking to an anxious knock on her door at about 2:30 that morning; it was Ingrid Schultz with the cruel news that SPEAR physicists had indeed discovered the same resonance, after all. Chen hurriedly drew up a histogram for release the next morning and then rushed back to MIT. Wu spent much of that morning on the telephone, calling physicists everywhere in Europe—including Italian friends at ADONE, advising them to look for the same resonance.

Next, Ting had to alert Rau of his results and plans to announce them, but the Brookhaven director was away at Los Alamos. Finally Ting found him at his motel, fast asleep. Having only learned sketchy details before this moment, Rau was "blown over like a candle" by the startling nature of the discovery.

Back at Brookhaven, Wu was desperately trying to locate a working Xerox machine to make copies of the histogram for general distribution. At six o'clock that morning A. J. McGreary, the AGS operator on shift, penned a note into his logbook:

> Experiment 598 has found a new particle!!! Ms's Wu and Schultz have been rushing around madly trying to make xerox copies of some data and calling outside. They are obviously very excited!

And on and on throughout the night. Sleep was forgotten in the desperate efforts to get word out about their discovery before it was too late. And Ting's next-door neighbor at the Flamingo had a hard time sleeping too. Robert Diebold, a physicist from Argonne National Laboratory, was in town for the same meeting; he remembers hearing the phone ringing incessantly from three o'clock on. He even overheard several of Ting's louder exhortations. "Now what could possibly be so important," thought Diebold, "that he needs to call a press conference?"

The wires were humming now all over the world. Particle physics ran up its biggest one-day phone bill ever.

Worn and haggard, Ting got to SLAC at 8 A.M. Monday and went straight to Panofsky's third-floor office to show him his discovery. Richter arrived moments later to discuss the agenda for the forthcoming advisory committee meeting.

"Burt," Ting greeted him, "I have some interesting physics to tell you about."

"Sam," Richter replied, still unaware of Ting's discovery, "I have some interesting physics to tell *you* about!"

Then followed "an astonishing conversation," according to Richter, in which he learned of the Brookhaven discovery and saw that it was the exact same resonance found at SPEAR. A bit later Richter went downstairs and found Schwitters. "Sam Ting's found the same damned thing!" he told Schwitters, who had already heard the amazing news in a phone call from Europe.

First on the agenda at the meeting, Schwitters told the members about the psi discovery. He recalls Ting sitting in the back of the room, ashen-faced and "looking like death itself." Ting seemed totally devastated by Schwitters's presentation. He got up to say his own group had already found the same thing, but stumbled badly in his delivery. For Schwitters, it was the only time he'd ever seen Ting lose his usual stern composure.

At a noon colloquium, Schwitters and Ting revealed their data in quick succession to an enormous, stunned audience. Schwitters remembers being surprised not so much by the size of the crowd, but by its *composition.* There were technicians, machinists, programmers, engineers, secretaries, and janitors jostling for standing room with the experimenters and theorists. Only then did he begin to grasp the truly historic dimensions of the discovery. Right on the spot, he made an impromptu change in his delivery, to reach this unanticipated audience. Following him Ting, having regained his composure, gave a short, crisp lecture about his own find. For those of us in the crowd, it was an unforgettable moment. Here were not one but *two* discoveries of the weirdest corpuscle we had ever encountered.

Frank Close was having lunch in the CERN cafeteria that Monday when he heard a remark that a new particle had just been discovered at SLAC. The young theorist, who had spent a few years as a SLAC postdoc after finishing up at Oxford, paid this rumor no heed. "One's always hearing rumors, they come and they go," he observed. "There was nice food to eat and digest, and if the rumor was true, we'd hear about it."

He was passing theorist Jacques Prentki's office on his way back from lunch, when he heard Prentki call out, "Hey Frank, come look at this!" Close wheeled and looked inside:

> It was a large room but there wasn't much space left in it because the room was full of people, all very excited, including Jack Steinberger. Now Jack is an eternal skeptic. But on this occasion he was looking distinctly non-skeptical, and I *knew* at that moment that something had happened.

Jean-Jacques Aubert, a French physicist who had been working as a postdoc on Ting's experiment, was standing in their midst showing them all a graph of a narrow pinnacle at 3.1 GeV. (He had drawn it up that morning, I learned, from numbers Chen read him over the phone.) "And when I saw it," Close recalled, "I knew it was true." The same peak had allegedly been seen at SLAC that weekend, too, though nobody at CERN had yet seen these data.

"There was tremendous excitement at CERN," Close recalled in a tape recording he made about a month later.

> It was indescribable. *Everybody* in the corridors was talking about this. And I think it was the first time in the history of CERN, if you were any place in the whole of the building at any time, the chances were [that] the people there were talking about one and the same thing.

The next afternoon an ad hoc meeting was hastily convened to discuss this striking object and hear Aubert recount the discovery of it. Close got there early, but already a hundred people were crowded on the stairs, waiting for a previous meeting to end. Steinberger finally brought the fever-pitched crowd to order. "It's nice," he began, "to have discovered something we can all believe in."

That same Monday morning the phone began jangling off the hook at *The New York Times,* too. Science reporter Walter Sullivan, who had long ago staked out particle physics as his personal territory, recalled the day vividly. First to ring him up was his old friend and trusted confidant, Victor Weisskopf, telling him about a startling particle discovery they'd be hearing a lot more about in the coming days.

Without explaining many details, Weisskopf implored Sullivan not to publish the story until a formal announcement had been made and scientific papers had been published in *Physical Review Letters.* To discourage physicists from scampering off to reporters with their latest results, Editor George Trigg had a stern policy of rejecting papers when the same results had already been announced in the newspapers. Sullivan agreed to hold back awhile but told Weisskopf he'd have to publish as soon as any other paper got wind of the story.

No sooner had Sullivan hung up than the phone rang again. This time it was a student at Columbia, reporting the Brookhaven find in greater detail. And a bit later that day, the patrician newspaperman learned more about the psi discovery in a phone call from SLAC. This particle really had these guys jumping. It would be hard to keep the lid on.

Meanwhile, Ting had skipped the rest of the advisory committee meeting and hurriedly finished his paper in a vacant SLAC office. After dictating his changes over the phone to MIT, he hopped an evening flight back to Boston. He rushed the typescript down to the journal offices on Tuesday afternoon, arriving there just before closing. Ting knew the SLAC-LBL paper was coming in the next day, hand-carried by Ted Kycia, and he wanted to get his paper in a day earlier, to help establish his priority. By now, he had dubbed his particle the "J," a letter that closely resembles the Chinese ideogram for Ting.

Still, despite heroic efforts, it would be three more weeks

before these papers appeared—in the December 2 issue. As the days went by and scientists all around the world heard the startling news over the grapevine, pressures began mounting on Sullivan to run the story. Convinced that the news was about to break anyway, Panofsky and Richter kept urging Trigg to let Sullivan publish it. The dike, they knew, was about to burst. Finally Trigg agreed, and Richter and Ting issued a joint press release on Saturday, November 16.

A Yale alumnus of long standing, Sullivan had garnered front-row tickets on the 50-yard line for the Yale-Princeton game that afternoon. But he spent much of the game in the press box, away from his family, on the phone to Stanford. Getting the go-ahead at halftime, he wrote up the story quickly and rushed it over the wire to New York. On Sunday morning, over a million *Times* readers awoke to front-page news of the dual discovery. "New and Surprising Type of Atomic Particle Found," announced the bold headline. Soon the entire world knew about the peculiar mote.

What, you may well ask, was all the excitement about? Why were grown men and women getting so worked up over a mere spike in some graph, or a computer scope lighting up with fireworks? "The suddenness of the discovery coupled with the totally unexpected properties of the particle are what make it so exciting," Richter and Ting announced in their press release. "It is not like the particles we know and must have some new kinds of structure."

The key to the frantic wonder was the extremely narrow *width* of the spire—less than 1 MeV, or about a part in three thousand. More detailed analyses, which removed effects that artificially *broadened* the peak, showed that its true width was only 0.05 MeV. According to the Uncertainty Principle, such a small width meant the resonance had a lifetime of 10^{-20} second, or 0.00000000000000000001 second. While this may seem like the splittest of split seconds to the average office worker trying to meet a deadline, it is a long, long time to high-energy physicists, who frequently deal with corpuscles disappearing a thousand times faster.

If this resonance were a hadron, which seemed as likely as any other possibility, it lived a thousand times longer than

Official portrait of Sam Ting and collaborators inside their Brookhaven trailer. Before him sits the finished plot of the J resonance, as eventually published.

normally expected. Why did it refuse to play along with its buddies, who disappeared far more quickly? *Something* had to be inhibiting its decay, and that something might well be a new property of matter never seen before. As Feynman and Gell-Mann often stressed, "Only conservation laws suppress reactions." A completely new characteristic of matter was probably being conserved in the "slow" disintegration of the new particle.

What was this characteristic? What was the new particle struggling so hard to conserve? At first, there were almost as many ad hoc proposals as there were theorists. Particle physics was off on a roller-coaster ride it had not witnessed in decades.

CHAPTER 13

A MATTER OF TASTE

They sought it with thimbles, they sought it with care;
They pursued it with forks and hope;
They threatened its life with a railway share;
They charmed it with smiles and soap.

—Lewis Carroll, THE HUNTING OF THE SNARK

IT is a short trip from Harvard and MIT to the campus of Northeastern University in the heart of Boston. A university worlds apart from its more famous Cambridge neighbors, Northeastern doles out a practical, no-frills education to legions of Boston-area commuters who hurry off after class to part-time jobs or suburban homes. It is hardly a place one would expect the esoteric field of particle physics to be taken seriously. But during the 1970s, Northeastern played host to regular gatherings of meson specialists. On April 26 and 27, 1974, about 200 physicists had gathered there for the Fourth International Conference on Experimental Meson Spectroscopy.

Spectroscopy has a respected, if not exactly glamorous, role in physics. In the nineteenth century, long before Bohr came along to interpret their work in terms of the energy levels of atoms, scientists were using devices called spectroscopes to measure and catalog the myriads of discrete spectral lines—the pure colors—emitted by glowing gases of hydrogen, helium, and other elements. A similar job was done by nuclear spectroscopists half a century later, when they classified the many hundreds of atomic nuclei, their excited states, and possible emissions. It is a tedious task far better suited for botanists or zoologists than for physicists.

When the meson spectroscopists met in 1974 to compare their latest bumps and resonances, the flora and fauna of their field, most of particle physics was looking the other way. During its heyday in the 1960s, spectroscopy had been one of the hottest subjects around. But with over a hundred creatures already catalogued, and others appearing every month, all in agreement with the quark model, who could get terribly excited?

The gauge theories of interparticle forces were grabbing the headlines that spring. Unity was in the air—and the hope that we might finally be able to understand the forces locking the putative quarks together. At MIT, our group was busy sifting old data and planning new experiments to determine if asymptotic freedom might explain the behavior of electron-nucleon scattering. Too busy to attend the meson conference, Jerome Friedman gave his invitation to Arie Bodek, who became one of the scant few MIT physicists to make the short cross-town trip. Samuel Ting was at Brookhaven with his entire group, shaking down their experiment.

The only physicist to come from Harvard, Sheldon Glashow delivered a paper titled "Charm: An Invention Awaits Discovery." As the case for neutral currents and gauge theories became stronger and stronger that spring, he began beating the drums harder and harder for experimenters to go search for evidence of a fourth quark—the charmed quark—that he and Bjorken had first proposed a decade before. If neutral currents really existed, as now seemed a good bet, a fourth quark "flavor" was sorely needed to explain why a few particular disintegrations of neutral kaons happened so rarely, if ever.

Forty-two years old that year, with his long graying hair falling helter-skelter across his forehead and his eyes framed by Coke-bottle glasses, Glashow gave his usual entertaining talk, bumbling good-naturedly through some calculations. "I would bet on charm's existence and discovery, but I am not so sure it will be the hadron spectroscopist who first finds it," he challenged his audience. "Charm will not come so easily as strangeness, yet no concerted, deliberate search has been launched." The way to find charm, he stressed, was to forget the bumps and resonances so dear to spectroscopy and look instead for rare decays involving muons and strange particles among the debris. A sudden onset of kaon creation, for example, was a sure sign of charm.

If spectroscopists didn't get going, however, Glashow warned his audience, charm would be discovered by "outlanders"—other kinds of physicists who did neutrino scattering or measured electron-positron collisions in storage rings. Then he offered his audience a bet. By the time the next meson conference rolled around, there were just three possibilities: "One, charm is not found, and I eat my hat. Two, charm is found by spectroscopists, and we celebrate. Three, charm is found by outlanders, and you eat *your* hats."

Robert Palmer and Nicholas Samios were two of Glashow's outlanders. Since the previous fall, he had been urging them to look for charm in the neutrino scattering experiments they were planning at Brookhaven. The group had just built a new, 7-foot bubble chamber expressly for use in neutrino and antineutrino beams. Though not much longer than the workhorse 80-inch chamber Samios and collaborators had used to discover the omega-minus, it contained far more liquid hydrogen or deuterium. In neutrino scattering experiments, mass is crucial.

Every once in a while, Glashow told Palmer and Samios, a high-energy neutrino impinging on the chamber might lob a W^+ at one of the target protons. The solitary down quark inside this proton could gobble up the W^+, becoming a charmed quark and transforming its parent into a charmed baryon for a brief instant. After less than a trillionth of a second, this charmed baryon would explode in a cascade of longer-lived particles, among them a lambda that served as the signpost for charm: a single strange particle among the debris.

Like his search for the omega-minus, this was the kind of

quest that intrigued Samios—hunting for a rare, exotic event. Filled with liquid hydrogen, the 7-foot bubble chamber sat chugging away in neutrino beams for the better part of 1974, logging hundreds of thousands of photographs for scanners to examine. They were to look for two bright rays zooming off from out of nowhere, making the characteristic "V" pattern indicating a lambda disintegration. If the V pointed back to another spray of debris, they were to call Palmer or Samios. This was a likely signature for charm.

One of the scanners found just such an event in late May, hardly a month after Glashow made his wager. Excited, Palmer and Samios were nevertheless cautious. It was not the kind of "gold-plated event" the first omega-minus event had been. Besides charm, there were other possible interpretations. Palmer worked through the detailed calculations that could tell him how else the lambda might have appeared, *without* the intercession of charm. Months passed before he was convinced.

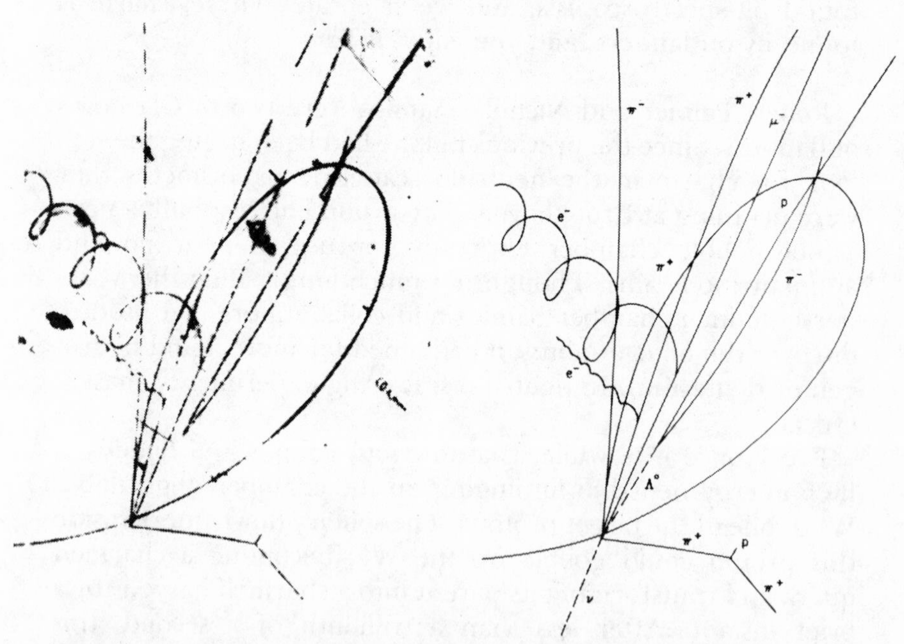

Photograph (left) of possible evidence for charm in Brookhaven's 7-foot bubble chamber plus a reconstruction of the decaying particle trajectories. A single strange particle, the lambda, emerges among the debris—suggesting the decay of a charmed baryon.

Meanwhile, their bubble chamber chugged contentedly along, logging one photograph after another in hopes of recording more such events. But the complex magnet surrounding their chamber failed on Sunday, October 20, and Ting became the beneficiary of the unused AGS beam. Three weeks later, high-energy physics was turned upside-down by the announcement of his momentous discovery.

Charm was in the air during the fall of 1974. Physicists talked about it in seminars, in corridors, and on experimental floors. The new kid on the block, it was the one we had all heard about and were hoping to meet someday soon.

Late that summer a lengthy review paper titled "The Search for Charm," by theorists Mary Gaillard, Benjamin Lee, and Jonathan Rosner, began making the rounds. There they discussed the reasons for a fourth quark and outlined the many possible ways experimenters might look for its evidence. Among their suggestions was the likelihood of creating a heavy neutral meson in collisions of electrons with positrons. They even proposed that such a particle might also have caused the curious "Lederman shoulder" between 3 and 4 GeV.

How much did the search for charm influence the choices made by Ting and company? Here the record is contradictory. According to Min Chen, Glashow came by at Brookhaven that August to recommend ways to find evidence of charm. They should not be looking for electron-positron pairs at all, he told them, but instead for pion-*kaon* pairs—strange particles in the debris.

Though a good friend of Glashow's, Ting vehemently denies that the search for charm had any influence on his experimental plans. But he was in fact one of the very first experimenters to be approached with the idea of charm. Way back in 1970, Iliopoulos recalled, Ting joined a going-away party for Luciano Maiani held aboard an ocean liner in Boston Harbor. It was just after Glashow, Iliopoulos, and Maiani had written their famous paper on charm, but well before it appeared in print. With the music playing and champagne flowing, the three Harvard theorists plied the lone MIT experimenter with repeated suggestions he start looking for charm.

On September 16, 1974, Ting met at Brookhaven with

Ronald Rau and Horst Foelsche, then director of experimental planning, to survey priorities for the fall. From the detailed notes Foelsche kept of that meeting, Ting stressed that a charmed particle search had high priority.

Ting likes to pretend that he never pays attention to theorists. As with many other experimenters, this is part of the public facade he presents to the physics community. But he is a master, one of his colleagues told me, at "pumping the theorists for information" that might help his team do good experiments. Separating the wheat from the theoretical chaff is a crucial talent for an experimental group leader.

Sau Lan Wu had read the "Search for Charm" paper. Several times that October she brought it to Ting's attention, pointing out its predictions of a sharp peak in proton-proton collisions. But her suggestion just did not seem to sink in. When Ting first revealed a discovery to Rau and Foelsche, on November 1, charm was proposed as one interpretation. "Sounds like he's onto a charmed particle, narrow width!" read Foelsche's personal notes.

Glashow thinks he first heard about the J discovery directly from Ting in a telephone call early that Monday morning, November 11, but Ting denies ever calling him. It was probably Chen who made the call from MIT. Glashow's initial impression was that it might be a "pseudo-Goldstone boson," because he then thought it had spin-0. But when word later arrived from SLAC that the new resonance probably had spin-1, he changed his mind. Here, at last, was concrete evidence for charm.

Late that same Monday morning, Iliopoulos was fast asleep in his Paris apartment. The apostle of charm liked to work on his calculations well into the morning hours and then sleep until noon. He was awakened suddenly by the phone ringing. It was Andre Lagarrigue, builder of Gargamelle and a good friend. "SLAC has discovered a narrow resonance near 3 GeV!" Iliopoulos snapped out of his slumber immediately. Might this be the evidence he'd hoped for?

He dressed hurriedly and rushed down to the Institut Henri Poincaré. Rumors flew, but there was little concrete information available. Then, late that afternoon, Bernard Jean-Marie, a French scientist working with the SLAC-LBL team, gave an ad hoc seminar to clear up the confusion. When Iliopoulos heard that the resonance probably had spin-1, he was convinced. Charm had turned up at last.

By luck or design, Jean-Jacques Aubert of Ting's group was sitting in the audience, too, having just rushed to Paris from Geneva. "We found the same particle a month ago!" he protested.

"Then why didn't you publish?" asked Jean-Eaudes Augustin, another SLAC-LBL collaborator who was listening as the priority battle erupted, and jumped up to defend his comrade.

"We had to make a few checks first," was all Aubert could offer.

For Iliopoulos and the rest of the physicists present, it was an astonishing performance they would not soon forget. Here were not one but *two* separate discoveries of the weirdest particle they had ever encountered.

Burton Richter had heard Iliopoulos beat the drums for charm at London, and even came up afterward to consider lodging a bet. He knew that additional quark flavors might help explain the "R-crisis"—the excess of hadrons created in electron-positron annihilation. But he was having none of this nonsense about charm, preferring to think that leptons were really hadrons at heart.

In the autumn of 1974, Roy Schwitters and other members of the SLAC-LBL team knew about the "Search for Charm" paper. But few of them had much opportunity to read it in any detail—what with all the work to be done upgrading the Mark I and preparing for the next experiments at higher energies.

If they possessed any idea of what to expect, it was just the vague general notion that strange particles should appear in much greater abundance if charmed quarks became involved. So when Gerson Goldhaber called that Friday morning to report observing an excess of kaons in those two fishy runs, it was the deciding factor that convinced Richter to go back and take another look. Though it later proved to be a red herring lacking any basis in fact, Goldhaber's "observation" had tipped the scales in charm's favor.

An avalanche of theorizing began almost immediately that Monday, as the first concrete information about the J and psi discoveries became available. At SLAC an ad hoc workshop of theorists met day and night, trying to interpret those peaks. "I remember the cathartic nature of the enterprise," Bjorken recalled:

> The lines of communication were very open. There was little thought about the usual kinds of priority and proprietary attitudes towards "ownership" of new ideas. The activity was extremely intense and very exciting. It was this way not only at SLAC, I am told. At many institutions it was somewhat the same way; the psi just electrified everybody.

A crucial first step was to clarify exactly how narrow this resonance was. The apparent widths of 1 to 2 MeV at SPEAR and 10 to 20 MeV at Brookhaven were just the smallest these detectors could resolve. Bjorken's more sophisticated analysis showed the intrinsic width had to be in the range of 50 to 100 thousand electron volts, an incredible factor of *twenty* narrower than observed in the Mark I. Such a narrow width meant, à la Heisenberg, that the lifetime of this object was about 10^{-20} second, an eon by nuclear standards.

Right from the starting gates, "hidden charm" led the theoretical pack. At the time, Bjorken gave it fifty-fifty odds. The favorite version held that the J and psi were the same vector meson composed of a charmed quark c plus its antiquark $\bar{c}$ with their spins aligned in parallel. That way their half-integer spins added to give the spin-1 expected of a vector meson, while charm plus anticharm canceled out to zero net charm. Harvard theorists Thomas Appelquist and David Politzer dubbed this marriage "charmonium" in a paper they sent around early the following week.

Other theorists, including Feynman, originally thought the psi to be a mating of an up quark with an anticharm quark. He said so in a letter to Richter dated Thursday, November 14. "Please send me any more facts, rumors, or ideas," he closes. "The world has opened up!" At the very bottom is a telltale postscript: "Murray didn't see this but he says it is all wrong, and he thinks that the charm-anticharm vector meson is more likely."

Another hypothesis that first frantic week held this resonance to be the lightest member of a new family of particles exhibiting "naked color." An octet of colored, spin-1 gluons was believed to carry the forces binding quarks together inside mesons and baryons. Perhaps one of them had finally emerged into the light of day.

Still others thought that this resonance might be the Z° itself, the carrier of the weak force responsible for neutral currents. Ting, who had conceived his experiment as a search for heavy forms of light, voiced such an opinion during the first few days. But this faction then had to explain why the Z° had appeared at a mass so far lower than the expected 80 to 100 GeV—and why it decayed so often into hadrons.

All this feverish activity could not go unnoticed by the press for long. When *The Daily Californian,* Berkeley's student newspaper, finally broke the story on Friday, November 15, Richter and Ting hurriedly issued a press release about their discovery. "The theorists are working frantically to fit it into the framework of our present understanding of the elementary particle," they said. "We hope to keep them busy for some time to come."

The experimenters spent that first week with SPEAR set at 3.1 GeV or close by, accumulating thousands of additional events, which helped them determine the properties of the psi—its mass, lifetime, and decay modes—more accurately. Meanwhile, "tourists" kept crowding into the SPEAR control room to gawk at the fireworks displays on the computer scope. To cut the confusion and get them out of the way, Harvey Lynch finally set up a second scope in a side room, and sent the sightseers there.

Late that week, Gell-Mann came to SLAC to get a first-hand look, and Panofsky took him down to SPEAR to witness the action. Thinking that *they* shouldn't have to use another scope, the two kept hanging around the control room, watching the main display. Finally Lynch could stand it no more. Nobel laureate or not, he politely but firmly escorted them into the side room. Panofsky took it good-naturedly, but Gell-Mann was duly miffed.

After about a week of running near 3.1 GeV, it was time to look elsewhere for more resonances. Both the charm and color ideas, the leading contenders, required entire families of new particles with higher masses. But nobody in the collaboration had any strong feelings about where to begin, except for

Breidenbach. Together with Terence Goldman, a SLAC theorist by way of Harvard, he concocted a crude model that took the psi to be composed of two objects orbiting one another—just like a hydrogen atom is built from an electron and a proton. If true, there should be another, heavier version that would show up as a narrow peak in the general vicinity of 3.7 GeV.

SPEAR had just been modified so that the energy of its two colliding beams could be changed in very small steps of about 1 MeV. And Breidenbach had established a direct link to SLAC's mainframe IBM computer so that he could immediately apply its full analyzing power to the events rolling in. The stage was set for a detailed energy scan that began at 3.6 GeV shortly after midnight on Thursday, November 21. Breidenbach drove wearily home to catch some much-needed sleep while others remained at the controls.

Hardly two hours later, he was awakened by a call. A promising spike had appeared near 3.7 GeV as suggested. Dead tired, he debated a few moments whether to return to sleep, but soon jumped into his car and sped off to SLAC. Upon arrival, he was greeted with graphs of *another* narrow peak, centered at 3.695 GeV and again only a few MeV wide.

At 6:30 that morning, Breidenbach called Panofsky to give him the good word, but got his wife instead—Panofsky was in the bathtub. "Well get him out!" Breidenbach insisted. Dripping wet and wrapped in a towel, Panofsky heard the great news. Richter learned about the discovery only upon arriving at SPEAR an hour later. Some of the group had gotten miffed at him for giving Schwitters so much of the credit in newspaper articles about the psi discovery, so they had deliberately avoided calling him.

Scheduled for maintenance that morning, the IBM computer was due to shut down soon, but Richter asked those in charge to keep it running so SPEAR could continue its analysis, and then swore them to secrecy. He wanted another day's worth of data before the crowds swarmed into the control room again. Group members were also told to keep it quiet. But at 9:45, a message appeared on the teletype linking them to the IBM computer: "DUE TO NEW PARTICLE DISCOVERY . . . THE SYSTEM WAS NOT TAKEN DOWN THIS MORNING." It also appeared on any other terminal hooked up to the computer.

Once again, throngs of happy tourists began to fill the control room. Once again, the champagne flowed while Richter and Goldhaber hastily wrote a short paper. Once again, the phone calls poured in from all over the world. On Friday morning, Breidenbach gave a short talk about the discovery of the ψ' ("psi-prime"), as it had been dubbed, to another auditorium crammed far beyond capacity with eager listeners.

It was his finest hour, and perhaps SLAC's, too. Breidenbach usually avoided giving talks or taking the spotlight, preferring to roll up his sleeves and work alongside the others while the leaders received much of the credit he deserved himself. But nobody else had played such a crucial part in *both* the MIT-SLAC experiments and the psi discoveries. He obviously savored his role now, showing only a few graphs and speaking hardly ten minutes before the audience showered him with applause. There was no other discovery to compete this time; the ψ' belonged entirely to SLAC.

"Those two weeks were the most exhilarating in my career as

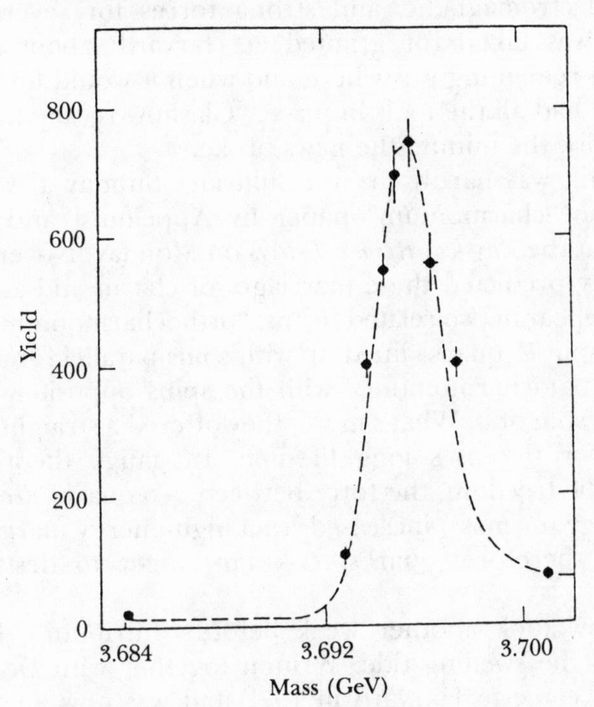

The second narrow resonance, the ψ', as first observed in electron-positron annihilation at SPEAR.

a physicist," observed Richter two years later, writing his personal memoir of the psi discoveries. "I have never worked at such a pace or pitch before, and I loved every minute of it."

That week a flood of theoretical papers began to clog the information channels of high-energy physics. Preprint shelves soon overflowed with every theorist's knee-jerk reaction to the recent surprises. Those still reluctant to recognize quarks imagined the new particles to be made from a heavy baryon—the omega-minus, say—paired with its antiparticle. Other die-hards hoped it might be a light version of the Z°, or maybe a heavy partner of the photon. But the majority of ideas invoked charm.

Nowhere was the charm hypothesis expressed more adamantly than at Harvard. Sheldon Glashow and Steven Weinberg, together with colleagues Thomas Appelquist, Sidney Coleman, Alvaro De Rujula, Howard Georgi, David Politzer, and Helen Quinn, had been elaborating the gauge theories of weak, electromagnetic, and strong forces for several years. Charm was taken for granted at Harvard; about the only question remaining was where and when it would finally show up. "We had all the tools in place," Glashow recalled, ready to go to work the minute the news broke.

The ink was hardly dry on Sullivan's Sunday *Times* article when the "charmonium" paper by Appelquist and Politzer arrived at the *Physical Review Letters* on Monday, November 18. In it they predicted these marriages of charm and anticharm would appear in two related forms: "orthocharmonium," where the two spin-½ quarks lined up with spins parallel for a total of 1, and "paracharmonium," with the spins oppositely aligned for a net spin of 0. What's more, they offered a straightforward reason for the psi's long lifetime. In gauge theories with asymptotic freedom, the force between two quarks *decreased* as the energy (or mass) increased. In a high-energy marriage like the psi, therefore, quarks took far longer to destroy one another.

Glashow took another week before contributing his own paper to the swelling tide. Written together with De Rujula, who had come to Harvard in 1972 and was now an assistant professor there, it sported the audacious title, "Is Bound

Charm Found?" No less audacious were its contents. With ample hindsight, one can say that this short, ten-page document foreshadowed much of the important work that would be done in particle physics over the next two years.

It is a pleasure to reread this paper and witness again the bold marks of two men at the very height of their careers, in possession of the truth and the courage to say so. Many articles in the physics journals are larded over with turgid prose and embroidered with caveats and qualifiers meant to cover one's flanks against carping colleagues. Not this one. With daring, forceful brushstrokes that reveal their certitude, De Rujula and Glashow deploy charmed quarks and asymptotic freedom to explain the J and psi properties, while clearing up an old puzzle about meson masses. The last half of the paper predicts how particles with "naked charm"—containing a *single* charmed quark, that is—will be found when electrons smash into positrons, neutrinos into nucleons, protons into protons, and photons into nuclei. Almost *all* of their prophecies eventually proved correct.

These two papers surfaced in late November amid the general proliferation of theoretical ideas fostered by the new particle discoveries. Awash in the tide, they had to battle for recognition with many other ideas and their endless variations.

Both papers hinted at still more narrow peaks with even higher masses. Indeed, the charmonium ideas emanating from Harvard probably influenced the discovery of ψ'. Goldman, who had helped Breidenbach figure out where to look, had been Appelquist's graduate student the year before. He had even furnished his old adviser with crucial information about the psi during the first hectic week after its discovery. "No, no," came the shocked reply. "It *can't* be that narrow!"

The *Physical Review Letters* published the J and psi discovery papers on December 2, hardly three weeks after their arrival, together with a paper from ADONE confirming the SPEAR find. It was already stale news to particle physicists. The very next week it brought out the ψ' paper, but by then it was old hat.

This was an exhilarating time for high-energy physics, a time when rumors, phone calls, memos, and preprints rebounded back and forth across oceans and continents with dizzying frequency and blinding speed. Established physicists fell back on personal information networks, built up over long years of practice, to help them sort fact from fiction—or reputable theory from wild speculation. Younger scientists lacking such networks banded together in study groups and tried to digest all the confusing information. Communication channels were open as never before.

Science reported the excitement in its December 6 issue. "Two New Particles Found: Physicists Baffled and Delighted," read the headlines. "The suddenness of the discoveries has left almost everyone baffled about what it all means," it observed. "Many physicists think that the discoveries may have opened a whole new dimension in the world of subnuclear particles, analogous to finding the first of many so-called strange particles in 1947." Sullivan tried to describe the great ferment in another *Times* article in mid-December. "Hardly, if ever, has physics been such an uproar as during the last three weeks, and the end is not in sight."

The flood of theoretical ideas finally hit the pages of *Physical Review Letters* a month later, on January 6, 1975. All told, nine papers were published there on successive pages, including the ones by Appelquist and Politzer and by De Rujula and Glashow. Others had distinguished authors like J. J. Sakurai, Julian Schwinger, and Chen Ning Yang. Out of the nine papers, five promoted the charm hypothesis and its variations.

It was not a time to keep data under your hat, either. In a bid for more time on the AGS, Palmer and Samios finally revealed their odd neutrino event, their single candidate for a charmed baryon, before the Brookhaven advisory committee. Word naturally leaked out to the rest of the physics community, so they published the event in early 1975. It was encouraging, but not convincing, evidence. Unlike their gold-plated omega-minus event, this one was ambiguous.

Ting had little difficulty in getting more time on the AGS. At Glashow's suggestion, his group now modified their spectrometer to look for a kaon in one eye and a simultaneous pion in the other. In their paper Glashow and De Rujula predicted that mesons containing a single charmed quark would have masses near 1.95 GeV and should show up as narrow peaks near that energy in pion-kaon mass plots. In early 1975, Ting's group made an intensive search for such peaks.

At Stanford the SLAC-LBL group finally began to crank SPEAR up to the higher energies for which it had been modified the previous summer, before all the commotion began. Stepping only 2 MeV at a time, lest they bypass another narrow peak, these experimenters gradually explored the entire region between 3.2 and 7.6 GeV—the highest energy then accessible. In January, they sent in two more papers showing early results. No more narrow peaks had appeared, but there was definitely a broad bump about 300 MeV wide centered at 4.1 GeV.

Things were looking very good for the charmonium idea by early 1975, while the other proposals were already falling back. The "naked color" hypothesis predicted a cluster of peaks, too, but it had difficulty explaining why any more than the first peak was narrow. In the Z° scheme, any extra peaks were an embarrassment. But from a pair of charmed objects bound together, a whole spectrum of peaks, narrow and broad, were expected—just as a hydrogen atom has a variety of energy levels it can jump to when excited by heat or light.

In fact, the term "charmonium" was coined by direct analogy with the simplest of aggregates, "positronium." This short-lived marriage between an electron and positron had been first isolated by Martin Deutsch in the 1950s. Essentially a hydrogen atom with the proton replaced by a positron, positronium has "ortho" and "para" states, too, with spins aligned in parallel and opposite directions—as well as higher-energy states that it can jump to. It provided a strikingly simple and increasingly powerful metaphor to help us visualize what might actually be going on inside these J and psi particles.

For hot-blooded experimenters, however, this was a time for action, not contemplation. We queued up at accelerators with hastily written proposals to seek all kinds of new particles and measure their properties in every collision imaginable. Lab directors and their advisory committees worked day and night trying to sort through the resulting mounds of paperwork. Other important experiments were abandoned or postponed while everybody rushed off to pan for gold in the vicinity of the recent discoveries.

* * *

DESY, too, had been building an electron-positron collider during the early 1970s. DORIS (for DOuble RIng Storage facility) had two oval racetracks, stacked one atop the other, crossing at two opposite points. By injecting many bunches of electrons and positrons into *two* separate storage rings like this, its designers had hoped to attain a much higher luminosity (more collisions per second) than SPEAR. The more racecars that speed through a crossroads per second, the better the chances of a smash-up.

But "the bunches talked to one another," Bjorn Wiik recalled, and the circulating beams deteriorated rapidly. So DORIS never came close to reaching its intended luminosity. DESY scientists were still grappling with this knotty problem when the startling news hit Hamburg on the morning of November 11, 1974.

Although DORIS was not working up to par, and its two detectors were not quite finished, Wiik, Gunther Wölf, and the other DESY physicists decided to go ahead and search for the new particle. Opportunities like this came knocking only once a decade. The psi offered a strikingly clear, strong signal with little background, so they might still be able to find it despite these dismal operating conditions. The Italians at ADONE, after all, had been able to push their machine beyond its 3 GeV limits and locate the tall, narrow peak.

Wiik and Wölf were SLAC alumni who had returned to Europe in 1971 to do electron-positron scattering at DORIS. Wiik, a quiet, hard-working Norwegian with hardly a hint of flamboyance, had worked with David Ritson and helped bail us out when our deuterium target began boiling in 1970. Wölf, a slender, handsome German who does everything with an air of great precision, had joined the SLAC bubble-chamber group in 1967. Together they designed and built the DASP (for Double Arm SPectrometer) detector for one of the two DORIS interaction regions. DASP had two large magnetic "eyes" sitting at opposite sides of this beam crossroads, staring across space at each other, just waiting to witness a smash-up.

Surrounding these eyes was another detector sensitive to photons, electrons, and positrons. As it was the only detector operating by November, Wiik, Wölf, and their collaborators began using it to look for electron-positron pairs produced in the vicinity of 3.1 GeV. At first they saw nothing because the DORIS energy calibration was offset slightly from SPEAR: they were looking in the wrong place. You could easily miss such a needle if you were only a few MeV off. Once this problem was licked, the DESY physicists found a huge peak at 3.09 GeV on Sunday, November 24—shortly after word hit town about the ψ' discovery.

Rather than scan up and down in energy, looking for more sharp features, these scientists decided to sit right on the two peaks and collect as much information about them as possible. That way they could also shake down their spectrometer and DORIS itself—and still collect meaningful data. "We were installing and replacing equipment almost every day," Wiik recalled.

Having worked at SLAC during the late 1960s, Wiik and Wölf had been indoctrinated into quark-parton thinking and were favorably disposed to the charmonium idea. What better way to explain the recent surprises than by adding another quark? And other members of the big DASP collaboration had worked on the MIT-SLAC experiments, too. Quarks and partons were everyday objects to this group, part of a conceptual framework exerting a powerful influence on their work. Thus many of them expected to find still other new particles by poking carefully among the debris left by the psis.

In May 1975, their strategy hit paydirt. Looking at particles emerging from the breakup of the ψ', DASP began to see pairs of photons shooting out with two well-defined energies. Perhaps the ψ' emitted a photon and became some intermediate particle; this new corpuscle might then decay immediately to the psi by emitting a second photon, in what was commonly known as a "cascade decay." Such dual emanations occur all the time with hydrogen and positronium, which can easily hop from one Bohr orbit to another by coughing up photons. A similar kind of behavior was expected from charmonium.

By the time of the European Physical Society Conference that June in Palermo, Sicily, DASP experimenters had witnessed only a few of these cascade decays, not quite enough yet to publish. Wölf sat nervously in the audience, hoping that the SLAC-LBL physicists were not about to scoop his team by reporting a similar find. Since November the Californians had been issuing paper after surprising paper, beating DASP to one important discovery after another. "I breathed a great sigh of relief," he recalled, when no such report occurred.

That August, Stanford again hosted the biannual Lepton-Photon Symposium. Because of the recent advances in muon and neutrino physics, the name and purview of the Symposium had been extended to cover *all* the leptons, not just the electron. These gatherings had come a long way in ten years, their growth reflecting the burgeoning importance of lepton scattering in the general scheme of particle physics.

From August 21 to 27, about 500 invitees crowded into spacious Kresge Auditorium in Stanford's new Law School building—along with the usual complement of gate-crashers from local physics communities. It was a not-to-be-missed event. The auditorium was packed from start to finish, despite a faltering air-conditioner that just could not cope with the heat being generated on the floor.

The schedule seemed to have been carefully structured to present the SPEAR results early, often, and in the best possible light. First up was Schwitters, presenting a detailed survey of the past year's discoveries on the Mark I, including a few surprises. The broad feature at 4.1 GeV had been scrutinized in finer detail; it now seemed to be breaking up into two and possibly *three* separate, narrower peaks—indicating at least a ψ'' and ψ'''. And the famous ratio *R*, which a year ago at London was headed for the stratosphere, had finally reached a plateau at energies above 5 GeV. Though still inexplicably high, the mere fact that *R* had finally leveled off was heartening news to the parton camp.

Next came Gerald Abrams from Berkeley and SLAC's Gary Feldman with more details about the first two psis. The energy difference between SPEAR and DORIS had been resolved, with the ψ now officially put at 3.095 GeV and the ψ' at 3.684 GeV. Feldman showed evidence for another three neutral particles, called the χ's (Greek *chi*) at SLAC, similar to the cascade particle seen by DASP and revealed earlier that summer at Erice. Finally, Wiik closed the morning session with a talk about the DESY findings, including possible evidence for the expected

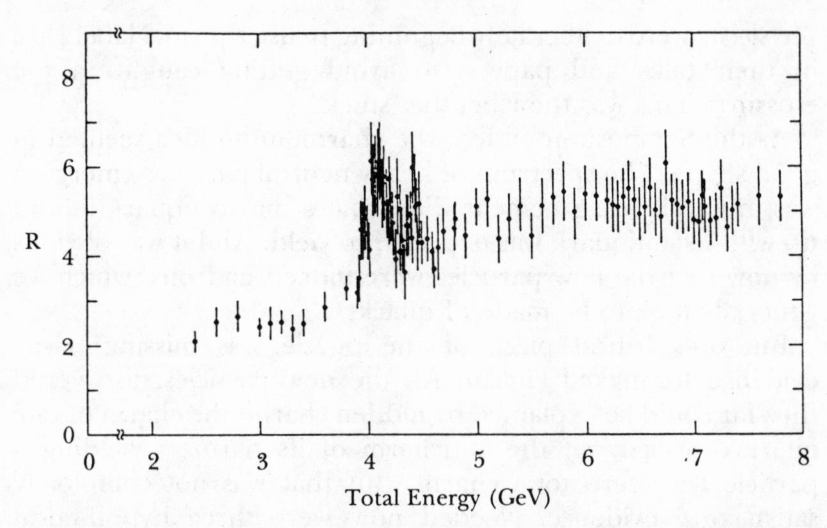

The graph of the ratio R *presented by Roy Schwitters at the 1975 Stanford Conference. After revealing more structure near 4.1 GeV, this curve hits a plateau above 5 GeV.*

*para*charmonium state near 2.8 GeV. Now there were *eight* heavy new particles!

It was an unforgettable performance for the storage rings, the coming of age of a radical new approach to particle physics that had only recently been struggling for funds. In a few short months, electron-positron colliders began to map out a brand-new spectroscopy of massive particles with a speed and clarity other kinds of machines could never hope to match. Comparable activity had not occurred since the early-sixties heyday of bubble chambers, which had spewed forth entire new families of mesons and baryons almost overnight. Many of us in the audience had come there expecting a good show, but we were absolutely dazzled by the actual performance.

Ting, who should have been granted a role in this opening session, if only out of respect for his truly pivotal contribution, had to sit quietly and listen while his J particle was repeatedly called the psi by one speaker after another. His own talk had been relegated to a session that afternoon covering production of the new particles in *other* kinds of collisions. The touchy issue of what to call the first particle, J or psi, had been smoldering ever since it became obvious the previous November that each group was sticking by its own name. Following DESY's lead, physicists were deliberately beginning to use the dual label J/psi in their talks and papers, to avoid getting caught in the crossfire. This was the label that stuck.

As the Symposium ended, the charmonium idea seemed in good shape. The spectrum of heavy neutral particles emerging was in excellent agreement with what a massive quark bound up with its antiquark was expected to yield. And it was obvious by now that the new particles were indeed hadrons, which we generally took to be made of quarks.

But one critical piece of the puzzle was missing: clear evidence for naked charm. All the new particles discovered thus far could be explained by hidden charm: the charm of one quark canceled out the anticharm of its partner, yielding a particle with zero total charm. But that was not completely satisfactory evidence. Needed now were three hypothetical motes called *D* mesons, in which charmed quarks were paired in mixed marriages with the familiar up and down quarks to exhibit net outward charm. Ting's group had searched extensively for narrow peaks in pion-kaon mass plots, a sure sign of a charmed *D* meson, but had found none. Nothing like them had been seen at SPEAR or DORIS either. In fact, the sudden surge of strange particles expected above energies of 4 GeV had completely failed to materialize. Charm left the Stanford Conference with only half a victory.

Since early 1975, ugly rumors had been drifting through the particle physics community that perhaps the psi discovery at SPEAR had not been so independent, after all. Perhaps Richter and company had learned of Ting's find over the grapevine—and then gone back for a closer look at some puzzling old data near 3.1 GeV. How else to explain why they would backtrack down to low energies when fresh new vistas were awaiting them above 5 GeV? Especially with DORIS close behind, as well. It was hard to identify a definite source for these rumors, but they did seem to emanate from the general direction of Boston.

The first hints of an emerging priority struggle between the two groups, what Schwitters calls "the geopolitics," came in mid-November 1974 when SLAC heard that Ting was naming his discovery the J, not psi. SPEAR then got one leg up on MIT by making the sole, undisputed discovery of the second particle. Ting had been looking in the same place and found nothing.

The speed and clarity with which SLAC-LBL physicists fleshed out the emerging spectroscopy lent enormous weight to their claims. The name psi was used far more often than J, except in Cambridge.

In the first few months the contest seemed to be fairly good-natured. Ting's troops began to offer and wear T-shirts with "J–3.1 GeV" emblazoned on the front. SLAC-LBL physicists kidded that the particles themselves preferred the ψ label. To prove their point they passed around photos of a computer display showing a ψ′ disintegrating into two pions and two muons. The four visible tracks had arranged themselves neatly into the four arms of a near-perfect letter psi.

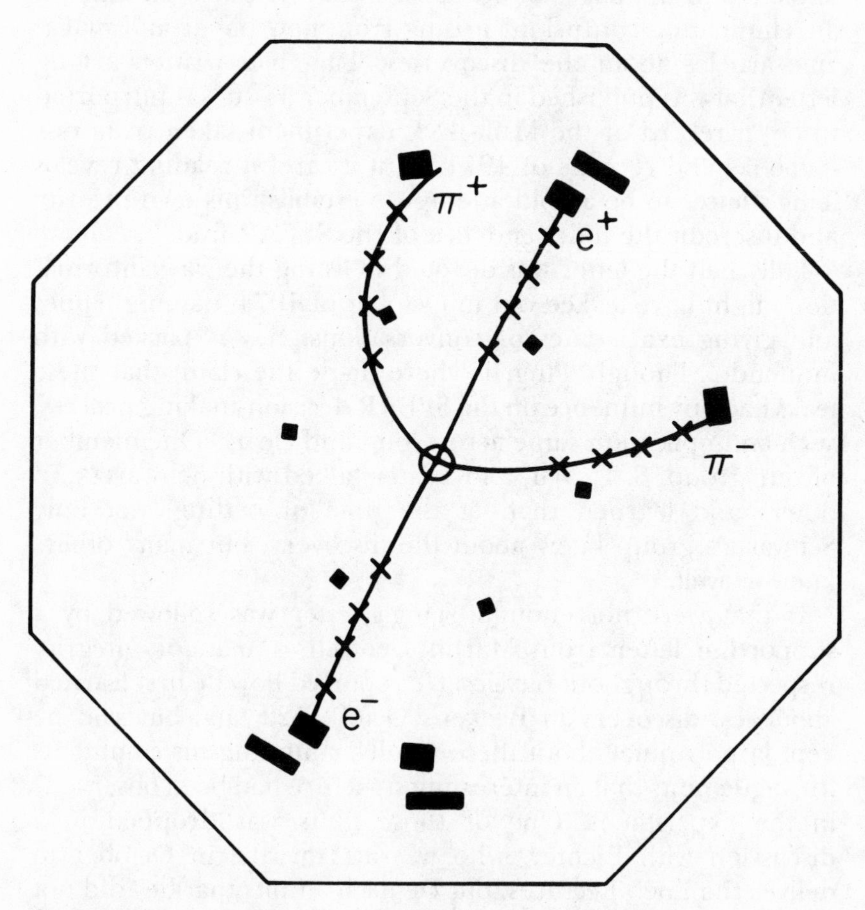

Computer reconstruction of a ψ′ decay in the Mark I detector, making a near-perfect image of the Greek letter psi.

In 1975 the nasty rumors began to circulate. They were the subject of countless discussions taking place at lunch, in offices, over beers—almost anywhere two physicists might get together. The rumor heard most often was that Schwartz, who obviously knew of Ting's find in late October, had carried this news back to California and told his SLAC cronies where to look.

But nothing concrete had appeared in print about these allegations. They just swam around in the great ocean of verbal discourse going on incessantly in high-energy physics. Though increasingly irksome to the SPEAR group, and particularly so to Richter, they could not be honored by any formal rebuttal.

Then, hardly a week after the Stanford Conference, a bomb exploded in the pages of *Science.* In what he called an attempt at "eliminating confusion" arising from newspaper and magazine articles about the discoveries, Ting had written a long letter that was published in the September 5 issue. It purported to be "a record of the MIT-BNL experiment taken from our logbooks and records of 1974." But a careful reading reveals Ting's letter to be a bold attempt to establish his own priority and discredit the independence of the SPEAR find.

Fully half the letter was devoted to listing the ways information might have leaked out in October of 1974, naming names and giving exact dates of conversations. It was packed with innuendo. Though Ting nowhere made the claim that these leaks had any influence on the SPEAR decision-making process, such an implication came across loud and clear. "One member of our group, S. L. Wu, and I later talked with Schwartz and others and learned that, at the time of betting, not only Schwartz's group knew about the discovery, but many others knew as well."

If that were not enough, Ting's letter was followed by a supporting letter from Martin Deutsch, a man of integrity respected throughout physics. He reported how he first learned about the discovery in Becker's October 22 talk, but said he kept largely quiet about these results, confining his comments to "vague hints that an interesting structure had been observed" in the experiment. One of those hints was dropped in a discussion with Richter, who was at Harvard in October to deliver the Loeb Lectures, but Deutsch admits that he "did not seem particularly impressed by my story."

The letters hit SLAC like a bombshell. Suddenly the nasty rumors were occurring in *print,* in black and white, for the entire scientific community to read. Richter was absolutely furious. He wrote a long, livid letter of reply, but was convinced by cooler heads not to publish it. Panofsky replied instead, reaffirming the independence of the two discoveries: "This should be a joyous occasion for all physicists."

It was pretty obvious right from the start that a Nobel prize was in the offing on the J/psi discoveries, and there seemed no question, at least at SLAC, that Richter and Ting would share the award when it eventually came. So many physicists there viewed Ting's letter as a brash attempt to lay claim to priority and with it an unshared prize.

People close to Ting, however, say that he seriously feared being bypassed by the Swedish Academy—that *Richter* might get an unshared prize—what with all the attention being given the follow-on discoveries at SPEAR. Ever since the earliest discoveries, in fact, the news media had lionized SLAC and generally ignored Brookhaven. Ting claims the two *Science* letters were an attempt to restore some sense of balance and to establish the importance of his own, prior discovery.

More than a decade later, the question is still asked. Did the rumors of Ting's discovery have any impact upon Richter's key decision to go back and take another look that fateful November weekend? Whenever physicists reminisce about "the good old days" of 1974, that touchy issue is always lurking in the background, refusing to roll over and die.

Of one thing I am absolutely certain. There *were* rumors of a Brookhaven discovery circulating around SLAC during the week of November 4, and possibly earlier. I remember them distinctly, especially my own reaction to Ting's presentation at SLAC on November 11: "So *that's* what those rumors were all about!" But the rumors I had heard were vague in the extreme, little more than, "There's been a big discovery at Brookhaven." You heard this kind of rumor all the time.

But Schwartz made a point when I spoke with him about those days. "What do you do when you hear a rumor like that?" he asked. "You get on the phone to your contacts at Brookhaven and ask questions until you find out." If the telephone activity after the psi discovery was any indication, he had a good argument.

So I tried to track down every possible point of contact. In his letter and two subsequent articles about the discoveries, Ting dropped a long list of names of people who knew about his find before November. Besides Deutsch and Schwartz, there were T. D. Lee, Ronald Rau, Willibald Jentschke, and Lee Grodzins of MIT—plus any other people they might have talked to. The options could easily have become endless.

As I dug into each possibility and talked to the people involved, however, I became impressed by how *little* they had known and how *late* in the game they had learned it. The most telling were the recollections of T. D. Lee, who had seen Ting's graph of a "rather unimpressive" bump on October 17. And the testimony of Ting's own group indicated that few of them knew much about any kind of peak before the previous Sunday group meeting.

That same week, however, Schwitters was already hot on the trail of the psi. The following Tuesday, October 22, he was asking his comrades to take a closer look at those two peculiar runs. So he was clearly interested in the 3.1 GeV region before any substantive rumors could possibly have reached SLAC.

By the end of that week, however, Schwartz had carried the news about Ting's narrow peak back to Stanford and told Jasper Kirkby about it. Both of them deny telling anybody else, particularly anyone in Richter's group. But it is difficult to believe the information simply stopped there, dead in its tracks. Where did the rumors I had heard come from?

Schwitters admits being aware of the same vague rumors about a Brookhaven discovery but says he paid them no heed, being far too engrossed by those two June runs. Brookhaven was then regarded as somewhat of a backwater by SLAC physicists—a torpid stream that had long ago been fished out. Any possible "discovery" there was probably a ho-hum affair, or so we thought before November 11. It was only in *hindsight* that the rumors assumed any significance.

In the final analysis, the overriding question is whether or not these rumors influenced Richter's decision during that fateful Friday meeting. After an exhaustive analysis, in which I interviewed many participants, my answer is a categorical *no*. Nor did the hints dropped by Deutsch at Harvard seem to have registered. By all accounts, Richter returned to SLAC on November 4 eager to push on to higher energies and extremely reluctant to waste any more time retracing his steps around 3 GeV. It took a mighty effort by Schwitters, Breidenbach, and others to convince him to go back, and even then he gave them just a weekend. Nobody I interviewed remembers any rumors being mentioned at that meeting. Everybody said it was Goldhaber's red herring that finally won Richter over.

Finally, I even pored over the SPEAR experimental logbook of that weekend, trying to find the slightest hint that somebody, anybody, had any such foreknowledge. There is no such hint, only genuine astonishment at such an incredible find. What impressed me most was how skeptical certain key people remained up to the very last moment. Harvey Lynch was writing detailed instructions for the weekend's efforts when he penned his gold-rush allegory (used as the epigraph for Chapter 12). As Schwitters's cospokesman, he was one of the three or four people in the best position to know. Only skepticism pervades his words: "Were those lumps gold or pyrite? He began to sift through the silt once again."

Throughout the fall of 1975 and into the following year, experimenters continued looking everywhere for evidence of naked charm, the one major piece missing from the puzzle. There were several tantalizing hints, like the bubble-chamber photograph from Brookhaven, other odd neutrino events turning up at CERN and NAL, and a cosmic-ray event found by Japanese physicists well *before* the November Revolution. Though encouraging, none of these could be taken as conclusive evidence for charm.

All eyes turned toward SLAC, waiting for the final proof. The best data on the psi and chi particles had come from the Mark I detector. If these were hadrons with hidden charm, then there also had to be at least a few cases in which charm came out of the closet. "What was needed," Richter later observed, "was simply the direct experimental observation of charmed particles, and the question was: Where were they?"

If Glashow's prediction that charmed *D* mesons would weigh in near 1.95 GeV was correct, then they should have appeared when the total electron-plus-positron energy climbed above twice that figure, or 3.9 GeV. Such a possibility was in keeping with the narrow widths of the first two psis. Because the 3.1 and 3.7 GeV particles did not have enough mass-energy to decay into a pair of *D* mesons, the quickest way to shed any hidden charm, they had long lifetimes and therefore very narrow widths, à la Heisenberg. But the other psis above 4.0 GeV *did* have sufficient energy; so they could decay rapidly, leaving the broader peaks observed. Some kind of "threshold" was obviously being crossed between 3.7 and 4.0 GeV, and many physicists expected it meant the creation of charmed mesons.

By early 1976, however, evidence for naked charm had still not shown up at SPEAR. Theorists were getting anxious. At an April conference in Madison, Wisconsin, Glashow took Goldhaber aside and tried to impress upon him the urgency of the search. Returning to Berkeley, Goldhaber took a closer look at some of the most recent data, reconstructing the masses of pion-kaon pairs he found among the debris. "I dropped what I was doing," he recalled, "and decided I'm going to look for charmed particles or find out why they're not there."

On May 3, after relaxing the conditions he had used to identify a kaon, Goldhaber finally began to see a narrow peak sprouting in his mass plot, at 1.87 GeV. Taking a different approach, François Pierre (another visiting French scientist working in the collaboration) saw a similar spike in the same kind of graph. Both peaks suggested the presence of a neutral *D* meson with a mass of 1.87 GeV, a bit below Glashow's prediction but close enough. On May 5, Goldhaber and Pierre sent around a joint memorandum to the whole collaboration. Naked charm had been found at last.

Three days later, Goldhaber phoned Glashow with the good news, asking him to keep it confidential. But Glashow called Walter Sullivan almost immediately, and *The New York Times* scooped the *Physical Review Letters* again. "It was all but impossible for me to keep the secret," Glashow apologized. "After all, John's wine and my hat had been saved in the nick of time!"

The 17th International Conference on High Energy Physics occurred that July in the Soviet city of Tbilisi, situated in the foothills of the Caucasus Mountains about halfway between the

Gerson Goldhaber (left) and Sheldon Glashow in the summer of 1976.

Black Sea and the Dead Sea. This Rochester Conference was dominated by the discovery of charm, just as John Iliopoulos had wagered at London two years before. SLAC and LBL physicists presented fresh new evidence for still more charmed particles—charged brothers and sisters of the neutral D mesons at 1.87 GeV, plus some spin-1 cousins, too. And the decay patterns of the D^+ followed the predictions of charm *exactly,* a real clincher. Solid evidence for charm surfaced in session after session. There was no longer any doubt.

But nobody came forth to pay off their bets, Iliopoulos

recalled, either then or afterward. Still, he could take pride in winning a victory of the spirit, if not the palate.

Three months after Tbilisi, on October 18, two similar telegrams arrived at Stanford and MIT. "The Swedish Academy of Sciences today awarded the 1976 year's Nobel Prize in Physics in equal shares to Professor Burton Richter, USA, and Samuel Ting, USA, 'for their pioneering work in the discovery of a heavy elementary particle of a new kind.' " It was not quite two years after that memorable November morning, one of the fastest awards on record.

The ever-cautious Academy omitted mention of the word "quark" in their telegram and in the full declaration of the 1976 physics prize, still refusing to acknowledge any obstreperous corpuscles that refused to show up in detectors. But it was part of a small and rapidly shrinking minority by then.

The award of a Nobel prize shared in equal measure by Richter and Ting helped quell the priority struggle and quiet the rumors that had circulated through the physics community

Steven Weinberg (left) and Sidney Drell (right) congratulate Burton Richter on October 18, 1976, the day his Nobel prize was announced.

that year. Tuxedoed and champagned, wined and dined, the two leaders were all smiles and graciousness in Stockholm that December. But the 3.1 GeV meson was still the J in Ting's Nobel speech and the psi in Richter's.

Glashow had to wait until 1979 for his prize, shared with Abdus Salam and Steven Weinberg for unifying the electromagnetic and weak nuclear forces. But he did get to see his own wager paid off at Northeastern the following April. Charm had indeed been found by outlanders, just as he had warned.

The 1977 conference on meson spectroscopy was a far cry from the 1974 affair, packed now with big-name physicists from around the world. Spectroscopy was fashionable again after the spectacular discoveries of the previous two years. Ever the iconoclast, Glashow gave a rabble-rousing talk titled "Charm Is Not Enough," trying to peer beyond the growing orthodoxy. Then they all had a big laugh as the spectroscopists present finally ate their hats—Mexican candy hats supplied by the organizers.

By the end of 1976, physicists the world over began to share a simple picture of the subatomic world, in which all matter was built of fundamental, pointlike quarks and leptons influencing one another by means of gauge forces. After all the confusion of the previous three decades, it was a great relief to have such a minimal, powerful theory. But we could not relax just yet. Nature still had a few more wrinkles to reveal.

CHAPTER 14

SIX OF ONE . . .

There are more things in Heaven and Earth, Horatio,
Than are dreamt of in your philosophy.

—Shakespeare, HAMLET

FOR physicists the world over, quarks finally emerged as "real" objects during the November Revolution and its aftermath. And after the discovery of naked charm, opposition to the quark idea collapsed—virtually overnight. The addition of a fourth quark flavor to clear up a nagging disease of the weak force, and its amazing success in predicting an entire forest of striking phenomena, had been the pivotal factors. As John Ellis remarked, "Charm was the lever that turned the world."

No latecomer to the quark idea, Sheldon Glashow described this mass conversion in a popular article he wrote for *The New York Times Magazine* in July 1976:

> The discovery of charmonium was an event of the utmost importance in elementary particle physics. Nothing so exciting had happened in many years. For believers in quarks, the new particle was the first experimental indication that a fourth quark existed. The successful interpretation of charmonium as a quark-antiquark combination, together with the difficulty in finding an attractive alternative hypothesis, led many doubters to see the error of their ways. As a result of the discovery of the J/ψ and its kin, the quark model has become orthodox philosophy.

A crucial element in this turnabout was the theory of quantum chromodynamics. Its key attribute of asymptotic freedom provided the only convincing explanation of why the J/psi was so incredibly long-lived. And the fact that the color force became stronger at large distances offered a genuine hope of resolving the mysterious absence of free quarks—the one major stumbling block remaining. The term *QCD* was used increasingly often after 1976, as theorists everywhere began working out its detailed implications.

The "Standard Model" of the universe, most of which John Iliopoulos had foreseen in his memorable 1974 London Conference talk, gained widespread acceptance after 1976. Leptons and quarks come in distinct "generations," each with four members. The first generation contains the electron and its neutrino, plus the up and down quarks; the second includes the muon, its own neutrino, plus the strange and charmed quarks. Quarks appear in three different colors, while leptons are colorless. All these spin-½ matter particles interact with one another by swapping another family of twelve spin-1, force-carrying particles that are intimately related by gauge field theory: a photon carrying the electromagnetic, the W and Z particles carrying the weak, and the eight colored gluons carrying the strong force.

So quarks are close kin to the very same leptons—the electron, muon, and their neutrinos—that were used to ferret them out in the first place. Apart from fractional charges, the only seeming difference is the fact that quarks carry color, which makes them cling so tightly together that they can never be torn asunder. Hadrons are *not* elementary particles; they are aggregates of quarks.

What is so remarkable about the conversions of the mid-1970s is how completely the new lepton-quark dogma won out. Prior revolutions in physics had not come so easily. Quantum mechanics, for example, encountered stiff opposition for many years from such stalwarts as Albert Einstein.

Part of the reason, at least, has to do with the fact that "atomism"—the breakdown of many complex systems into a few simple parts—has always been a familiar strain in Western scientific thought. All physicists are thoroughly comfortable with the idea, if a little worried that it may go on endlessly. To

Fermions, or matter particles

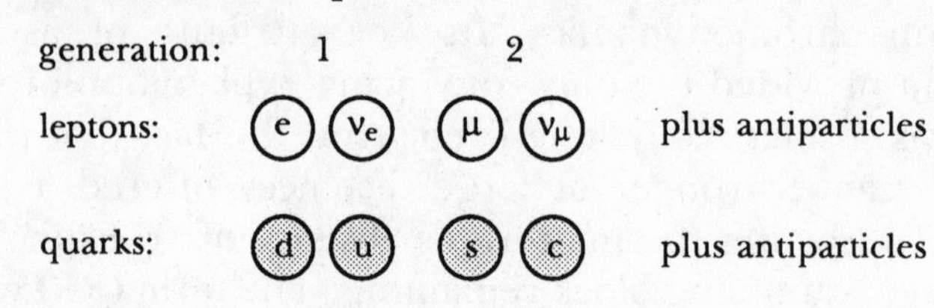

Bosons, or force particles

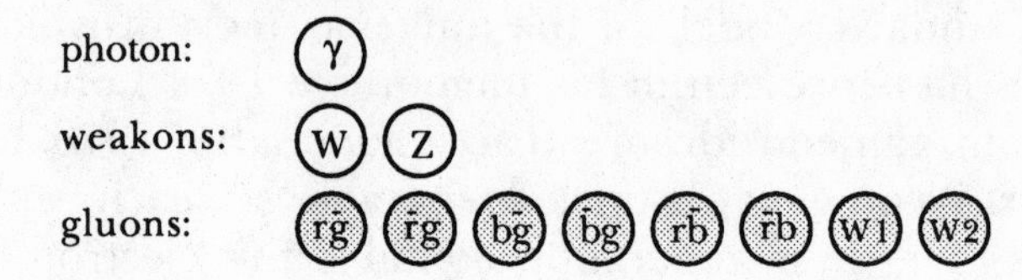

The elementary particles of the Standard Model, circa 1976.

return to it now, for the third time this century, required no great leap of imagination, only some thoroughly convincing evidence. That had been provided by the discovery of charm.

So the triumph of the lepton-quark world view was perhaps not a true scientific "revolution" in the classic sense of a sharp break from prior thought patterns into totally new ways of conceiving reality. But it was certainly an upheaval of major proportions. And it led to a grand resolution of many baffling problems that had been nagging particle physics since the late 1920s, when theorists first began trying to accommodate relativity and quantum mechanics under the same roof. The floodgates now stood wide open to the great tide of theorizing that followed.

There were a few pockets of resistance, to be sure, the most noteworthy being some bootstrap holdouts cloistered mainly at Berkeley. These infidels usually took refuge in the fact that quarks had still not been "seen" in a naked state, leaving tracks in some particle detector. But they found themselves treated by their colleagues with increasing disdain—especially when their arguments began to rely too heavily on mysticism. Fritjof Capra's *The Tao of Physics,* which appeared in early 1976, espoused this bootstrap philosophy and found an enormous, devoted audience among those laypersons completely out of touch with the most recent discoveries of high-energy physics. From the moment it hit the bookstores, however, it was ten years out of date.

Heresies and renegade sects arose, too. By far the commonest were theories that simply added new quarks and leptons to those already established. The Standard Model set no clear-cut limits on such activities, so theorists took quick advantage of this license. "There could be other heavy quarks as well and new hadron quantum numbers other than charm," Glashow and De Rujula had observed in early 1975, referring to certain "fancy" quarks they had proposed the previous year. By the 1977 Northeastern Conference, Glashow was convinced. Additional quarks and leptons were needed to explain all the experimental anomalies cropping up.

But the Standard Model could easily accommodate these kinds of heresy—as long as they had solid experimental backing. Adding more quarks was not much of a problem. Thus, the remainder of the 1970s was a period of gradual consolidation for the new world view.

Experimenters had been looking for heavy leptons since the early sixties. Why should Nature stop at four leptons—the electron and muon, plus their two associated neutrinos? The muon itself is merely a heavy version of the electron, about 200 times as massive. Why not an even more ponderous cousin? So whenever a brand-new atom smasher opened up virgin territories, experimenters were often there searching for heavy leptons at the new frontiers. It was by no means a popular pastime, but there were usually a few hardy souls willing to make the effort. Like the quark-hunters, however, they always came away emptyhanded. Until the mid-1970s, that is.

Martin Perl had been Ting's thesis adviser at Michigan before coming to SLAC in 1963. A tall beanstalk of a man with an unruly shock of curly, graying hair, Perl formed a small group there that did muon experiments in End Station B. "My hope was to find some difference between the electron and muon," he recalled, "a difference not predicted by quantum electrodynamics and the different masses."

But five years of difficult experiments had produced few noteworthy results—and no discernible differences. So in 1972 he joined forces with Richter as part of the burgeoning Mark I collaboration. From the beginning, his pet project was to search for heavy leptons using this revolutionary detector.

"Most active physicists did not believe in heavy leptons in a serious way," Perl remembered of those years. "There was almost no useful literature; indeed, that is why I began to work in that field. I have never liked to work in crowded fields."

Perl and his colleague Gary Feldman figured that heavy leptons might be produced in *pairs* at SPEAR—one with a positive charge, one negative. As any such corpuscles would often disintegrate into other leptons, there occasionally might be events with only two visible tracks in the Mark I, one of them

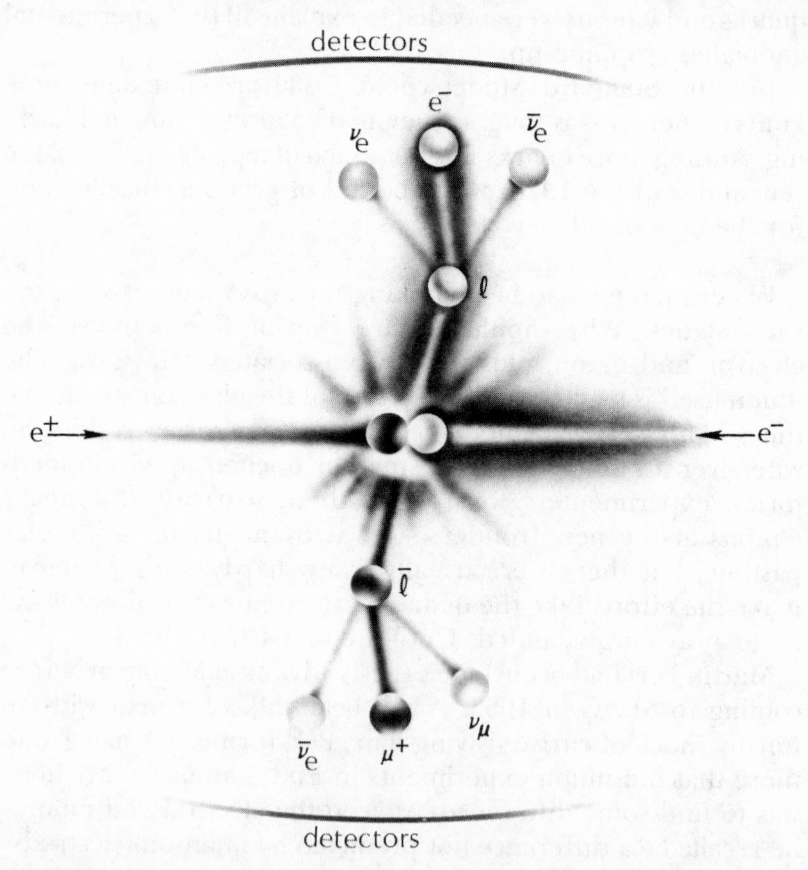

One possible mechanism for "anomalous e-mu events" at SPEAR. An electron and positron annihilate, forming a heavy lepton and its antiparticle, which disintegrate almost immediately to an electron, a muon, and four invisible neutrinos.

an electron or positron and the other a muon of the opposite charge. Four ghostly neutrinos supposedly released in such outbursts would escape the detector unnoticed, spiriting away plenty of energy.

Perl and Feldman sought such "anomalous e-mu events" during 1973 and early 1974, when SPEAR ran at low energies, but without success. Trouble was, there were other ways to get these kinds of events, too. The scant few examples they found were dismissed as these backgrounds. They were noise, not signal.

But when SPEAR started to run at higher energies in the fall of 1974, just after the earthshaking psi discoveries, these scientists began to see a fair number of anomalous e-mu events—at about 4 GeV and above—that they could not so readily dismiss. By the spring of 1975, Perl and Feldman were composing a paper, which was completed and shipped off to the *Physical Review Letters* in August. Feldman mentioned these anomalous events at the Stanford Conference later that month, suggesting they arose from the decay of a new corpuscle he called the "*U*" particle (for "unknown"). Could it be a heavy lepton?

At the time the key question mark in everybody's mind was the puzzling absence of naked charm at SPEAR. Many physicists thought the anomalous e-mu events were not really the offspring of heavy leptons at all but of the mysteriously recalcitrant charmed mesons, the *D*'s. These curious lepton pairs had begun to show up, after all, very close to the energy expected for charmed meson creation.

Perl and Feldman stuck doggedly to their guns, however, arguing for the heavy lepton interpretation over stiff opposition coming even from other members of the collaboration. In May 1976, shortly after Goldhaber spotted his first *D*'s, they became absolutely convinced. With over a hundred anomalous events now in hand, they began a paper claiming the discovery of a new heavy lepton, calling it the τ (the Greek letter *tau*), with a mass somewhere between 1.6 and 2.0 GeV. But their complex arguments failed to sway the entire collaboration. Several dissenters even took their names off the paper before it went off to *Physics Letters* that July.

When DESY scientists reported finding no evidence for a heavy lepton at DORIS, it seemed as if the skeptics might have been right. But anomalous events of all sorts continued to show up at SPEAR, bolstering Perl's confidence. The very next spring, DESY came in with positive results, putting any remaining doubts to rest.

The tau lepton eventually weighed in at about 1.78 GeV, a little less than the charmed *D* mesons. Through a freak of Nature, it had popped up at almost exactly the same energy, complicating the interpretation of electron-positron annihilation there. In retrospect, naked charm had been so difficult to find at SPEAR because so damn much new physics was erupting between 3 and 4 GeV.

With unit electric charge, the tau lepton added one unit to the ratio *R* in this region. Coupled with the additional 4/3 contribution of the charmed quarks, this meant that R had to jump from a value of 2 to well above 4—pretty much what had been measured at the electron-positron colliders. The *R*-crisis, which had baffled experimenters and theorists to no end, was solved.

Like the electron and muon, the tau has its own neutrino, the tau neutrino, which is emitted whenever a tau disintegrates. Later experiments at SPEAR could only set an upper limit of 250 MeV on its mass—about as much as two pions. Quite possibly, this ghost has no mass at all.

But with the discovery of the tau and its neutrino, we now had *six* leptons instead of four, disrupting the hard-won symmetry between quarks and leptons. "If there are more than four leptons," Glashow declared, "there must almost certainly be more than four quarks." Were there two more quark flavors lying around somewhere, as yet undetected? And whatever the answer, Nature was obviously getting more complex again. Could the Standard Model cope?

Twice Leon Lederman had narrowly missed the J/psi—in 1968 and 1972. In his first encounter, watching the creation of muon pairs at Brookhaven's AGS, his Columbia group had noticed a funny enhancement—the famous Lederman shoulder—between 3 and 4 GeV. "This 'shoulder' excited us," he wrote. "We wondered if it could represent a sharp resonance that was smeared by our crude apparatus but marked the presence of some new particle." But rather than repeat the experiment with better equipment, as some of his coworkers had urged, Lederman chose to forge ahead to higher energies and explore the frontiers about to open up elsewhere. "Why do it again at Brookhaven," he recalls thinking, "when you could do it better at Fermilab?"

Lederman had his second shot at the J/psi on the CERN Intersecting Storage Rings, but unexpected backgrounds swamped his equipment and obscured his vision. Largely because of this tremendous noise, his group triggered their detectors *above* 3.1 GeV, making them effectively blind to the J/psi.

Undaunted, Lederman returned to Fermilab (as NAL was renamed in 1974) to continue the experiments he had pioneered at Brookhaven—the creation of lepton pairs in collisions of protons with nuclei. An outgoing, whimsical man then in his early fifties, fond of fine cigars and good company, he is widely recognized as one of the world's best experimenters. Having convinced Robert Wilson to become the laboratory Director, Lederman enjoyed plenty of clout there. Where others often had to make do with severely limited budgets, his group usually had adequate funding. So it was able to build a good double-eyed spectrometer a lot like Ting's Brookhaven device, one with the extremely fine resolution needed to hunt for narrow peaks.

Unfortunately for Lederman, however, Ting and Richter beat him to the decade's greatest prize. But early in 1975, Lederman's group, including physicists from Columbia, Stony Brook, and Fermilab, finally began using its new spectrometer to search for particles like the J/psi with masses above 5 GeV.

In 1976, they thought they had found one. Observing electron-positron pairs produced in the collisions of 400 GeV protons with beryllium nuclei, they found only twenty-seven at high mass. But twelve of these pairs clustered around a mass of 6 GeV, suggesting the existence of a neutral, spin-1 particle. Lederman and company published their find in May, dubbing their particle the Y (the Greek letter *upsilon*).

When they repeated the experiment that summer, however, the peak withered away. Studying emerging muons this time, they witnessed a total of 159 high-mass pairs, without finding any unusual feature near 6 GeV. The earlier peak must have been just a "one-in-fifty" statistical fluctuation that had fooled the group. Jokes began to circulate around Fermilab that they had only discovered the "oops leon."

But now the experimenters noticed another small bump sprouting at 9.5 GeV. Once burned and twice shy, they checked their analysis and began preparing yet another experiment. The anticipation was high, Lederman recalls. "John Yoh of our group labeled a bottle of champagne '9.5' and placed it in the group refrigerator."

Using vastly improved equipment, they began searching once again in May of 1977. During the very first week the bump reappeared. "The following week we doubled our data and still the bump remained," wrote Lederman. Soon there was little doubt left—and no champagne, either. But before they could study the region in greater detail, a fire broke out in their experimental area, and the electronics became drenched with hydrochloric acid. The physicists waited anxiously while a salvage expert flew in from Europe to help clean up the mess. Eight precious days elapsed before they got back on line.

In June, they mailed a paper to the *Physical Review Letters* announcing their discovery. Among a total of 9000 muon pairs, they had uncovered 770 between 9.0 and 10.5 GeV—an excess of 420. There was clearly a peak at 9.5 GeV and the hint of yet another just above 10.

By the time of the biannual Lepton-Photon Symposium, held that August at DESY, the group had another 18,000 muon pairs in hand. Lederman presented the combined data there, revealing not one, not two, but *three* narrow peaks less than 100 MeV wide—at 9.4, 10.0, and 10.4 GeV. He recommended calling these heavy particles the Υ, Υ', and Υ'', as this label was obviously no longer needed for the spurious 6 GeV peak.

The interpretation of the upsilons as evidence for yet another quark flavor was not long in coming. With the tau lepton a virtual certainty by mid-1977, theorists were clamoring for two additional flavors to flesh out another fourfold generation of pointlike fundaments. Strong similarities between the upsilon family of narrow peaks and the charmonium states suggested that these neutral, spin-1 particles were combinations of a new heavy quark, with mass about 5 GeV, together with its antiquark.

About the only question left was the charge of this quark.

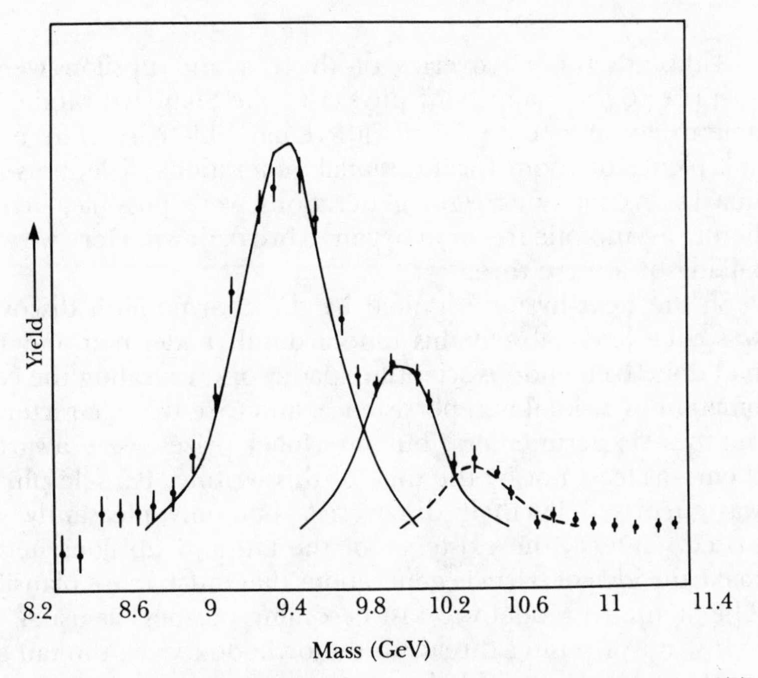

The three upsilon peaks Lederman revealed at DESY in August 1977, evidence for a new family of heavy neutral particles like the psis.

Was it ⅔ or −⅓? On other grounds, theorists had already predicted such a pair with these charges, naming them the "*t*" and "*b*" quarks for "top" and "bottom"—variants of the labels "up" and "down." The more whimsically inclined favored "truth" and "beauty," so that experimenters might begin the search for naked truth and beauty. But for once the more prosaic choice seemed to win out. It was still possible, however, for avid experimenters to go looking for "bare bottom," the combination of a bottom quark with an up or down antiquark to form a heavy meson.

Indeed, Lederman's data seemed to favor the bottom quark with a charge of −⅓. More precise measurements made in 1978 at DORIS, which had been driven to its limits to reach this 10 GeV region, confirmed this interpretation, and that was that. Now the only quark missing was the top quark, presumably lurking at still higher mass. With powerful colliders being built at SLAC and DESY, able to explore the 10 to 40 GeV region, it seemed only a matter of time before it would be found.

* * *

Although the discoveries of the tau and upsilon were a surprise to the majority of physicists, the Standard Model was able to accommodate them with remarkable ease. There was still plenty of room for additional generations of leptons and quarks. As many as eight generations were possible, in fact, before asymptotic freedom began to break down. Here we were talking of a mere three.

So the near-hysteria ignited by the charmonium discovery was completely absent this time around. Lederman and Perl had done tremendous scientific spadework, revealing the combination of risk-taking, persistence, and care that characterizes our best experimenters, but no Nobel prizes were awarded them—at least not by the time of this writing. Particle physics was surprised by their discoveries, but only pleasantly surprised. Indeed, the existence of the tau and upsilons helped make the idea of particle generations that much more plausible. The Standard Model was fast becoming business-as-usual.

Instead of posing threats to the orthodox view, the tau and upsilon particles provided important new laboratories for testing its implications at extremely high energies and small distances. The upsilon family was especially significant in this regard, presenting high-energy physics with what Lederman termed "an embarrassment of riches."

With *two* separate systems of heavy quarks, the psis and the upsilons, one could make far more detailed studies of the forces tethering quarks together. Because the charmed and bottom quarks were so ponderous and slow-moving within the parent mesons, their interactions could be calculated without all the bewildering relativistic complications of speedier motes. So the good old Schrödinger Wave Equation, on the shelf now for many years since its great successes in atomic physics, could be dusted off and deployed in this entirely new realm. The nonrelativistic quark model of Dalitz and Morpurgo (see Chapter 4) was back in vogue a decade after its demise. Hadrons are atoms composed of quarks.

During these same years, DESY and SLAC were racing to complete their new electron-positron colliders. After the recent overwhelming success of SPEAR, funded on the sly out of the

SLAC operating budget, Panofsky had little trouble getting government approval to build a much bigger ring. But prying the necessary $78 million out of a penny-pinching Congress, burdened by the costs of Vietnam and a worldwide recession, was a separate issue altogether. Even though the new SLAC machine was proposed in April 1974, six months before DESY's plans went in, it was late 1976 before Congress authorized the actual money and the project could begin. The Bundestag, by contrast, reacted to the recession by pumping money into the West German economy, and DESY was one of the lucky beneficiaries.

Housed in a tunnel over a mile around, the PEP (for Positron-Electron Project) ring was designed to store rotating beams of 18 GeV positrons and electrons fed it by the linear accelerator. With DESY's PETRA (for Positron-Electron Tandem Ring Accelerator) already underway and aiming for 19 GeV per beam, some scientists wondered aloud why SLAC didn't shoot higher. But given past experiences with SPEAR and DORIS, Stanford's machine builders were confident they could beat the Germans again.

Getting finished first meant having first crack at the search for the next prize, "toponium." This hypothetical spin-1 meson, composed of the putative top quark and its antiquark, was expected to appear at the high-energy frontiers these colliders would be opening up. This time the German physicists won the race, completing PETRA in November 1978, and had almost an entire year to explore these territories all by themselves.

As PETRA scanned up and up in energy, however, its four detectors revealed no evidence for the top quark. Starting at a total energy of 13 GeV, and reaching more than 30 GeV by the summer of 1979, DESY scientists found no hint of any narrow peaks like the J/psi or the upsilons. PETRA eventually hit 46 GeV without striking any peaks. A few seeming structures evaporated under a second, closer examination. All that remained was just a broad, featureless desert devoid of new quarks or leptons.

So entrenched had the Standard Model become by then, however, that the absence of toponium at PETRA caused no great alarm. There had never been any hard-and-fast prediction of its mass, after all, only several guesses and hunches that it might appear in this energy range. When measurements proved differently, it was no great problem to admit that the top quark had to be heavier than 23 GeV—and was thus invisible at PETRA and PEP energies.

The hard-won consensus of this lepton-quark world view was not about to be shattered so quickly. The bottom quark needed a partner to fill out the third generation of spin-½ fundaments. It would be found someday, once sufficient energy could be concentrated in one spot to create it.

But the growing unity of particle physics had been bought at the price of increasing complexity, a price physicists are generally loath to pay. We prefer a minimum of arbitrary parameters and unwarranted assumptions. Now there were at least six leptons and probably an equal number of quarks—plus all their antiparticles, and the three quark colors, too. And yet we had no reasonable explanation for the very different masses of these pointlike motes. It was all a bit unsettling.

CHAPTER 15

THE POWER OF GLUE

Pure logical thinking cannot yield us any knowledge of the empirical world. All knowledge of reality starts from experience and ends in it.

—Albert Einstein

On the surface, the Deutsches Electronen Synchrotron seems a far cry from the other major accelerator centers. Where CERN, SLAC, Fermilab, and Brookhaven are all situated in expensive rural or semirural settings, the better to accommodate their enormous facilities, DESY is crammed onto a compact urban site close to Hamburg, West Germany's busy North Sea port. Driving southwest on Notkestrasse, an unassuming suburban street lined with tidy homes and comfortable apartments, you come to a long driveway with only a small, unobtrusive sign announcing the laboratory's presence.

Once past the guard gate, you find a cluster of low, undistinguished brick buildings laid out in an orderly fashion along several spacious, well-kept streets. Everything about the place speaks efficiency. It could just as well be a bustling electronics factory as a laboratory for high-energy physics. The only hint of a particle accelerator is the low mound of earth occasionally visible toward the site boundaries. This is PETRA, DESY's electron-positron collider.

Before PETRA, DESY had toiled in the long shadow cast by her American counterpart, SLAC. The great discoveries of scaling, the psi particles, and the tau lepton had all occurred at the Stanford machine, only to be confirmed later at Hamburg. DESY did solid, but rarely earth-shaking, work. With their new collider, however, its scientists finally had first crack at virgin territories. Although toponium obstinately refused to turn up there, an equally valuable prize lay waiting in the wings: PETRA found the first visible evidence for gluons.

These carriers of the color force were proving to be almost as elusive as the quarks they supposedly glued together. Indeed, the odds-on bet during the late 1970s was that gluons would *never* appear outside the hadrons, just like quarks. Any possible evidence for them had to be indirect.

But gluons were absolutely mandatory for the success of quantum chromodynamics and the quark-parton model. They were the missing link that might explain, once and for all, why no quarks had ever turned up in particle detectors.

Quark confinement was still somewhat of a question mark for QCD. The theory *did* require that the color force between two quarks increase at large separations, after all, even if it could not prove explicitly that quarks remained forever trapped. That the proof could eventually be done, once the appropriate techniques were finally developed, was fast becoming a widespread belief among particle theorists, an act of *faith.* Quark confinement was gradually demoted to the lowly status of an "engineering problem," according to John Ellis, to be solved whenever enough computing power was finally available.

Here was yet another area where experimenters could make a valuable contribution, by "finding" these putative gluons or at least measuring some of their proclivities. It became an active area of particle physics research during the late 1970s.

At the time, scaling violations were viewed as the premier test of quantum chromodynamics—of the nature of the gluons. As early as 1974 we had already seen hints of such behavior in the structure functions measured at SLAC. But the ambiguities about the appropriate path to take in attaining higher energies precluded any stronger statement.

After the 1975 Stanford Conference, those ambiguities evaporated. No matter what path was taken, both SLAC and Fermilab reported significant departures from the flat behavior originally predicted by Bjorken. Coming at a time when asymptotic freedom was back in favor because it provided a cogent explanation for the long lifetime of the J/psi, these violations of scaling were soon hailed as additional support for QCD.

But the most convincing test was the exact *pattern* of scaling violation. That was hard work—for both experimenters and theorists alike. The former had to make incredibly precise measurements over vast expanses of electron, muon, or neutrino energy; the latter had to ferret out the many theoretical ambiguities cropping up along the way. Although it offered great precision, SLAC lacked the necessary high energies for the job and dropped out of the picture. Fermilab had plenty of energy, but it was too busy chasing imaginary anomalies to devote enough precious accelerator time to meticulous studies of nucleon structure.

So European physicists picked up the football late in the decade, after their Super Proton Synchrotron (SPS) came on line at CERN. Originally delivering protons at 300 GeV, this 4-mile ring eventually reached 450 GeV, slightly less than the Fermilab synchrotron. Almost from the day the SPS was turned on late in 1976, large international teams of experimenters were using its intense beams of muons and neutrinos to measure the structure functions in excruciating detail.

Jack Steinberger had been studying neutrino scattering since its infancy. During the fifties and early sixties he had worked shoulder-to-shoulder with such Columbia stalwarts as Lederman, Samios, and Schwartz before returning to Switzerland later in the decade. A trim, compact man of piercing, ice-blue eyes and firm, deliberate manner, he had little patience for a weak scientific argument. With a reputation as the most skeptical of skeptics, "Jack the Knife" had lost more than his share of fine wine in various wagers, but he had been a good sport about it.

A latecomer to quark-parton thinking, Steinberger became spokesman for a collaboration of CERN, Dortmund, Heidelberg, and Saclay physicists building a colossal neutrino detector for the SPS. A remarkably simple but effective design, the CDHS detector was a long series of 2-inch, magnetized iron slabs interleaved with planes of taut wires and thin plastic strips that recorded any tracks left there by charged particles. All told, there were 1200 tons of iron playing the dual roles of massive target for the impinging neutrinos and magnetic spectrometer to locate and identify the various bits of matter emerging from a smash-up.

One of the first things the CDHS collaboration went stalking was Carlo Rubbia's "high-y anomaly." In the mid-1970s the HPW (Harvard-Penn-Wisconsin) team at Fermilab had reported seeing an overabundance of low-energy muons emerging from the collisions of antineutrinos with nuclei. Too large to be readily accommodated within the Standard Model, this anomaly had already spawned a vigorous theoretical effort. The commonest explanation required very new types of quarks quite unlike the four or five then established.

Steinberger remembers sitting with Rubbia on the patio in his garden, sometime in 1976, discussing the upcoming CDHS experiment. "You're not going to find very much, Jack," Rubbia challenged him. If they confirmed the anomaly, it would only redound to HPW's credit; if they instead disproved it, well that was no great discovery and would soon be forgotten, too. Undeterred, Steinberger and his group forged ahead with their measurement and found no such excess. The Standard Model was still in fine shape, after all, and the high-y anomaly was soon a footnote in history.

Over the next few years, as the CDHS collaboration logged more and more neutrino and antineutrino collisions in the mammoth detector, a consistent pattern began to emerge. At low values of the Feynman parameter x (the fraction of a nucleon's momentum carried by the struck quark), the structure functions rose ever so slightly. At high values of x they *decreased* almost as slowly. It was exactly the pattern predicted by QCD, once all the theoretical ambiguities had been resolved. One could look at these graphs and actually "observe" how the color force between quarks gradually became weaker at small distances.

To many particle physicists the scaling violations seen by the CDHS physicists, and similar effects soon measured by the

European Muon Collaboration using muons as probes, were the most convincing evidence for gluons. Other groups like our own had reported subtle hints or partial glimpses of such behavior. Or they had combined the results of two or more different experiments before drawing their conclusions. That was a chancy approach, at best. These two European collaborations were the first to make such detailed studies over wide enough ranges of energy to prove or disprove QCD convincingly. Painstaking, unglamorous work that would win nobody a Nobel prize, it became one of the firmest cornerstones of the Standard Model.

An intuitive picture of the color force between quarks emerged during the late 1970s. Based on close analogies between QCD and QED, this approach treated the force as if caused by multiple exchanges of colored gluons—just as the electromagnetic force between two charges is due to multiple photon exchanges. Equivalently, you can imagine field lines extending from one quark to another, transmitting the color force between them. Gluons are just the physical manifestations of these imaginary lines. Like the photon, they are spin-1 particles with no mass.

Using such a picture, theorists imported many calculation techniques directly from QED—especially at high energies or short distances, where the analogy works best. By doing so, they could predict experimental results like the scaling violations just mentioned. The more venturesome theorists even tried to figure what happens at low energies or large distances, too, where the analogy is on a weaker footing—the true domain of really new and different physics.

Imagine what happens when you pull two charged particles apart. The field lines extending from one to the other begin to thin out, because there is more space opening up and there are just so many lines available. Now in this picture the force between two particles depends on the *density* of the field lines between them. The closer the lines, the bigger the force. Because they thin out as you yank the charges apart, the force becomes weaker. It is a natural consequence that seems to arise out of pure geometry itself, little more.

But consider the case of two quarks—for instance, the quark and antiquark in a meson. Here the field lines, which emanate

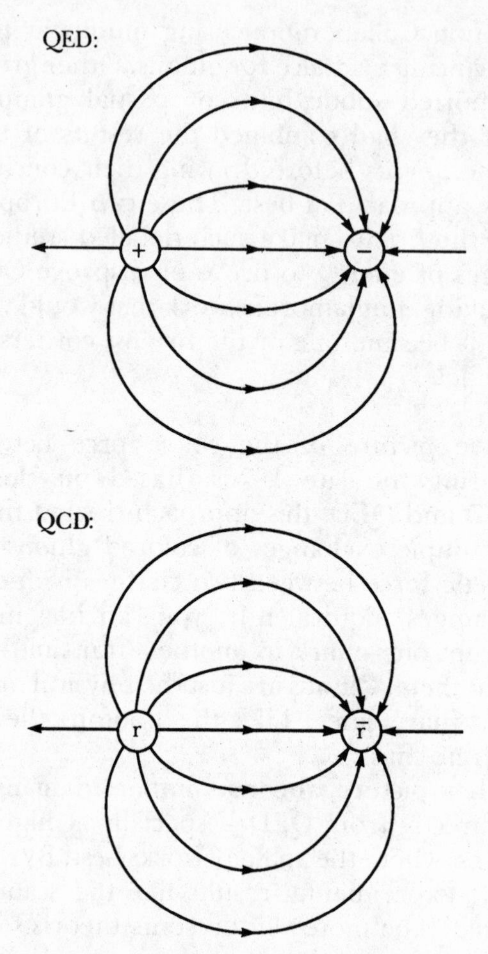

Field analogy between electric and color forces, which works extremely well at small distances.

from one and end on the other just as before, also *carry color.* That is the one true difference here, the crucial difference. Though it seems so trivial at first glance, it is very, very important. Because the field itself carries color, the analogue of electric charge, it can act as the source of new field lines and thereby exert a force on *itself.* This a photon cannot do. Pictorially, the field lines extending between two quarks attract *each other* into a tighter, denser bundle. Hence the interquark force grows as the quarks separate.

Now if you try to pull the two quarks farther and farther apart, you will stretch this bundle into a narrow tube—called a

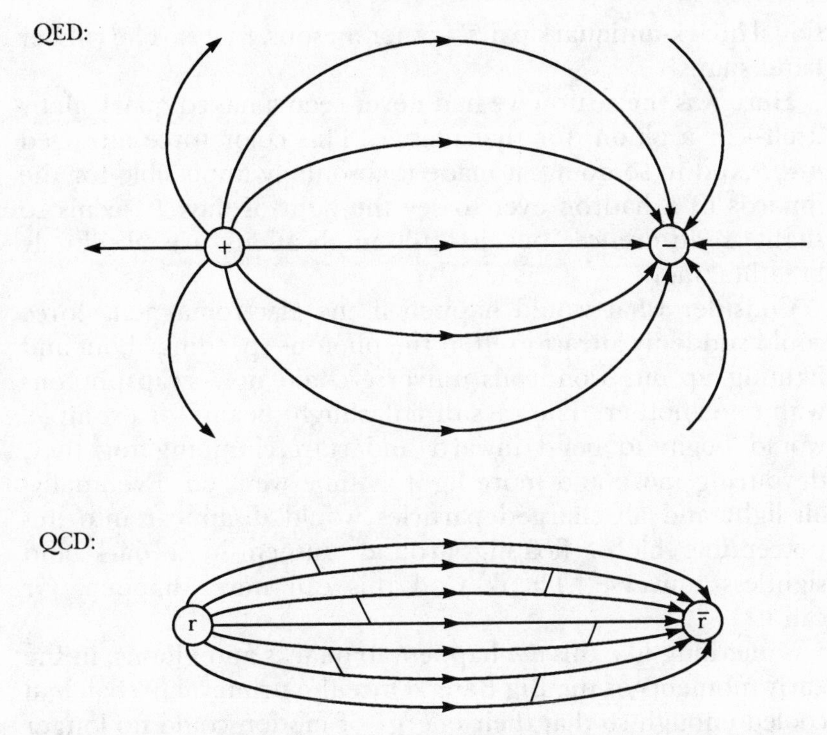

What happens to the field lines at larger separations, in quantum electrodynamics as compared to quantum chromodynamics. The color field lines between two quarks attract each other, *forming a "flux tube" and binding them tightly together.*

flux tube—where the density of field lines remains essentially the same, no matter what the separation. So the force you now must apply is independent of the distance between the quarks, once they are more than about a proton radius apart. Another way to describe this effect is to say that more and more gluons pop up out of the vacuum to fill the widening gap between the quarks. Color begets color, which begets more color, ad infinitum. Unlike gravity or electromagnetism, this color force is forever. It never lets go.

Trying to separate two quarks is a lot like pulling on two ends of a broken rubber band. You pull and pull, injecting more and more of your energy into the stretched band. To pull two quarks a foot apart, it would require the same energy as needed to lift an automobile a foot off the ground. Long before you can ever reach such a separation, the energy materializes as additional quark-antiquark pairs—other mesons, that is. The rubber band snaps!

Here was the reason we had never seen a naked quark all by itself—or a gluon, for that matter. This color force attracted *itself!* And in so doing, it made it absolutely impossible for the innards of a hadron ever to see the light of day. It seems so slight a difference, but its full implications are absolutely breathtaking.

Consider what would happen if the electromagnetic force could suddenly attract itself, if the photons speeding about and lighting up our wondrous universe could now swap photons with one another. The rays of a flashlight beam, for example, would begin to bend inward and start clumping together, devouring more and more light as time went on. Eventually, all light and all charged particles would disappear into imperceptible blobs, floating around unseen in a dark and sightless universe. Thank God this can never happen. Or can it?

Something like this *did* happen, to quarks and gluons, in the early moments of the Big Bang. Once the primeval fireball had cooled enough so that their energy of motion could no longer counter the inexorable pull of the color force between them, the remnant quarks and antiquarks began clumping together in a great frenzy of mating, forming the pairs and threesomes that could neutralize this relentless tug. The polychromatic early universe full of colorful characters quickly became a staid, ordinary world populated only by colorless leptons, hadrons, and photons. Color had no existence, no meaning whatsoever, beyond the cramped confines of hadrons. Only when two of these blobs came very close together could the residual impact of this color force be felt.

There matters rested until, billions of years later, a few intelligent but colorblind apes began groping around in the dark with special canes—quantum mechanics, gauge field theory, and high-energy beams of pointlike leptons. With these penetrating tools we peered into the relics remaining from those glorious early moments and caught a fleeting glimpse inside these tiny microcosms of what a beautiful world it must have once been.

* * *

In the late nineteenth century, despite the wealth of indirect evidence for atoms, a few major scientists refused to believe in their existence as distinct material entities. Led by the physicist and philosopher Ernst Mach, the positivist school of thought insisted that reality was to be found only in direct sensations, in observational experience. Although the atomic theory had successfully explained chemical combination and the behavior of gases, nobody had really *seen* a single atom yet. To the positivists, these "atoms" had only a mathematical existence like that which Gell-Mann ascribed to quarks many years later. They were not "things."

It was Einstein's 1905 explanation of Brownian motion—the abrupt, jerky movements of pollen grains and dust particles seen in a microscope—that finally won even the positivists over. These motions are the result, he argued, of impacts with very fast-moving atoms or molecules. Scientists could at last see, with their very own eyes, the effects of a single atom.

To establish their reality once and for all, quarks still required the same kind of direct, observational evidence of their individual existence. But finding a single quark in a particle detector seemed out of the question by the late 1970s; it would have been a big embarrassment for QCD. William Fairbank of Stanford, who claimed to be finding fractional charge on balls of superconducting niobium metal, had to endure severe criticism when nobody could confirm his "discovery."

Even the electron scattering experiments at SLAC were ambiguous on this point; we could observe only the effects of large ensembles of quarks, not one or a few. Soon after the early MIT-SLAC experiments were reported in 1968, other scientists began trying to find single quarks among the debris of electron-proton collisions. If the electron was indeed striking individual quarks, they reasoned, then these fragments should be seen flying off in other directions, away from the ricocheting electron, leaving distinctive tracks. But only the usual, familiar hadrons ever turned up.

Those who still wanted to believe in quarks had to go out on a limb and insist that they somehow metamorphosed into hadrons before emerging. Somehow, by some kind of mysterious sleight of hand, these wily quarks "dressed up" and appeared as normal hadrons by the time they reached our

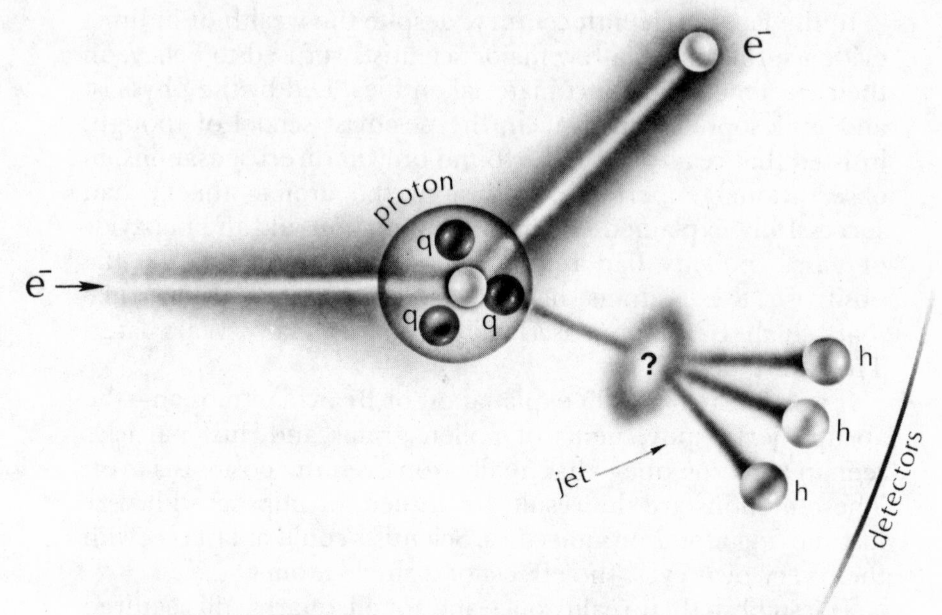

The anticipated formation of "jets" in electron-proton collisions. The quark-parton struck by an electron somehow metamorphoses into a spray of hadrons before reaching particle detectors.

detectors. But even so, Bjorken and Feynman suggested, we might still witness the behavior of individual quarks by watching for "jets" of debris flying away from a collision. If these shifty quarks were indeed disguising themselves as hadrons, so the argument went, we should see closely packed sprays of particles emerging in a few specific directions—not just a random scatter of debris.

Evidence for jets would be the same kind of visual proof for the quarks' existence as Brownian motion had been for atoms. Though we might never see a single quark itself, we should still be able to witness its effects upon the immediate surroundings. If they were indeed trapped in some kind of hadronic cocoon, you should find pieces flying off in a single direction when one of their more impetuous brothers kicked this cocoon hard enough.

But intensive searches at SLAC and elsewhere turned up no convincing evidence for jets—at least not immediately. Trouble was, we were all looking in the wrong place at first. By shooting high-energy particle beams into fixed targets, we were in effect slamming trailer trucks into stationary Volkswagens, and almost *all* of the debris was being thrown forward in a tight cone. To pick a few well-defined jets out of this morass defied the best efforts of experimenters. It was well nigh impossible.

A much better place to find jets happened to be the storage rings. Here, scientists were slamming two trucks into each other almost head-on, and the debris could fly away at a great variety of possible angles. With detectors like the Mark I at SPEAR watching the collision on many sides, it became a lot easier to tell whether or not the emerging particles were packed into a few distinct jets.

Gail Hanson of the SLAC-LBL collaboration was the first to uncover hard evidence for jets. One of the few women experimenters in particle physics, she had earned her Ph.D. at MIT and worked at the CEA Bypass before coming west in 1974 to join Richter's group. Her own special interest at SPEAR became the search for jets in electron-positron annihilation.

If a quark-antiquark pair was created after the annihilation, each had to zoom off back-to-back, in order to conserve momentum. If they subsequently dressed up as hadrons, not one but *two* jets would emerge in opposite directions. So if quarks had anything to do with this process, Hanson expected to find—on the average—that hadrons flew off in two back-to-back jets, not randomly.

The most difficult part was identifying two distinct jets in the first place. As always, there was plenty of scatter in the data to confuse matters. Jets did not simply leap off the page. But by a sophisticated computer analysis performed in early 1975, before the Stanford Conference, Hanson was able to establish that hadrons were indeed emerging in two back-to-back jets, especially at the highest SPEAR energies. And the directions preferred by these jets were characteristic of spin-½ particles. Whatever was producing them looked very much like two quarks.

The discovery of jets was a crucial finishing stone in the quark-parton edifice. "Such behavior is possible without assuming the existence of quarks, but any other explanation seems difficult and cumbersome," announced Richter at Stockholm. "In my view the observations of these jet phenomena in electron-positron annihilation constitute one of the very stron-

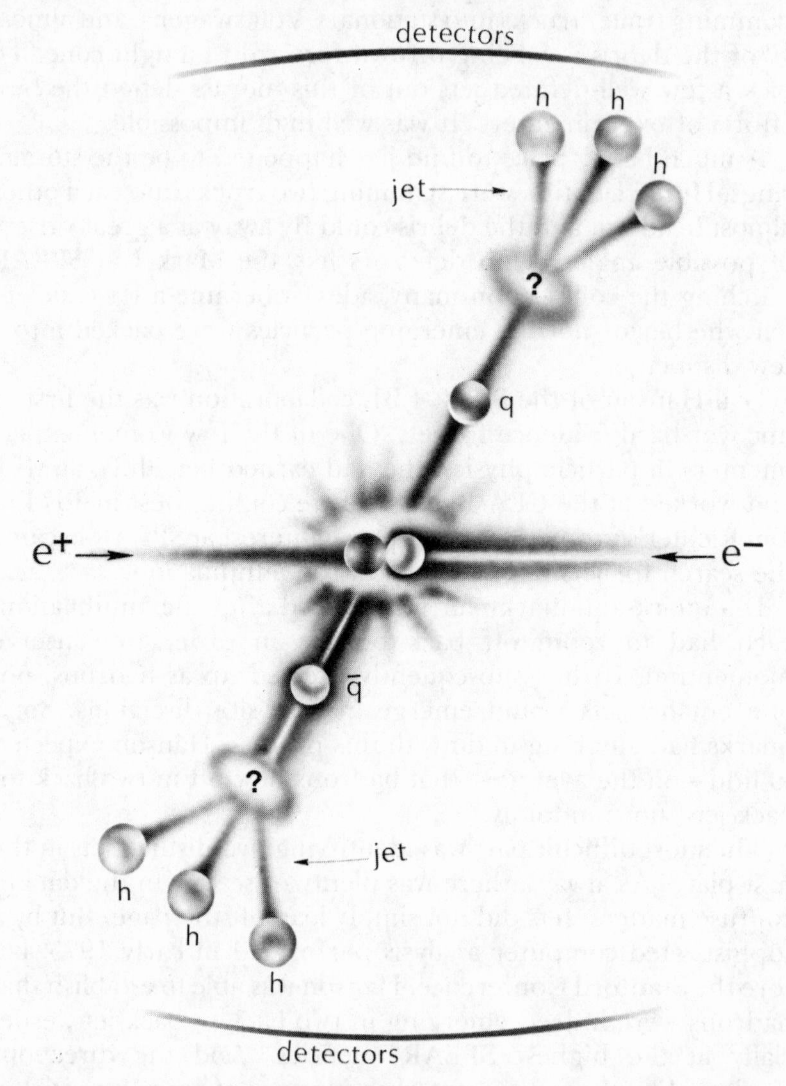

How two back-to-back jets might occur in electron-positron annihilation.

gest pieces of evidence for believing that there really is a substructure to the hadrons."

If there were any lingering doubts about these jets, they evaporated when data began rolling in from DESY's new collider in early 1979. Whereas a sophisticated computer analysis had been necessary for SPEAR physicists to discern jets within the random scatter of debris, at PETRA they literally jumped into your lap, one by one. At these higher energies, the quark-antiquark pair created in electron-positron annihilation zoomed apart with great vigor. So their hadron remnants were packed together in much narrower jets that could be identified immediately. "On PETRA you could literally *see* the jets appearing on the screen," Albrecht Böhm recalled, "one after another."

What a thrill it must have been for the experimenters on shift. They could finally observe the very "track" of a single, individual quark. What had sprung into being inside the minds of two men and been thought only a mathematical curiosity at first, what had then made its presence felt through many indirect tests during the late 1960s and throughout the 1970s,

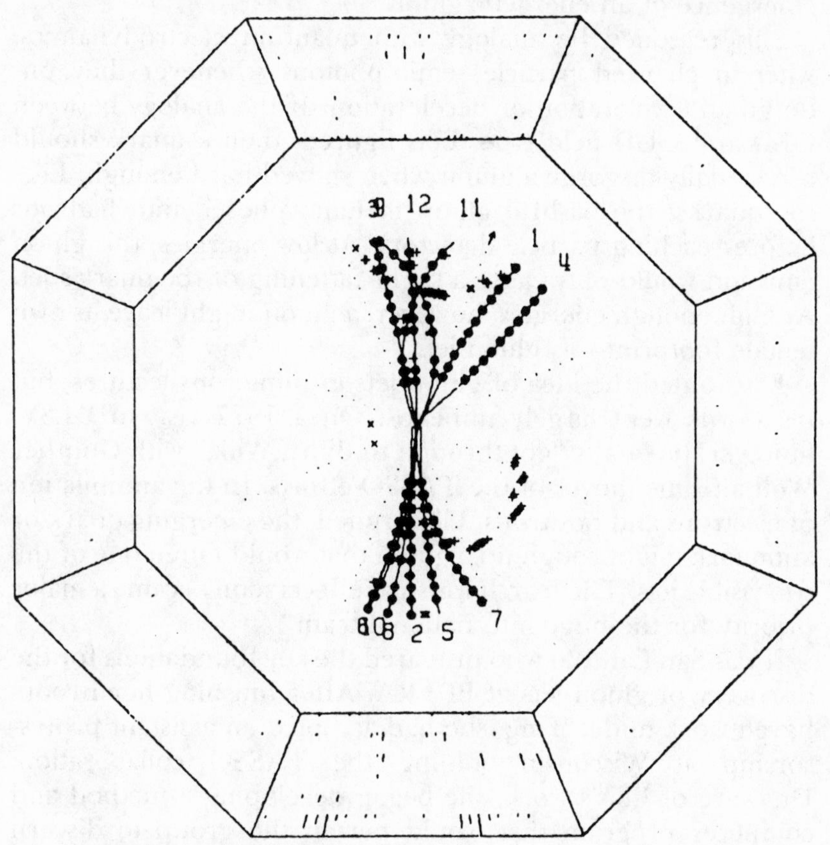

A two-jet event witnessed at PETRA.

was now leaving actual *footprints,* two by two, in their particle detectors. The word had become flesh. Quarks were "things," in every sense of the word.

But there were other insights to be had at PETRA, where four separate experimental groups operated independent detectors like the Mark I, each surrounding one of the four available interaction regions. One of these groups, including over seventy physicists from nine institutions in Germany, Great Britain, Israel, and the United States, had been watching a large detector named TASSO (for Two-Armed Solenoidal SpectrOmeter) for events with *three* jets emerging, not just two. Such three-pronged events had been first predicted in 1976 by John Ellis and two other CERN theorists. Two of the three jets would be footprints of quarks, but a third might arise from the emergence of an energetic gluon.

Ellis reasoned by analogy with quantum electrodynamics, wherein charged particles emit photons whenever they undergo an acceleration or deceleration. If the analogy between QED and QCD held true, Ellis figured, then a quark should occasionally disgorge a gluon when shoved hard enough. Like the quarks, the bashful gluon metamorphosed into hadrons before reaching particle detectors. At low energies, this gluon emission would only cause a slight fattening of the quark's jet. At high enough energies, however, a gluon might leave its own telltale footprint—a "gluon jet."

Ellis touted the idea of gluon jets in numerous lectures, but his words went largely unheard. On a 1977 visit to DESY, however, he finally "got through to Bjorn Wiik," with Gunther Wölf a prime mover of the TASSO efforts. In the annihilation of electrons and positrons, Ellis argued, the emerging quark or antiquark might cough up a gluon that would fatten one of the two visible jets. The search for such effects soon became a major priority for the huge international team.

It was Sau Lan Wu who prepared the key foundations for the discovery of gluon jets at PETRA. After finishing her Brookhaven work under Ting, she had accepted an assistant professorship at Wisconsin, joining the TASSO collaboration. Unaware of Ellis's work, she began developing a method and computer program that would permit the group to discern three distinct jets amid the surge of debris hitting its detector.

Color seems an integral part of this woman's life. As a young girl growing up in Hong Kong, she dreamed of becoming a painter, but opted for physics at Vassar because of "practical reasons." Her tiny office at DESY is stacked with the usual piles of books, reports, and computer output, but gaily colored drapes frame the small window and a bright red and green coverlet adorns the couch.

Inspired by Hanson's work at SPEAR, Wu began asking herself, "Wouldn't it be nice to find a way to *see* all three jets distinctly?" That way they could leapfrog the whole messy business of trying to tell whether one quark jet was fatter than the other. After a year of hard work and little success, she hit on a method in 1978 and published it early the next year.

The earliest runs on PETRA had come at total energies of 13 and then 17 GeV, high enough to see two-jet events though not quite enough to produce three. But after a shutdown in March 1979, the collider began operating above 27 GeV during April and May. With the aid of Wu's program, which in effect allowed TASSO physicists to rotate the events around and view them from the most convenient angles, they now began to find events with three distinct prongs, all in the same plane.

By June, TASSO had about forty events at this energy, including a few with some kind of extra appendage. Scheduled to review the recent PETRA findings for a gathering of mostly neutrino physicists that month in Bergen, Norway, Wiik was reluctant to say much about gluon jets. But Wu arrived at Wiik's country cottage, where he was hosting Ellis, bringing with her the latest candidates. Among them was one that Ellis called a "gold-plated event" he was absolutely convinced contained a gluon jet.

Thus emboldened, Wiik revealed this event at Bergen, suggesting that TASSO was possibly beginning to witness the footprint of the gluon itself. But most of his listeners remained unconvinced. One single event with three prongs could easily be a fluke—a fortuitous event in which all the visible hadrons had merely *happened* to line up in three directions. "C'mon Bjorn, do you really believe this one event?" challenged Donald Perkins. "I want to see all forty!"

As the weeks passed, the TASSO collaboration found more and more three-jet events. With a distinctive geometry—three

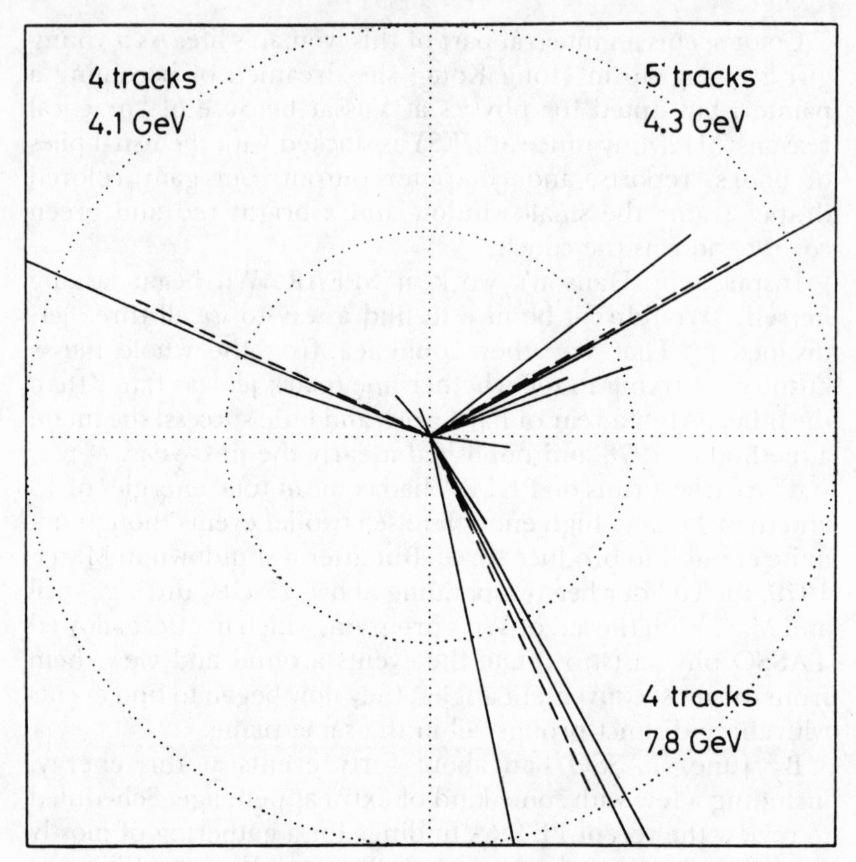

The first three-jet event, as revealed by Bjorn Wiik at Bergen. Solid lines represent individual particles; the dashed lines indicate the central axis of each jet.

separate prongs, all in the same plane—resembling the hood ornament of a certain expensive German automobile, they were soon nicknamed "Mercedes events." At the July meeting of the European Physical Society at CERN, TASSO physicist Paul Söding showed another four examples and presented a much more thorough analysis. Many things pointed toward gluons, but Söding stopped short of making an unambiguous claim.

Stimulated by the highly suggestive TASSO reports, the other PETRA collaborations had begun searching in earnest for similar effects. Working with exotic detectors named JADE,

PLUTO, and MARK-J, they pored over data, old and new, looking for events with the same planar, three-pronged geometry being witnessed at TASSO.

Events finally converged upon the biannual Lepton-Photon Symposium, held in August 1979 at Fermilab. There all four PETRA collaborations reported evidence for three-jet events, with the favorite interpretation being the emergence of gluons. It was the hit of the conference. Leon Lederman, recently named Director of Fermilab to succeed the retiring Robert Wilson, was extremely enthusiastic about the new European findings. To publicize the discovery, and perhaps to underscore how America was falling behind in particle physics, he called a press conference on the last day of the Symposium.

Under the glare of the klieg lights, a priority struggle began to erupt. Led by Samuel Ting, the MARK-J collaboration had claimed the "*Discovery* of Three-Jet Events," in disregard of the prior TASSO reports. The MARK-J collaboration certainly

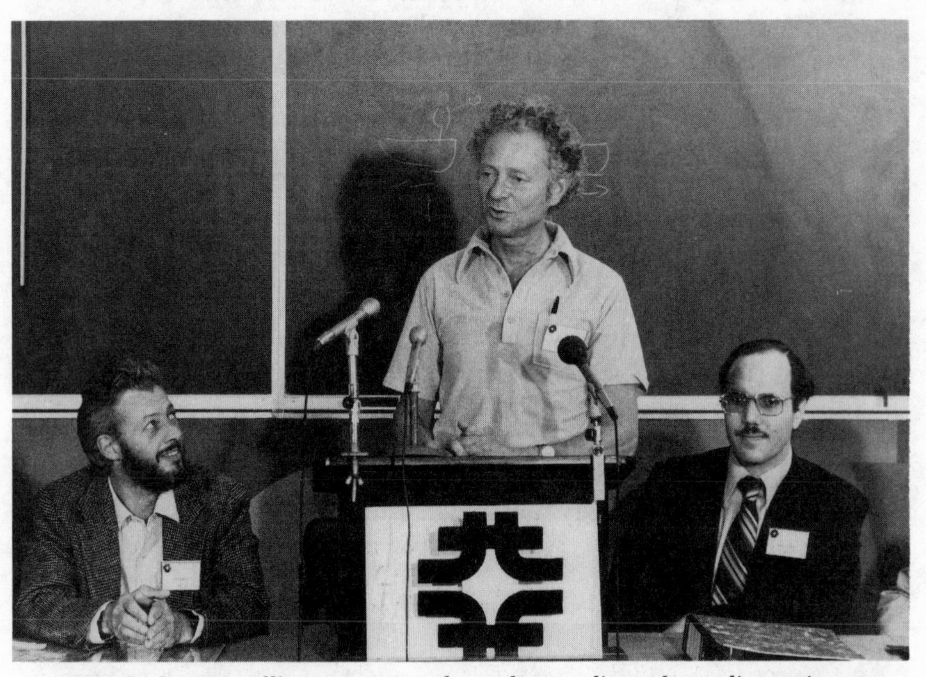

Leon Lederman telling reporters about the startling gluon discoveries at PETRA. Flanking him are Gunther Wölf of TASSO, left, and Harvey Newman from the MARK-J group.

made an important contribution, but could hardly claim its own "discovery." To make matters worse, the paper they subsequently sent to *Physics Letters* contained no reference to the published talks of Wiik and Söding. When pressed about this oversight, Ting replied, "Traditionally, we have always respectfully declined to refer to and comment on preliminary conference reports with small statistics and uncertain conclusions."

TASSO physicists had a completely different view of the matter. Most felt that Ting had tried to grab the credit for the three-jet discovery by engineering what Wiik termed "a massive PR campaign" in American newspapers. A Nobel laureate has far greater access to reporters, noted Wu, and they could not be expected to comprehend the detailed story behind such a complex scientific discovery.

Walter Sullivan remembers Ting's PR campaign well. On the day after the press conference, this veteran reporter, dean of American science writers, was relaxing with his wife at a secluded lake in northern New Hampshire. The only telephone for miles was the one at the boat landing across the lake from his cottage. Suddenly it rang. After learning there was a call for him, Sullivan rowed over to the landing only to find Ting on the other end talking excitedly about his "discovery of the gluon" at DESY. Unimpressed, Sullivan suggested Ting call Malcolm Browne, who was covering the science beat in his absence.

But Ting insisted it was too important to trust to anybody else. Several hours after returning to his cottage, Sullivan glanced out the window and discovered a young man rowing toward his dock through the pouring rain. Soaking wet by the time he arrived, the oarsman handed him an envelope containing Ting's press release. "We took pity on the poor chap," Sullivan recalls, "and invited him to warm up over a cup of hot coffee." But their soggy messenger hurried off into the gloom, apologizing that "Professor Ting told me to hurry right back to work" at MIT.

Browne finally wrote up the story, based largely on Ting's account, and it appeared on the front page of the September 2 issue. *Times* readers across the country learned that the gluon had been discovered at DESY by an international collaboration including Ting and twenty-seven physicists from the People's Republic of China.

Getting an article published in the scientific journals proved only a bit more difficult. When Klaus Winter, editor of *Physics Letters,* insisted that he drop the word "discovery" and reference the earlier TASSO work, Ting took the paper instead to the *Physical Review Letters.* According to Editor George Trigg, who was away on vacation, his stand-in Eugene Wells decided the "discovery" was so important that it should be published quickly as is—bypassing the customary refereeing process. A solid-state physicist, Wells knew nothing about the earlier TASSO reports and may have been unduly swayed by a Nobel laureate's pressing him for immediate publication. He had the typesetting staff work on it over the Labor Day weekend and published in only eighteen days. So even though this paper had been submitted two days after the TASSO article arrived at *Physics Letters,* it appeared in print a full week earlier.

By and large, European physicists took a dim view of the American theatrics. Though extremely suggestive, the evidence for gluon jets was not yet thoroughly convincing. These premature newspaper announcements only subverted the aims of the patient investigators then working hard at PETRA. And TASSO's Söding felt a lot of credit belonged not to experimenters but to Ellis and company for predicting such jets in the first place, three years earlier.

Others felt that this attempt to grab the spotlight in what was basically a European discovery betokened a faltering American high-energy physics program. "So why have these results been hailed as so remarkable, particularly in the US?" asked the British science weekly, *New Scientist.* "The only conclusion seems to be that American particle physicists are trying hard to keep up the momentum for federal funding for their expensive form of research."

After another year of running, little doubt remained at PETRA about gluon jets. With over a thousand events in hand, TASSO experimenters even managed to prove that the third jet possessed spin-1, not spin-0, just as required by QCD. Along with the detailed proofs of scaling violations, these gluon jets became one of the principal showpieces of the theory, another cornerstone of its empirical foundations. The very "glue" binding quarks together had itself appeared in the flesh.

* * *

Although quark confinement had still not been proved by the decade's end, few physicists doubted that QCD was the true theory of the interquark force. Not only was it a beautiful idea that fit well with all the revolutionary new insights of gauge theory, but the experimental evidence was also getting awfully strong. The proof of confinement would come, in due time, when powerful enough computers were finally developed to do the arduous calculations.

The presumed "reality" of quarks still rested on such an explanation of their complete absence from particle detectors. They were somehow imprisoned within hadrons by forces we could easily imagine—using "bags," "strings," "flux tubes," and their ilk—but not fully calculate, at least not yet.

CHAPTER 16

THE ROAD TO UNITY?

We shall not cease from exploration
And the end of all our exploring
Will be to arrive where we started
And know the place for the first time.

—T. S. Eliot, Four Quartets

A decade's end is often just a handy placemark, a convenient way of punctuating history that does not necessarily correspond to any signal trends or important events. But there was definitely a *fin de siècle* air about 1979, as if physics were witnessing the end of one era and the beginnings of another. Speaker after speaker at the major conferences that summer talked with a sense of solid accomplishment and resolution. At the heart of the widespread contentment stood the "new orthodoxy" of the Standard Model, triumphant after weathering five years of stormy challenge.

"We have now reached a quiet time," observed Bjorken at the 1979 Bergen Conference on neutrino physics. "The situation really is remarkably satisfactory. It is no wonder that, after the false alarms and relatively extravagant theoretical responses of the past few years, there is at present such a mood of minimalism and complacency."

He was referring to the confusion spawned in the particle physics community by a number of spurious experimental results. Theorists more prone to ambulance-chasing, as Gell-Mann would call it, had responded to these alarms with all sorts of exotic variations on the gauge theory scheme, heresies from the Standard Model. But these new sects disintegrated after further experiments failed to bear out their radical prophecies. The simplest possible path—the one proposed by Weinberg and Salam over a decade earlier—had proved to be the chosen path. "Nature seems to do things economically," observed Glashow at Bergen.

A major heresy had erupted when experimenters initially failed to find any violations of parity in the interactions of electrons, as required by the Glashow-Salam-Weinberg theory unifying electromagnetic and weak nuclear forces. With the admission of the Z° to the select circle of force messengers, electrons should have been able to tell left from right—just as can neutrinos—by occasionally swapping a Z° instead of the customary photon. But scientists at Oxford and the University of Washington had found no such left-right asymmetry in atomic physics experiments of the mid-1970s.

Like so much else that had happened in that remarkable decade, the parity-violating electrons finally surfaced at SLAC. Prodded by Charles Prescott, who had come to SLAC via Caltech and the nearby University of California at Santa Cruz, Group A and a team of Yale scientists had been doggedly pursuing an extremely difficult experiment in which "polarized" electrons caromed off deuterium nuclei. If electrons could actually tell left from right, then their chances of scattering had to be slightly smaller when their spins pointed *along* their direction of motion rather than the opposite way. Left-handed electrons, that is, should ricochet a bit more often than right-handed ones. But the expected effect was only 1 part in 10,000, and a first experiment in 1976 had been unable to prove anything, one way or another.

With a far more intense source of polarized electrons built by Edward Garwin and Charles Sinclair, these physicists tried again in early 1978. End Station A, where the measurement was performed, was in danger of becoming deserted by then, with only an occasional group stopping by to use the dusty old spectrometers inside this historic concrete cavern. This experiment was its crowning glory. Using a vastly improved beam of polarized electrons, the SLAC-Yale collaboration finally discovered a tiny but significant difference between left- and right-handed electrons, in close accord with what the Glashow-Salam-Weinberg theory predicted. A major victory for the Standard Model, this experiment dissolved many of the remaining heresies.

By the summer of 1979, the atmosphere of consensus was so thick it had become almost suffocating. Here were many of the most argumentative men and women on the face of the earth, all lining up to be counted among the elect, true believers in the new physics. Experiment after experiment solidly confirmed the unification of weak and electromagnetic forces. A key parameter in the theory, called the "Weinberg angle," came up virtually the same no matter how it was measured. With it one could predict the W and Z° masses fairly accurately—about 80 and 90 GeV respectively, as heavy as a hundred protons, far too massive to appear at existing atom smashers. But few of us now doubted these ponderous messengers would show up eventually, when enough energy became available to create them.

So it was, too, with the missing top quark. Though PETRA

Fermions, or matter particles

generation:	1		2		3	
leptons:	e	ν_e	μ	ν_μ	τ	ν_τ
quarks:	d	u	s	c	b	t

Bosons, or force particles

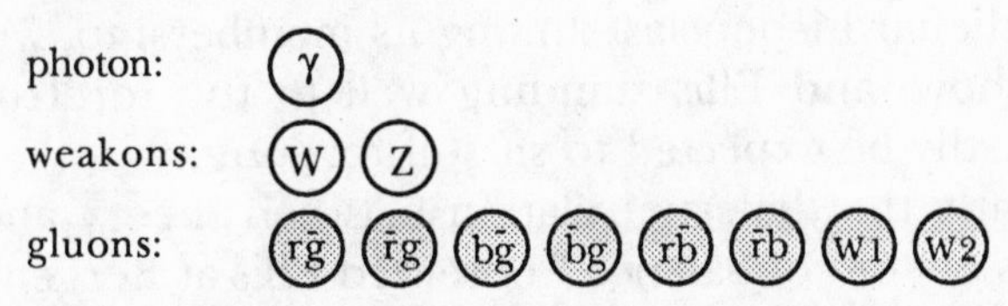

Elementary particles of the Standard Model, circa 1980, with the top quark not yet confirmed, but expected to be found. Leptons and quarks come in at least three generations with four distinct members; in addition, each quark can appear in three colors.

had reached energies well above 30 GeV by the summer of 1979, it gave absolutely no evidence for toponium, the anticipated marriage of a top quark and its antiquark that should have given narrow peaks like the J/psi and the upsilon particles. Theorists just concluded that the top quark had to be heavier than previously imagined, and hence invisible at PETRA energies. The Standard Model was unspecific about quark masses, so it was far easier to push the problem aside for the moment than to abandon the new faith entirely. Only a few renegades dared to suggest that maybe a sixth quark wasn't so necessary, after all.

Asked to summarize the current state of particle physics at Bergen, Glashow offered his audience an anecdote:

> On holiday in Jamaica, we noticed that there is only one kind of cheese. It is called "standard cheese" and is used in sandwiches, with salads, and on pizzas. So it is in particle physics. We have no real alternative to the standard gauge theory. It is used for strong interactions, weak interactions, and electromagnetism.

The Swedish Academy of Sciences agreed with his assessment, at long last. It finally awarded Glashow the Nobel prize that fall, to be shared with Abdus Salam and Steven Weinberg for their pivotal roles in the unification of the weak and electromagnetic forces. Ever cautious, the Academy had held back while heresies were still raging during the previous few years. But now that the dissenters had returned to the flock, the three theorists could be canonized at last into the sainthood of their scientific profession.

Still, there were those physicists who could grumble aloud about the widespread complacency. The profession counts many a diehard iconoclast among its membership, with Bjorken, Glashow, and Ellis standing well to the forefront. They could hardly be expected to sit still for long.

"Recently the dialectic relation between theory and experiment has got out of balance," observed Ellis at Bergen, "and we have a tendency to regard everything as solved, at least in principle. Our main sin is now that of hubris, and we should be more humble in our ignorance, remembering previous occasions when Nature unleashed upon us unexpected surprises." He looked to the act of measurement for new inspiration. "The most important experiments will probably be those which fly in the face of orthodoxy. Nature always has some surprises up her sleeve."

Glashow agreed, for once. His lecture at Stockholm that December was as much a critique of the "premature orthodoxy" he now perceived crystallizing as it was an overview of his past twenty years' work. "We are still far from Einstein's truly grand design of unifying *all* known forces," he admonished.

> Physics of the past century has been characterized by frequent great but unanticipated experimental discoveries. If the standard theory is correct, this age has come to an end. Only a few important particles remain to be discovered, and many of their properties are alleged to be known in advance. Surely this is not the way things will be, for Nature must still have some surprises in store for us.
>
> Nevertheless, the standard theory will prove useful for years to come. The confusion of the past is now replaced by a simple and elegant synthesis. The standard theory may survive as a part of the ultimate theory, or it may turn out to be fundamentally wrong. In either case, it will have been an important way-station, and the next theory will have to be better.

The "ultimate theory" that Glashow had in mind was one of the so-called "grand-unified" theories that had sprung up during the mid-1970s, in the wake of the gauge theory revolution of those hectic years. Jogesh Pati and Salam had proposed an early version, the idea about electrons changing into quarks that so enamored Richter during the famous R-crisis. Glashow and his Harvard colleague Howard Georgi advocated yet another, based on the symmetry group called SU(5) for *S*pecial *U*nitary group in *5* dimensions.

Grand-unified theories, or GUTS for short, took the next logical step of unifying the strong force with the "electroweak" force, a unity suggested by the fact that both were successfully described by gauge field theories. There were many possible grand-unified theories, but they all shared a few common features. In GUTS the unity of leptons with quarks is absolute. A quark can become a lepton, for example, by disgorging an unbelievably heavy particle called an "X boson," just as a neutrino changes into an electron by emitting a W^+. Such theories also purport to explain the seemingly arbitrary masses of the leptons and quarks. And GUTS require that the strong, weak, and electromagnetic forces have the exact same strength whenever the interaction energy exceeds the mass of these ponderous X particles. These forces only *appear* different at lower energies, where the symmetry is supposedly broken.

Unfortunately for experimenters, the mass of the X's had to be about 10^{15} GeV in order to explain certain well-known features of the subnuclear world. That's 1,000,000,000,000,000 —or a million billion—times as heavy as a proton, about the mass of a good-sized bacterium! Far too heavy to be created at any atom smasher we could ever build, at least on Earth. And if Georgi and Glashow's SU(5) scheme were correct, there was nothing interesting to be found between the 100 GeV mass scale of the W or Z° and the 10^{15} GeV masses of these X's. The vast, barren, unpopulated wasteland in between became known, half-jokingly, as "Glashow's Desert."

There was one conceivable way to test these grand-unified theories, but to do so physicists had to go underground. Because a quark might occasionally become a lepton, protons might not be absolutely stable, as previously thought. The heavy particles at the heart of the nucleus might actually *decay*. "Perhaps diamonds are not forever," joked Glashow. But they had an exceedingly long, long life. Because the X particles causing the purported decay were so incredibly massive, they were awfully hard to produce, and so a proton should have lived about 10^{31} years, on the average, before coughing one up and disintegrating. That's about 1,000,000,000,000,000,000,000 times the present age of the universe (about 15 billion years).

The only way to detect such a rare event is to gather a lot of protons together, say a few thousand tons' worth, and watch them all very closely for a year or more, hoping for a few to expire. To shield out most of the cosmic rays that could easily mimic a proton decay, these tremendous volumes had to be buried deep in the earth. This is the kind of experiment physicists started planning in the late 1970s. Large tanks of water or piles of steel were about to be buried in a railroad tunnel under Mont Blanc, in a salt mine below Cleveland, Ohio, and in a lead mine two miles beneath Kamioka, Japan. There, experimenters had to wait with great patience for gold-plated evidence of a single proton decaying spontaneously into a pion and a positron. Only then could we truly believe in grand unification.

There were many who thought this unification business more grandiose than grand. Sure, there were still far too many unexplained masses and arbitrary parameters left lying around after Glashow, Salam, and Weinberg had finished unifying the weak and electromagnetic forces. But why not try to explain this disquieting complexity by seeking a still deeper, simpler level of matter beneath the quarks and leptons? "Does the beginning proliferation of flavors," asked Victor Weisskopf in the summer of 1979, "indicate an internal structure of the quark, a new spectroscopy, a higher rung of the quantum ladder?" After all, this more plebian approach had history on its side, having already worked well three times in this century.

As far as physicists could tell, quarks and leptons still looked like mathematical points by the end of the decade. If they had any "size" at all, it was less than 10^{-15} cm—a hundredth the size of a proton and less than a millionth that of an atom. So it was stretching the imagination just a bit to start talking about their "structure" so soon. And how could we even begin to think seriously about the structure of something we could "see" only indirectly?

But you had to admit there were an uncomfortable number of supposedly "fundamental" entities to cope with. If you granted the existence of the top quark, then there were six flavors of quarks and six leptons—plus each quark came in three colors, too. Not even counting their antiparticles or any possible future generations, that made at least twenty-four fundamental fragments of matter. And that number did not even include the force particles—the photon, the two W's, and the Z°, plus eight gluons. Things were indeed getting a bit cumbersome.

So it was inevitable, I suppose, that soon after quarks had been accepted as "real" particles, physicists began trying to dissect them. We began to hear the word *preon,* a general name that Salam coined for the imaginary constituents of quarks and leptons—the next rung of the quantum ladder.

In early 1979, Haim Harari, an Israeli theorist who spent much of his time at SLAC, suggested an idea along these lines. He needed but two fundamental, spin-½ "rishons"—a "*T*" with charge ⅓ and a neutral "*V*"—to build up each generation of leptons and quarks, including all antiparticles and colors. The positron, for example, was a *TTT* combination (with charge +1), while the up quark was a *TTV, TVT,* or *VTT* (with charge +⅔) and the down antiquark a *TVV, VTV,* or *VVT* (with charge +⅓), the three different permutations representing the three colors. It was a remarkably economical idea, the kind that appeals to physicists.

To test these preon ideas, however, we had to find some kind of internal structure to leptons and quarks, which as far as we could tell were still mathematical points. To study such tiny features meant, à la Heisenberg, that we needed probes with far higher energies than available at existing atom smashers—even *more* than necessary to create a ponderous Z°. Though tremendous, such energies were not inconceivable, and mammoth new accelerators were soon being planned to deliver them. Ever sharper knives to cut the supposedly "uncuttable." The cycle seemed about to repeat itself, once again.

So despite the upheavals and insights of the past decade, as the 1980s began the age-old debate between idealists and materialists seemed in excellent health. On the one hand, Nature was to be reduced to pure symmetry groups, the modern equivalent of Plato's triangles. On the other hand, all one might need to explain its enduring complexity was yet another deeper, simpler level of matter—the present-day manifestation of Democritus's atoms. Two roads led off to unity, and physicists still had plenty of work to do before getting there.

There were really two profound triumphs of high-energy physics during the seventies. One was of course the unification of the weak and electromagnetic forces, the first time such a unification of two forces had occurred in over a century. The other was the discovery of another layer of matter, "the reasoned evolution of the quark hypothesis from mere whimsy to established dogma," as Glashow put it so eloquently in his Nobel lecture. One triumph could never have occurred without the other. Each provided the context for working out its complement. Nature is a multifaceted gem we can see in many ways.

EPILOGUE

QUARKS are taken for granted today, as if little more problematical than pebbles or stones, inanimate objects at the very base of reality. As I finish this book in early 1987, they are being used almost unthinkingly throughout particle physics—and are even making their way into nuclear physics and cosmology, too. But those of us who struggled so long and hard to understand whatever was happening inside hadrons, and who finally nurtured these oddest of fundaments into "existence," can never take them so lightly. Quarks will always have a special place in *my* heart, at least.

How future generations will regard quarks I cannot know. Perhaps they will be seen as just another way station, one more stopover on the road to unity. Perhaps this whole search for a "deepest" level of matter is ultimately as futile as the hunting of Lewis Carroll's Snark. Or maybe there *is* something truly unique about quarks and leptons, and genuine unity is almost at hand. I have no easy answers, just my own prejudices. Only the Gell-Manns and Glashows, Richters and Tings of the future can answer these questions.

Not much seems to have changed in high-energy physics since the Standard Model took a firm hold about 1980. Governments continue spending billions on this esoteric research, but with few major new results to show for it. There have certainly been a large number of discoveries *claimed,* but these have either found an explanation within the Standard Model or, more often, have simply disappeared upon closer examination. Compared to the lively seventies, this has been a remarkably uneventful decade so far.

In 1983, Carlo Rubbia indeed found the long-sought W's and

the even more weighty Z°, with masses very close to what had already been predicted by the Standard Model; he shared the 1984 Nobel prize for this truly important work. But his other "discoveries" at CERN's powerful new proton-antiproton collider either have been completely discredited or remain extremely questionable, like his purported 1984 discovery of the top quark. Physicists who have been fooled once too often by his sketchy results now prefer to wait for much better evidence before they consider the top quark "found."

Grand unification continues a tantalizing possibility, sorely needed by particle theorists and cosmologists, but still singularly lacking in experimental proof. Despite five years of waiting and watching, underground detectors around the world have not witnessed the decaying protons anticipated by grand unified theories. Protons seem to live a lot longer than theorists had originally hoped, and Glashow's SU(5) theory is in deep decline. Perhaps diamonds *are* forever, after all.

Supersymmetry and the more recent superstring theories are the fashion today, particularly the latter, which consider fundamental particles to be one-dimensional "strings" instead of mathematical points. This is an important new idea with much to recommend it, including the wonderful possibility of unifying gravity with all the other forces—Einstein's fondest dream. But a convincing experimental proof of the superstring idea seems far in the distant future, if possible at all.

The less fashionable "preon" ideas, which consider quarks and leptons to be aggregates of still smaller entities, remain viable because atom smashers powerful enough to study such tiny distances have simply not been available. A new generation of colliding-beam machines just now coming on line at Fermilab and SLAC may have more to reveal about this question. And colliders under construction at CERN and DESY, scheduled for completion in a few years, may even provide some answers.

So far, it has been a remarkably productive decade on the theoretical front—if variety is a valid criterion. There seem to be at least as many ways to get beyond the Standard Model as there are theorists. But like the grand-unified theories, many of these proposals make predictions that are all but inaccessible to observation or experimental verification. Physics begins to look like metaphysics.

No, there is nothing today with solid experimental proof that cannot be included within the broad purview of the Standard Model, even as it stood in 1980. Sure, there are many troubling complexities that cry out for simpler solutions and no shortage of sweeping proposals about how to begin. But these remain hopeful speculations in the continuing absence of any concrete evidence. Incomplete though it may seem, the Standard Model reigns supreme throughout the subatomic world—at least for now.

More and more, it appears that the ten-year period from 1967 through 1976 represents a genuine watershed in the history of science. The vast tide of theoretical work coming in the last ten years is merely the natural fallout normally expected after such a period of upheaval—similar to what occurred during the late 1920s when physicists began deploying the powerful new techniques of quantum mechanics. We have entered a period of "normal science" again.

Any great upheaval in physics raises important questions of epistemology, of how we can even begin to know what we know. The recent establishment of the Standard Model is no exception. Did we actually "find" the one true way Nature "is," a way that was lying there all the time waiting to be "discovered"? Or did we *construct* this new reality using the sense impressions available from experiments, the metaphoric imagination intimately tied to the larger culture, and the mathematical structures of pure thought? Answers to these questions must come to grips with the modern idea of "objectivity" and what it really means to science.

Virtually all practicing physicists deny that their science harbors any cultural bias whatsoever; they will assert that its statements are absolute. Such influences may guide the progress of a new idea in its original stages, they agree, but these are gradually winnowed out of the final product by the unmerciful rigor of the scientific method—the exposure of new ideas to the withering glare of experiment. Physicists speak the exact same language here on Earth as on Sirius or Alpha Centauri.

But there is solid merit to the opposing viewpoint, too. As argued by Andrew Pickering in his recent book, *Constructing Quarks,* scientists in effect *choose* the image of reality dominant in any particular era. That choice reflects, in part, the social and cultural norms of the age. They cannot be winnowed out completely by experiment, because experimenters share these very same biases and employ them in the interpretation of their data. Objectivity is really a social choice made by a scientific community acting communally; it is not just "the way things are."

According to this second viewpoint, the Standard Model won out over alternative pictures of subatomic reality because it allowed the greatest degree of social accord among practitioners of high-energy physics. For the time being, at least, this theory is the accepted norm against which alternatives are compared, to be replaced by a better theory only when one comes along good enough to unseat it. Scientific truth is temporal and culturally conditioned.

So what do physicists mean when we claim that quarks "exist"? Are there really pointlike entities "out there" imbued with colors and fractional charges and spin-½? Or is this peculiar image merely the way our twentieth-century scientific culture has found it convenient to organize a lot of otherwise confusing experimental data?

I find merit in both viewpoints and ultimate truth in neither. Those who claim that physicists discover the way things are forget that we can only make *representations* of what is "out there." We cannot know the quark-in-itself. And anyone familiar with quantum theory has to admit that no representation is ever the one unique description of a subatomic entity. There is always an element of choice—the act of a willing subject—in every such picture.

But those who take advantage of quantum arbitrariness to claim that the only reality is "in here," in the offspring of the human mind, leave me absolutely cold. Anybody who has repeatedly banged his head against Nature, as experimenters do day in and day out, knows that physical reality is not some wild, fantastic concoction of an overworked mind. There *are* regularities in the subatomic world "out there" that by some wondrous miracle we can at least begin to describe in ordinary human language.

Like artists, physicists paint suggestive pictures of the reality we see using familiar materials and pigments available from the larger culture. But, unlike artists, we have a standard way of telling good art from bad art: we do experiments. We insist that our theories have observable consequences—and accept only those that satisfy this criterion. Although cultural biases can and do remain, at least we make certain we have *successful* descriptions of Nature.

Physical reality is *neither* some objective truth "out there" nor a purely subjective experience "in here." It is a uniquely human description of the *interaction* between humanity and Nature, involving elements of both inner and outer realms.

In retracing the steps taken by particle physics during the sixties and seventies, what impressed me most was the role of language in creating this reality. A large part of the power of a physical idea lies in its ability to *communicate,* to involve a great variety of listeners in the understanding process. Thus current algebra remained the esoteric practice of a few oddball theorists until Bjorken and Feynman came along to interpret its tenets in terms of quarks and partons, the "atoms" of our era. Then the fires began to spread in earnest, leading eventually to the Standard Model in power today.

Although particle physicists may work in one of the most abstract realms known to humanity, we still have to talk to one another. And to do so we often have to fall back on the everyday language of the larger culture. We have little choice. So we borrow words like "quarks" or "flavor" or "charm" or "color" and apply them in completely new ways to describe our weird, Lewis Carroll world. Other than having some intrinsic metaphoric content, these are words plucked from the thin air of everyday discourse and redeployed in this subatomic realm. Any "meaning" these words carry comes mainly from their repeated use and elaboration by the whole community of practicing physicists.

Nothing illustrates the evolution of a word better than the history of quark "color." The threeness of individual quarks was a concept that languished virtually unnoticed from 1964 until 1971, when Gell-Mann started using this very fortunate term to describe it. So apt was the word "color" that it caught on almost immediately. By the decade's end, color was recognized as the source of a fundamental new force of Nature, and was even beginning to make a kind of appearance in particle detectors—as the gluon jets seen in electron-positron collisions.

A similar evolution occurred with the idea of quarks. When Gell-Mann introduced the concept in early 1964, it was a mathematical whimsy useful mainly in organizing a plethora of different subatomic particles. But by the end of the 1970s, quarks were hard physical entities as "real" as the electrons used to ferret them out. Form had become substance.

This key role of language in creating physical reality gives an important hint about the nature of objectivity. The reality scientists know is the one that *communicates:* it is above all a *shared* reality. Scientists escape the realm of the purely subjective by projecting their individual ideas and insights into the great ebb and flow of their discipline. Those ideas that capture the allegiance of a scientific community become—at least for a time—the objective reality of that field. This objectivity is not "the way things are," but the way they are understood and discussed by the great majority of its practitioners.

In its relentless pursuit of objective truth, a scientific community resembles, if anything, a vast collective "mind." Of paramount importance to its vigorous growth is the free and rapid exchange of information among its many parts. At the heart of this exchange sits language. So the truth scientists speak about is not some rigid, static entity "out there." It is a living, growing, evolving body of successful ideas they share with their colleagues.

Can science, then, provide ultimate answers? This is a difficult question whose answer, if it exists at all, lies beyond the proper domain of science. There are certainly elements of physics, like quantum mechanics and relativity, that now have an air of eternity about them. But we can never be completely sure. We can only keep on probing and theorizing—as if someday we might actually reach a final answer.

What science does remarkably well, again and again, is to build surprisingly successful images of the universe—in all its wondrous variety and complexity. Somehow, in some mysterious way, the human mind can come to grips with the remotest realms of experience, like the interior of a proton or the first split seconds of the Big Bang. It is not at all obvious to me that this should even be possible. Yet it happens again and again.

NOTES

The sources of direct quotations used in the text are given in the notes below. The numbers listed to the left of these notes indicate the page or pages where the corresponding source material is quoted.

Page

19 Democritus quoted from Carl Sagan, *Cosmos* (Random House, 1980), p. 181.

20 Titus Lucretius, *De Rerum Natura,* Book I: lines 498–501.

Isaac Newton, *Opticks* (Dover Publications, 1952), pp. 375–6.

21 Dalton quoted from Sheldon Glashow, "The Hunting of the Quark," in *The New York Times Magazine,* 18 July 1976, p. 9.

24 Rutherford quoted in John G. Taylor, *The New Physics* (Basic Books, 1972), p. 54.

25 Ernest Rutherford, *Philosophical Magazine,* vol. 21, p. 669 (1911).

27 Banesh Hoffmann, *The Strange Story of the Quantum* (Dover Publications, 1959), p. 54.

28 Niels Bohr, "On the Constitution of Atoms and Molecules," *Philosophical Magazine,* vol. 26, p. 2 (1913).

33 Einstein and Schrödinger quoted by John Archibald Wheeler, "Bohr, Einstein and the Strange Lesson of the Quantum," in Richard Q. Elvee, editor, *Mind in Nature* (Harper & Row, 1982), pp. 4 and 6.

Werner Heisenberg, *Physics and Philosophy* (Harper & Row, 1962), p. 42.

Ibid., p. 42.

38 *Ibid.,* pp. 46 and 54.

39 Alfred North Whitehead, *Science and the Modern World* (New American Library, 1948), p. 4.

48 Hideki Yukawa quoted from Robert A. Millikan, *Electrons* (+ *and* −), *Protons, Neutrons, Mesotrons and Cosmic Rays* (University of Chicago Press, 1947), p. 501.

51 Carl Anderson quoted from Millikan, *op. cit.*, p. 505.
Niels Bohr letter quoted from Millikan, *op. cit.*, p. 509.

52 Yukawa quoted from Hideki Yukawa and Chihiro Kikuchi, "The Birth of the Meson Theory," *American Journal of Physics*, vol. 18, p. 154 (1950).

56 "Feynman is a splendid . . ." from C. P. Snow, *The Physicists* (Macmillan London Limited, 1981), p. 143.

57 "Like today's silicon chip . . ." from Julian Schwinger, "Renormalization Theory of Quantum Electrodynamics: An Individual View," in Laurie M. Brown and Lillian Hoddeson, eds., *The Birth of Particle Physics* (Cambridge University Press, 1983), p. 329.

68 Pauli quoted by Frederick Reines, in lecture presented at Fermilab, Batavia, Illinois, 3 May 1985.

83 Chew quote taken from Geoffrey F. Chew and S. C. Frautschi, "The Principle of Equivalence for All Strongly Interacting Particles Within the S-Matrix Framework," *Physical Review Letters*, vol. 7, p. 394 (1961).

84 "Each particle helps . . ." from October 22, 1963, speech given at the National Academy of Science. Published as Geoffrey Chew, "Elementary Particles?" *Physics Today*, April 1964, p. 34.
"Twentieth century physics . . ." quoted by Alexander W. Stern, "The Third Revolution in 20th Century Physics," *Physics Today*, April 1964, p. 43.

85 Chew quoted from Geoffrey F. Chew, Murray Gell-Mann, and Arthur F. Rosenfeld, "Strongly Interacting Particles," *Scientific American*, February 1964, p. 92.

90 Buddha's aphorism quoted from Chew et al., *op. cit.*, p. 89.

95 Gell-Mann quoted from *Proceedings* of the International Conference on High-Energy Nuclear Physics, Geneva, 1962 (Geneva: CERN Scientific Information Service, 1962), p. 805.

98 "The great unifying invention . . ." from Chew et al., *op. cit.*, p. 74.

101 Gell-Mann quoted by Robert Serber, interview, 8 November 1983.
"a terrible idea" from T. D. Lee, telephone interview, 5 August 1985.

"If the bootstrap . . ." quoted from Murray Gell-Mann, "Particle Theory from S-Matrix to Quarks," Caltech Report No. CALT-68-1214, p. 29.

"those funny little things" from Gell-Mann interview, 13 February 1985.

102 "Please, Murray, . . ." quoted from Gell-Mann, "Particle Theory from S-Matrix to Quarks," *op. cit.*, p. 32.

"Three quarks for Muster Mark! . . ." from James Joyce, *Finnegans Wake* (The Viking Press, 1959), p. 383.

"probably would have been rejected" from Gell-Mann interview, 13 February 1985.

103 M. Gell-Mann, "A Schematic Model of Baryons and Mesons," *Physics Letters*, vol. 8, p. 214 (1964).

George Zweig, "Origins of the Quark Model," in Nathan Isgur, editor, *Baryon 80: Proceedings of the Fourth International Conference on Baryon Resonances* (University of Toronto Press, 1981), p. 448.

104 "to eat one another," from George Zweig, telephone interview, 4 January 1985.

Both Zweig quotes from Zweig, "Origins of the Quark Model," *op. cit.*, p. 450.

105 "It is fun . . ." from Gell-Mann, "A Schematic Model of Baryons and Mesons," *op. cit.*, p. 215.

"We construct a . . ." from Murray Gell-Mann, "The Symmetry Group of Vector and Axial Vector Currents," *Physics*, vol. 1, p. 73 (1964).

106 "We will work . . ." from George Zweig, "An SU3 Model for Strong Interaction Symmetry and Its Breaking," CERN Report No. 8419/TH.412, published in Don B. Lichtenberg and S. Peter Rosen, eds., *Developments in the Quark Theory of Hadrons* (Nonantum, Massachusetts: Hadronic Press, Inc., 1980), p. 27.

108 "the concrete block model" from Gell-Mann interview, 13 February 1985.

"In February . . ." from *CERN Courier*, vol. 4, p. 26 (March 1964), Geneva, Switzerland.

112 Lederman quoted from "Search for Quarks Using Fermi Motion and Discovery of the Anti-Deuteron," *Adventures in Experimental Physics*, vol. 2, pp. 300 and 301 (1972).

113 Zichichi quoted from interview, 2 May 1984.

114 "I have discarded . . ." from R. A. Millikan, "A New Modification of the Cloud Method of Determining the Elementary Electrical Charge and the Most Probable Value of That Charge," *Philosophical Magazine*, vol. 19, p. 209 (1910).

115 "If the present ideas . . ." from G. Morpurgo, "Is a Non-Relativistic Approximation Possible for the Internal Dynamics of 'Elementary' Particles?" *Physics,* vol. 2, p. 103 (1965).
"I simply thought . . ." from G. Morpurgo, letter of 17 March 1983 to Andrew Pickering, p. 13.

117 Murray Gell-Mann quoted from "Elementary Particles?" *Proceedings of the Royal Institution,* vol. 41, p. 160 (1966).

118 Murray Gell-Mann, "Current Topics in Particle Physics," *Proceedings of the Thirteenth International Conference on High Energy Physics,* Berkeley, California, 1966, p. 5.

120 Stephen Gasiorowicz, *Elementary Particle Physics* (John Wiley & Sons, Inc., 1966), p. 278.
"It is clear . . ." from Plato, *Timaeus,* 22:1.

124 Robert Hofstadter quoted from interview, 20 January 1984.
Edward L. Gintzon, "An Informal History of SLAC, Part One," *SLAC Beam Line,* Stanford, California, April 1983, pp. 13 and 15.

128 Taylor quoted from interview, 6 March 1984.

132 "Why do you want . . ." from interview with David Coward, 26 September 1983.

134 Panofsky quoted from "Electromagnetic Interactions—Experiment," *Proceedings* of the 14th International Conference on High-Energy Physics, Vienna, August 1968, p. 37.

136 "standing in front . . ." from Robert Jaffe interview, 19 October 1983.

139 Bjorken quotes taken from J. Bjorken, "Theoretical Ideas on Inelastic Electron Scattering," *Proceedings* of the 1967 International Symposium on Electron and Photon Interactions at High Energy, Stanford, California, 5–9 September 1967, pp. 109 and 117–118.

140 "I would also . . ." from *ibid.,* p. 126.
"What can we do . . ." from interview with Richard Taylor, 5 July 1984.
"The kind of logic . . ." from Bjorken interview, 7 September 1984.

143 Barish quoted from interview, 15 March 1984.

148 Kendall quoted from interview, 24 October 1983.
Taylor quotes from interview, 6 March 1984.

148–52 Feynman comments on these pages taken from interviews, 14 and 15 March 1984, except as noted below.

150 "The experimenters have . . ." paraphrased by Feynman, interview, 14 March 1984.
"All my life . . ." quoted from interview with Paul Tsai, 3 April 1984.

151 "I've really got something . . ." quoted by Jerome Friedman, interview, 24 October 1985.

153 Bjorken quoted from interview, 7 September 1984.

154 "It should be . . ." taken from G. Morpurgo, "The Quark Model," *Proceedings* of the 14th International Conference on High-Energy Physics, Vienna, 1968, p. 225.

161 "The data on . . ." from J. D. Bjorken, SLAC Report No. SLAC-PUB-571, Stanford, California, March 1969, p. 2.

162, 164 Sakurai quoted from J. J. Sakurai, "Vector-Meson Dominance and High-Energy Electron-Proton Scattering," *Physical Review Letters,* vol. 22, p. 981 (1969).

164 Sakurai quoted from his talk, "Vector Meson Dominance—Present and Future Prospects," *Proceedings* of the 1969 International Symposium on Electron and Photon Interactions at High Energy, Liverpool, 1969, p. 91.

165 "Sakurai had always . . ." from Friedman interview, 1 February 1985.

167 "I have difficulty . . ." from Richard P. Feynman, "Very High-Energy Collisions of Hadrons," *Physical Review Letters,* vol. 23, p. 1415 (1969).

"... to a gnat's whisker" from Feynman interview, 14 March 1984.

168 Taylor quotes from interview, 6 March 1984.

169 Perkins quote recounted by Frank Close in interview, 17 April 1984.

171 "Inelastic lepton-nucleon . . ." from J. D. Bjorken, "Asymptotic Sum Rules at Infinite Momentum," *Physical Review,* vol. 179, p. 1547 (1969).

176 "the measurements come . . ." from *CERN Courier,* Geneva, Switzerland, vol. 10, p. 271 (September 1970).

179 Walter Sullivan, "Subatomic Tests Suggest a New Layer of Matter," *The New York Times,* 25 April 1971, p. 1.

"just been talking . . ." from Taylor interview, 6 March 1984.

180 "to design and build . . ." from Panofsky letter of 5 May 1971 to *The New York Times.*

Kendall quoted from his "Deep Inelastic Electron Scattering in the Continuum Region," *Proceedings* of the 1971 Symposium on Electron and Photon Interactions at High Energy, Ithaca, New York, 1971, p. 255.

181 Feynman quoted in *ibid.,* p. 260.

186 Perkins comments from his talk, "Neutrino Interactions," *Proceedings* of the XVI International Conference on High-Energy Physics, Batavia, Illinois, 1972, vol. 4, p. 189.

187 Perkins quoted in *ibid.*, p. 217.

189 Feynman quoted from M. Bander, G. L. Shaw, and D. Y. Wong, eds., *AIP Conference Proceedings No. 6,* Particle Physics, Irvine, California, 3–4 December 1971, pp. 178–80.

191 "You know that Gell-Mann . . ." from Feynman interview, 14 March 1984.
"Feynman's 'put-ons' . . ." from Gell-Mann interview, 14 March 1984.

192 "I have a principle . . ." taken from Bander et al., *op. cit.*, p. 170.

195 Yang quoted from his 1959 Vanuxem Lecture, Princeton University, published as *Elementary Particles* (Princeton University Press, 1962), p. 59.

198 Exchange between Yang and Pauli recounted by Yang, 2 May 1985, at the International Symposium on the History of Particle Physics, Batavia, Illinois, 1–4 May 1985.

199 Glashow quoted from Arthur Fisher, "Glashow's Charm," *Science 84,* March 1984, p. 55.

201 "At first sight . . ." from Sheldon Glashow, "Partial Symmetries of Weak Interactions," *Nuclear Physics,* vol. 22, p. 579 (1961).

202 "As theorists sometimes do . . ." from Steven Weinberg, Nobel lecture, 8 December 1979, reprinted in *Science,* vol. 210, p. 1212 (12 December 1980).
Lear quote from *ibid.*, p. 1213.

203 Coleman's quote from "The 1979 Nobel Prize in Physics," *Science,* vol. 206, p. 1290 (14 December 1979).

205 "revealed Weinberg and . . ." from *ibid.*, p. 1291.

209 Schwartz quoted from interview, 30 May 1985.

210 "We called our construct . . ." from Sheldon Glashow, "The Hunting of the Quark," *The New York Times Magazine,* 18 July 1976, p. 9.

211 "The solution was trivial" from Iliopoulos interview, 24 May 1984.
"We argued a lot . . ." from Iliopoulos interview, 4 June 1985.
"totally unaware" from Glashow interview, 18 May 1984.
"Steve, why didn't . . ." from Iliopoulos interview, 4 June 1985.

214 "He was at MIT . . ." from Rubbia interview, 13 August 1985.
"We might now stand . . ." from letter of 14 March 1972 to Robert R. Wilson, as quoted by Peter Galison, "How the First Neutral Current Experiments Ended," *Reviews of Modern Physics,* vol. 56, p. 494 (April 1983).

216 Faissner letter quoted in Galison, *op. cit.,* p. 487.
219 "They had been vaccinated . . ." from Rubbia interview, 13 August 1985.
"If you've got results . . ." quoted by Wesley Smith, interview, 2 May 1985.
220 Exchange between Musset and Salam recounted in Salam's Nobel lecture, 8 December 1979, reprinted in *Reviews of Modern Physics,* vol. 52, p. 530 (July 1980).
221 "It is perhaps . . ." from S. Weinberg, "Recent Progress in Gauge Theories of Weak, Electromagnetic and Strong Interactions," *2nd International Conference on Elementary Particles,* Aix-en-Provence, France, September 1973, p. c-47.
223 Rubbia quoted from interview, 13 August 1985.
228 Jaffe quoted from interview, 11 October 1983.
229 "a natural choice" from Gell-Mann interview, 13 February 1985.
232 "a garden where . . ." quoted by V. Silvestrini, "Electron-Positron Interactions," *Proceedings* of the XVI International Conference on High-Energy Physics, Batavia, Illinois, 1972, vol. 4, p. 10.
238 "The simplest and . . ." from H. Fritzsch, M. Gell-Mann, and H. Leutwyler, "Advantages of the Color Octet Gluon Picture," *Physics Letters,* vol. 47B, p. 366 (1973).
242 Weinberg quoted from his "Recent Progress in Gauge Theories of the Weak, Electromagnetic and Strong Interactions," *Reviews of Modern Physics,* vol. 46, p. 273 (April 1974).
245 "This was a marvelous theory . . ." from Burton Richter, "Colliding Beams at Stanford," *SLAC Beam Line,* Stanford, California, November 1984, p. 2.
247 Richter's comments taken from *ibid.,* p. 6.
248 "I've had a soft . . ." from *ibid.,* p. 7.
250 Exchange with Richter recounted by Bjorken in interview, 7 September 1984.
254 "These results . . ." from J. D. Bjorken, "A Theorist's View of Electron-Positron Annihilation," *Proceedings* of the Sixth International Symposium on Electron and Photon Interactions at High Energies, Bonn, West Germany, 27–31 August 1973, p. 25.
Rees quoted from lecture given at SLAC, 14 November 1984.
257 Bjorken and Drell quotes taken from "Particle Physics: Is the Electron Really a Hadron at Heart?" *Science,* vol. 184, pp. 783–4 (17 May 1974).
259 Ellis quoted from his "Theoretical Ideas About $e^+e^- \rightarrow$

Hadrons at High Energies," *Proceedings* of the XVII International Conference on High-Energy Physics, London, July 1974, pp. IV–30.

Richter quotes from his "Plenary Report on $e^+e^- \rightarrow$ Hadrons," *Proceedings* of the XVII International Conference, *op. cit.*, pp. IV–37, 41, and 54.

260 Iliopoulos comments from his "Plenary Report on Progress in Gauge Theories," *Proceedings* of the XVII International Conference, *op. cit.*, pp. III–91, 97, and 100.

"It was all presented . . ." from James D. Bjorken, "The November Revolution—A Theorist Reminisces," lecture given at SLAC, 14 November 1984, reprinted in the *SLAC Beam Line*, Stanford, California, Special Issue Number 8 (July 1985), p. 5.

263 Ting quote from author's recollection of this seminar.

264 "a little desperate" from Wit Busza interview, 15 May 1984.

267 "The total cross-section . . ." from first draft of paper circulated to SLAC-LBL collaboration, 5 July 1974, private collection of Roy Schwitters, p. 7.

Rau quoted from his letter to Sam Ting, 26 September 1974.

268 Chen quotes from 2 November 1983 interview with Ulrich Becker and Min Chen.

270 Becker quoted by Min Chen in interview, 9 August 1985.

Lee quoted from telephone interview, 5 August 1985.

271 Comments by Schwitters from interview, 17 November 1984.

272 "had a very narrow . . ." from Schwartz interview, 30 May 1985.

Exchange between Schwartz and Ting recounted by Schwartz in interview, 30 May 1985.

"I owe Mel . . ." from Samuel Ting, "One Researcher's Personal Account," *Adventures in Experimental Physics*, vol. 5, p. 125 (1976).

Schwartz quoted by Rau in interview, 18 May 1984.

273 Exchange between Lee and Schwartz recounted by Y. Y. Lee in interview, 4 June 1984.

"I just won . . ." and "And I know . . ." from Schwartz interview, 30 May 1985.

"squirming in his seat" from Becker interview, 2 November 1983.

274 "Here were . . ." from Gerson Goldhaber, "One Researcher's Personal Account," *Adventures in Experimental Physics*, vol. 5, p. 135 (1976).

Schwitters comments from interview, 17 November 1984.

277 "I was *very* . . ." from Ting interview, 26 July 1986.
Ting comments from Samuel Ting, "One Researcher's Personal Account," *op. cit.*, p. 126.

278 "What shall we do?" quoted by Y. Y. Lee, interview, 4 June 1984.
"The storage ring boys . . ." quoted in 1 November 1974 meeting notes taken by Horst Foelsche, read by telephone, 22 May 1985.
Weisskopf comment from interview, 12 December 1984.

279 "remarkable increase" from Goldhaber, *op. cit.*, p. 136.

280 "It looks like the Messiah . . ." quoted by Schwitters in interview, 17 November 1984.

281 Schwitters comments from entry in SP-II Experimenters' Logbook, 11 November 1974, Stanford, California (original kept at the Smithsonian Institution, Washington, D.C.).
"Oh my God! . . ." recounted by Breidenbach, interview, 31 August 1984.

282 Bjorken quotes and "BJ, I think you'd . . ." from James D. Bjorken, "The November Revolution," *op. cit.*, p. 5.

285 Breidenbach quoted from interview, 31 August 1984.
"Have you heard . . ." quoted by Sau Lan Wu in interview, 8 May 1984, and personal notes recorded in December 1974.

286 Rau reaction from interview, 18 May 1984.
"Experiment 598 has found . . ." from AGS Operators' Logbook, 11 November 1974.

287 Diebold quoted from telephone interview, 20 June 1985.
Conversation between Richter and Ting quoted from Burton Richter, "One Researcher's Personal Account," *Adventures in Experimental Physics,* vol. 5, p. 147 (1976).
"Sam Ting's found . . ." quoted by Roy Schwitters in interview, 2 August 1984.
Schwitters comments from interview, 2 August 1984.

288 Frank Close comments and Steinberger quote from tape recording made by Close in early December 1974, transcribed 21 April 1984.

290 Richter and Ting quoted from joint MIT/SLAC press release dated 16 November 1974, p. 1.

295 Glashow quoted from Sheldon Glashow, "Charm: An Invention Awaits Discovery," *Proceedings* of the Fourth International Conference on Experimental Meson Spectroscopy, Boston, 26–27 April 1974, p. 392.

298 "Sounds like he's onto . . ." from personal notes of Horst Foelsche, read by telephone, 22 May 1985.

Glashow quoted from interview, 18 May 1984.

Lagarrigue quoted by John Iliopoulos, interview, 24 May 1984.

299 Exchange between Aubert and Augustin recounted by Iliopoulos, interviews of 24 May 1984 and 4 June 1985.

300 Bjorken quoted from his article, "The November Revolution," *op. cit.,* p. 5.

Richard Feynman quoted from letter to Burton Richter dated 14 November 1974.

301 Richter and Ting quoted from press release, 16 November 1974.

302 Breidenbach quoted from interview, 31 August 1984.

"DUE TO NEW PARTICLE . . ." from SP-II Experimenters' Logbook, *op. cit.,* 21 November 1974, p. 103.

303 Richter quoted from his article, "One Researcher's Personal Account," *op. cit.,* p. 149.

304 "We had all . . ." from Glashow interview, 18 May 1984.

305 Applequist quoted by Goldman, telephone interview, 18 November 1985.

306 "The suddenness of the . . ." from William D. Metz, "Two New Particles Found: Physicists Baffled and Delighted," *Science,* vol. 186 (6 December 1974), p. 909.

"Hardly, if ever . . ." from Walter Sullivan, "Physics in a Ferment of Theories Over Discovery of Two Particles," *The New York Times,* 16 December 1974.

308 "the bunches talked . . ." from Wiik interview, 5 May 1984.

309 Wiik comment from interview, 5 May 1984.

310 Wölf quoted from interview, 7 May 1984.

312 "the geopolitics" from Schwitters interview, 3 August 1984.

314 Ting and Deutsch quotes taken from their letters published in *Science,* vol. 189 (5 September 1975), p. 750.

315 Panofsky quoted from "Particle Discoveries at SLAC," letter to *Science,* vol. 189, p. 1045 (26 September 1975).

"What do you do . . ." from Schwartz interview, 30 May 1985.

316 "rather unimpressive" from Lee interview, 5 August 1985.

317 Harvey Lynch quoted from SP-II Experimenters' Logbook, *op. cit.,* 9 November 1974, p. 11.

"What was needed . . ." from Burton Richter, "From the Psi to Charm: The Experiments of 1975 and 1976," Nobel lecture of December 1976, reprinted in *Reviews of Modern Physics,* vol. 49 (April 1977), p. 262.

318 Goldhaber quoted in Andrew Pickering, *Constructing Quarks* (Edinburgh University Press, 1984), p. 177.

“It was all . . .” from Sheldon Glashow, “The Hunting of the Quark,” *The New York Times Magazine,* 18 July 1976, p. 9.

320 Telegram quoted from text reprinted in *SLAC Beam Line,* Stanford, California, November 1976, p. 2.

322 Ellis quoted from Andrew Pickering, *op. cit.,* p. 254.

“The discovery of . . .” from Glashow, “The Hunting of the Quark,” *op. cit.,* p. 9.

325 “There could be other . . .” from A. De Rujula and S. L. Glashow, “Is Bound Charm Found?” *Physical Review Letters,* vol. 34 (1975), p. 46.

Perl comments from letter to author dated 10 December 1985.

328 “If there are more . . .” from Sheldon Glashow, “Charm Is Not Enough,” *Proceedings* of the 5th International Conference on Experimental Meson Spectroscopy, 29–30 April 1977, (Northeastern University Press, 1977), p. 421.

“This ‘shoulder’ excited . . .” from Leon Lederman, “The Upsilon Particle,” *Scientific American,* vol. 239 (October 1978), p. 75.

329 “Why do it again . . .” quoted from Lederman interview, 7 September 1984.

330 Lederman quoted from his article, “High Energy Experiments,” in M. K. Gaillard and R. Stora, eds., *Gauge Theories in High Energy Physics* (Amsterdam: North Holland Publishing Company, 1983), p. 849.

332 “an embarrassment of riches” from Lederman, “The Upsilon Particle,” *op. cit.,* p. 79.

336 “engineering problem” from Ellis interview, 12 July 1984.

338 “You’re not going . . .” recounted by Jack Steinberger in interview, 3 May 1985.

345 “Such behavior is possible . . .” quoted from Burton Richter, “From the Psi to Charm,” *op. cit.,* p. 264.

347 Böhm quoted from interview, 4 May 1984.

348 Ellis comments taken from interview, 12 July 1984.

349 “practical reasons” quoted from Wu interview, 8 May 1984.

Wu quoted from letter to author dated 29 December 1985.

Perkins quoted by Sau Lan Wu, interview, 8 May 1984.

352 Ting quoted from letter to G. Wölf dated 13 September 1979.

“a massive PR campaign” from Wiik interview, 5 May 1984.

Exchange between Sullivan and MIT student quoted from Sullivan interview, 1 June 1984.

353 “So why have these . . .” quoted from “Do Gluons Really Exist?” *New Scientist,* 6 September 1979, p. 709.

355 "We have now reached . . ." from James D. Bjorken, "The New Orthodoxy: How Can It Fail?" *Proceedings* of Neutrino 1979, Bergen, Norway, 18–22 June 1979, p. 9.

356 Glashow quoted from his summary talk at Bergen, in *Proceedings* of Neutrino 1979, *op. cit.*, p. 519.

358 "On holiday in . . ." excerpted from *ibid.*, p. 518.

"Recently the dialectic . . ." quoted from John Ellis, "Status of Gauge Theories," *Proceedings* of Neutrino 1979, *op. cit.*, p. 451.

359 Glashow quoted from his Nobel lecture, "Towards a Unified Theory: Threads in a Tapestry," reprinted in *Reviews of Modern Physics,* vol. 52 (July 1980), p. 539.

360 Glashow quoted from his summary talk at Bergen, *op. cit.*, p. 520.

361 Weisskopf quoted from his lecture, "Personal Impressions of Recent Trends in Particle Physics," published in A. Zichichi, editor, *Pointlike Structures Inside and Outside Hadrons* (Plenum Press, 1982), p. 2.

362 Glashow quoted from Nobel lecture, *op. cit.*, p. 539.

INTERVIEWS AND CONVERSATIONS

William Atwood, SLAC. At Stanford University, Stanford, California, 1 August 1984; at SLAC, 4 March 1985.

Raymond Arnold, American University. At SLAC, Stanford, California, 14 November 1984.

Jean-Jacques Aubert, University of Marseille. At University of California, Santa Cruz, 22 & 23 August 1984.

Barry Barish, Caltech. At Caltech, Pasadena, California, 15 March 1984; at Fermilab, Batavia, Illinois, 10 September 1984.

Ulrich Becker, MIT. At MIT, Cambridge, Massachusetts, 25 October 1983; 2 November 1983; 15 & 17 May 1984; 4 June 1985 (by telephone); 25 July 1986.

Mirza Beg, Rockefeller Institute. At Rockefeller, New York City, 9 November 1983 (by telephone); 24 May 1984.

Bruce Beron, Stanford Research Institute. At Palo Alto, California, 29 May 1985.

Jeremy Bernstein, Stevens Institute. At New York City, 6 December 1984 (by telephone).

Hans Bienlein, DESY. At DESY, Hamburg, West Germany, 7 May 1984.

Peter Biggs, Massachusetts General Hospital (formerly of MIT). At Boston, Massachusetts, 11 December 1984.

James Bjorken, Fermilab (formerly of SLAC). At Fermilab, Batavia, Illinois, 7 & 12 September 1984; at SLAC, Stanford, California, 14 November 1984; at Fermilab, 2 & 3 May 1985, 11 December 1985 (by telephone); at SLAC, 28 & 29 May 1986.

Eliott Bloom, SLAC. At Stanford University, Stanford, California, 1 August 1984.

Arie Bodek, Rochester University (formerly of MIT). At La Honda, California, 10 & 11 August 1984, and 23 May 1985; at Fermilab, Batavia, Illinois, 11 September 1984, and 1 & 4 May 1985.

Albrecht Böhm, DESY and Technische Hochschule, Aachen, West Germany. At DESY, Hamburg, West Germany, 4 May 1984.

Adam Boyarski, SLAC. At SLAC, Stanford, California, 8 & 10 July 1985.

Martin Breidenbach, SLAC (formerly of MIT). At Menlo Park, California, 31 August 1984; at SLAC, Stanford, California, 13 May 1985.

Witold Busza, MIT. At MIT, Cambridge, Massachusetts, 15 & 17 May 1984.

Min Chen, MIT. At MIT, Cambridge, Massachusetts, 25 October 1983 and 2 November 1983; at DESY, Hamburg, West Germany, 4 May 1984; at MIT, 17 May 1984; at Erice, Italy, 9, 12 & 13 August 1985.

Frank Close, Oxford University and Rutherford Laboratory. At Oxford, England, 17 April 1984; at London, England, 11 May 1984.

David Coward, SLAC. At SLAC, Stanford, California, 26 September 1983; 9 July 1985.

Robert Crease, Columbia University. At Fermilab, Batavia, Illinois, 4 May 1985; at New York City, 21 & 28 May 1985, 2 & 25 June 1985 (by telephone).

Donald Cundy, CERN. At CERN, Geneva, Switzerland, 26 April 1984.

Richard Dalitz, Oxford University. At Oxford, England, 17 April 1984; at Fermilab, Batavia, Illinois, 2 May 1985.

Robert Diebold, Argonne National Laboratory. At Berkeley, California, 20 June 1985 (by telephone).

Sidney Drell, SLAC. At SLAC, Stanford, California, 29 March 1984.

Max Dresden, State University of New York, Stony Brook. At Fermilab, Batavia, Illinois, 1 May 1985.

David Dorfan, University of California, Santa Cruz. At CERN, Geneva, Switzerland, 24, 26 & 29 April 1984; at Santa Cruz, 9 May 1985.

John Ellis, CERN. At SLAC, Stanford, California, 12 July 1984.

Glen Everhart, formerly of MIT. At Morristown, New Jersey, 16 December 1985 (by telephone).

Gerald Feinberg, Columbia University. At Columbia, New York City, 17 October 1983.

Richard Feynman, Caltech. At Caltech, Pasadena, California, 14 & 15 March 1984; 8 February 1985 (by telephone).

Jean Flanagan, MIT. At MIT, Cambridge, Massachusetts, 24 June 1985 (by telephone); 27 August 1985.

Horst Foelsche, Brookhaven. At Brookhaven, Upton, New York, 21 & 22 May 1985 (by telephone).

Jerome Friedman, MIT. At MIT, Cambridge, Massachusetts, 11, 18, 19, 21 & 24 October 1983; 29 March 1984 (by telephone); 17 & 18 May 1984; 2 January 1985 and 1 February 1985 (by telephone).

David Frisch, MIT. At MIT, Cambridge, Massachusetts, 12 December 1984.

David Fryberger, SLAC. At SLAC, Stanford, California, 5 & 10 June 1985.

Murray Gell-Mann, Caltech. At Caltech, Pasadena, California, 12 & 14 March 1984 (by telephone); 20 & 30 November 1984 (by telephone); 13 February 1985; at Fermilab, Batavia, Illinois, 2 & 3 May 1985; at Caltech, 4 November 1985 (by telephone).

Frederick Gilman, SLAC. At SLAC, Stanford, California, 22 October 1984; 12 November 1984; 17 January 1985.

Sheldon Glashow, Harvard University. At Harvard, Cambridge, Massachusetts, 18 May 1984.

Gerson Goldhaber, University of California, Berkeley. At SLAC, Stanford, California, 25 July 1985; 26 July 1985 (by telephone).

Terence Goldman, Los Alamos (formerly of SLAC). At Los Alamos, New Mexico, 18 November 1985 (by telephone).

Lee Grodzins, MIT. At MIT, Cambridge, Massachusetts, 17 May 1984.

Robert Hofstadter, Stanford University. At Stanford, California, 20 January 1984.

John Iliopoulos, Ecole Normale Supérieure. At Rockefeller Institute, New York City, 24 May 1984; at SLAC, Stanford, California, 4 June 1985.

Maurice Jacob, CERN. At CERN, Geneva, Switzerland, 25 April 1984.

Robert Jaffe, MIT. At MIT, Cambridge, Massachusetts, 11 & 19 October 1983; 16 May 1984.

Willibald Jentschke, DESY and CERN. At Stanford University, Stanford, California, 1 & 2 August 1984; at Fermilab, Batavia, Illinois, 3 May 1985.

Kenneth Johnson, MIT. At MIT, Cambridge, Massachusetts, 17 May 1984.

Henry Kendall, MIT. At MIT, Cambridge, Massachusetts, 21 & 24 October 1983; 16 May 1984.

William Kirk, SLAC. At SLAC, Stanford, California, 17 September 1985.

Jasper Kirkby, CERN (formerly of SLAC). At CERN, Geneva, Switzerland, 25 April 1984; at SLAC, Stanford, California, 8 November 1985.

Ted Kycia, BNL. At Brookhaven, Upton, New York, 18 & 19 June 1985 (by telephone).

Donald Lazarus, BNL. At Brookhaven, Upton, New York, 5 June 1984.

John Learned, University of Hawaii. At Honolulu, Hawaii, 21 March 1984; 29 April 1985 (by telephone).

Leon Lederman, Fermilab (formerly of Columbia University). At Fermilab, Batavia, Illinois, 7 & 10 September 1984.

T. D. Lee, Columbia University. At New York City, 5 August 1985 (by telephone).

Y. Y. Lee, BNL. At Brookhaven, Upton, New York, 4 June 1984.

Luigi di Lella, CERN. At CERN, Geneva, Switzerland, 20 April 1984.

Christopher Llewellyn-Smith, Oxford University. At Oxford, England, 17 & 18 April 1984.

Gloria Lubkin, *Physics Today*. At New York City, 29 July 1986.

Harvey Lynch, SLAC. At SLAC, Stanford, California, 14 May 1985; 21 June 1985.

Thomas Lyons, MIT. At SLAC, Stanford, California, 21 June 1985.

Robert Messner, SLAC (formerly of Caltech). At La Honda, California, 18 March 1984; at SLAC, Stanford, California, 15 November 1984, 18 April 1985, and 14 May 1985.

Giacomo Morpurgo, University of Genoa. At CERN, Geneva, Switzerland, 24 April 1984.

Heinz Pagels, Rockefeller Institute. At New York City, 17 October 1983.

Robert Palmer, BNL. At Brookhaven, Upton, New York, 5 June 1984; at SLAC, Stanford, California, 26 & 29 July 1985.

Wolfgang Panofsky, SLAC. At SLAC, Stanford, California, 31 January 1984; 17 January 1985; 5 February 1985.

Donald Perkins, Oxford University and CERN. At CERN, Geneva, Switzerland, 26 & 27 April 1984; at Fermilab, Batavia, Illinois, 4 & 6 May 1985.

Martin Perl, SLAC. At SLAC, Stanford, California, 16 December 1985.

Robert Phillips, BNL. At Brookhaven, Upton, New York, 4 June 1984 and 16 May 1985; at Stanford, California, 2 & 3 August 1984.

Andrew Pickering, University of Illinois. At MIT, Cambridge, Massachusetts, 11 December 1984.

Helen Quinn, SLAC. At SLAC, Stanford, California, 10 June 1985.

Petros Rapidis, Fermilab. At Fermilab, Batavia, Illinois, 10 September 1984.

Ronald Rau, BNL. At MIT, Cambridge, Massachusetts, 18 May 1984; at Brookhaven, Upton, New York, 3 December 1985 (by telephone).

Frederick Reines, University of California, Irvine. At Fermilab, Batavia, Illinois, 1 & 3 May 1985.

Jean and Terence Rhoades, Fermilab. At Fermilab, Batavia, Illinois, 27 April 1985 (by telephone); 1 May 1985.

Burton Richter, SLAC. At Palo Alto, California, 2 May 1986; at SLAC, Stanford, California, 15 August 1986.

Alan Rothenberg, SLAC. At Stanford, California, 2 August 1984.

Carlo Rubbia, CERN and Harvard University. At Erice, Italy, 13 August 1985.

Nicholas Samios, BNL and Columbia University. At Brookhaven, Upton, New York, 5 June 1984.

Herwig Schopper, CERN. At Berkeley, California, 22 July 1986.

David Schramm, Fermilab and University of Chicago. At Fermilab, Batavia, Illinois, 31 July 1984 (by telephone); at Chicago, Illinois, 5 May 1985 (by telephone).

John Schwartz, Caltech. At Erice, Italy, 13 August 1985.

Melvin Schwartz, Stanford University. At Palo Alto, California, 30 May 1985; 31 May 1985 (by telephone).

Roy Schwitters, Fermilab and Harvard University. At Fermilab, Batavia, Illinois, 27 July 1984 (by telephone); at Stanford, California, 2 & 3 August 1984; at Fermilab, Batavia, Illinois, 6 September 1984; at SLAC, Stanford, California, 17 November 1984; at Cambridge, Massachusetts, 28 December 1985 (by telephone).

Robert Serber, Columbia University. At Columbia, New York City, 8 November 1983; at New York, 5 April 1984 (by telephone).

Wesley Smith, Columbia University. At Fermilab, Batavia, Illinois, 2 May 1985.

Paul Söding, DESY. At DESY, Hamburg, West Germany, 7 May 1984; at Berkeley, California, 18 July 1986.

Jack Steinberger, CERN. At Stanford, California, 2 August 1984; at Fermilab, Batavia, Illinois, 1 & 3 May 1985.

Lynn Stevenson, University of California, Berkeley. At Stanford, California, 2 August 1984.

Karl Strauch, Harvard University. At Harvard, Cambridge, Massachusetts, 10 May 1985 (by telephone).

Walter Sullivan, *New York Times*. At New York City, 1 June 1984.

Richard Taylor, SLAC. At SLAC, Stanford, California, 17 January 1984; 6 March 1984; 5 July 1984; 12 November 1984.

Samuel Ting, MIT. At MIT, Cambridge, Massachusetts, 2 November 1983; at SLAC, Stanford, California, 14 November 1984 and 23

May 1985; at MIT, Cambridge, Massachusetts, 23 May 1986 (by telephone), 24 & 26 July 1986.

Walter Toki, SLAC. At Stanford, California, 1 August 1984; at SLAC, Stanford, California, 14 August 1986.

George Trigg, *Physical Review Letters.* At Brookhaven, Upton, New York, 4 June 1984.

Y. S. Tsai, SLAC. At SLAC, Stanford, California, 2 & 3 April 1984; 9 January 1985.

Victor Weisskopf, MIT. At MIT, Cambridge, Massachusetts, 12 December 1984; at Fermilab, Batavia, Illinois, 2 May 1985; at Cambridge, Massachusetts, 29 May 1986 (by telephone) and 13 August 1986 (by telephone).

Bjorn Wiik, DESY. At DESY, Hamburg, West Germany, 5 May 1984.

Frank Wilczek, University of California, Santa Barbara. At SLAC, 1 April 1985.

Gunter Wölf, DESY. At DESY, Hamburg, West Germany, 7 & 8 May 1984.

Sau Lan Wu, DESY and University of Wisconsin. At DESY, Hamburg, West Germany, 8 May 1984; at Stanford, California, 2 & 3 August 1984; at Madison, Wisconsin, 5 December 1985 (by telephone); at San Francisco, California, 9 December 1985; at DESY, 30 December 1985 and 30 July 1986 (by telephone).

Antonino Zichichi, University of Bologna and CERN. At CERN, Geneva, Switzerland, 2 May 1984, 20 November 1984 (by telephone); at Erice, Italy, 16 August 1985.

George Zweig, Caltech and Los Alamos National Laboratory. At Los Alamos, New Mexico, 4 January 1985 and 1 February 1985 (by telephone).

Key to Acronyms:

BNL—Brookhaven National Laboratory, Upton, New York.

CERN—European Organization for Nuclear Research, Geneva, Switzerland.

DESY—Deutsches Electronen Synchrotron, Hamburg, West Germany.

MIT—Massachusetts Institute of Technology, Cambridge, Massachusetts.

SLAC—Stanford Linear Accelerator Center, Stanford, California.

ACKNOWLEDGMENTS

I wish to thank all the men and women of elementary particle physics who made this book possible by the generous amounts of time they spent with me in conversations and interviews, recalling the events I have attempted to describe in this book. Too numerous to cite here, their names are listed on pp. 382–87.

I am indebted to many of these people and to others for reading and commenting upon drafts of various sections and chapters. They include Luis Alvarez, Ulrich Becker, Arie Bodek, Min Chen, David Coward, Richard Dalitz, Sidney Drell, Richard Feynman, Henry Kendall, Leon Lederman, Giacomo Morpurgo, Wolfgang Panofsky, Donald Perkins, Martin Perl, Ronald Rau, Frederick Reines, Burton Richter, Nicholas Samios, Emilio Segrè, Robert Serber, Karl Strauch, Walter Sullivan, Victor Weisskopf, Gunther Wölf, Antonino Zichichi, and George Zweig. A few went far beyond necessity and spent untold hours with me, helping to shape the book with their insights, recollections, and detailed criticisms of large portions of the manuscript. For this invaluable aid I owe my deep gratitude to James Bjorken, Jerome Friedman, Murray Gell-Mann, Roy Schwitters, Samuel Ting, and Sau Lan Wu. Any errors that remain in the book are my fault alone.

My visits to the national and international laboratories of high-energy physics were highly informative and enjoyable thanks to the efforts of Neil Baggett and Robert Phillips at Brookhaven, Roger Anthoine and Gordon Fraser at CERN, Petra Harms and Pedro Waloschek at DESY, Margaret Pearson at Fermilab, and Nina Adelman and William Kirk at SLAC.

Many of these people helped me enormously with photo research, too, as did May West at Fermilab. In addition to helping me find many SLAC photos, Walter Zawojski prepared a number of truly exquisite airbrush drawings of subatomic processes.

I owe an intellectual debt of the highest order to Martin Gardner, Gerald Holton, and Thomas Kuhn, whose ideas about the process of science have guided my own thinking and inform these pages. Visits to the Niels Bohr Library at the American Institute of Physics in New York provided information and photographs about early twentieth-century physics. The historical insights of Andrew Pickering, communicated over many pints of ale, have been much appreciated—as are his friendship and hospitality.

John Brockman and Tobi Saunders helped to place the manuscript with Simon and Schuster, where Henry Ferris, Veronica Johnson, and Ursula Obst have worked untiringly to render it into English. Russell Tkaczyk added his fine pen-and-ink drawings. My editor, the incomparable Alice Mayhew, deserves my deepest thanks for her wise counsel and unflinching support throughout this project, even when it fell well behind schedule.

I have benefited from the encouragement of many friends and colleagues, particularly Richard Merrill, Robert Messner, and Stephen Rock, who lent a ready ear to my musings and speculations. Arie Bodek deserves a special thank you for his support during the later stages of the writing.

Finally, *The Hunting of the Quark* would not exist were it not for the advice, encouragement, and support of my wife, Linda Goodman. She gave the book this title when it was still only a hazy idea and made invaluable criticisms and suggestions. To her I owe my deepest and warmest gratitude.

Michael Riordan
Menlo Park, California
March 1987

INDEX

Abbadessa, John, 248
Abrams, Gerald, 310
"ace" theory, 104–10
ADONE, 232, 247, 249–50, 252, 254–55, 286, 305, 308
AGS (Alternating Gradient Synchrotron), 73–74
alpha particles, 24–27, 42–45, 124
Alvarez, Luis, 61, 75–76, 81, 89, 95, 118
Ampère, André, 214
Anderson, Carl, 49–51, 74
angular momentum, 65–67
anomalous e-mu events, 326–28
antibaryons, 69–72
anticharm, 304
antideuteron, 112
antilambdas, 81–82
antileptons, 70–72
antineutrinos, 68, 186–87, 195
antinucleons, 81
antiparticles, 49; particles vs., 66
antiprotons, 66–67, 77
antiquarks, 110, 184
Appelquist, Thomas, 300, 304–5
Aristotle, 19–20, 38
Aspen Center for Physics, 237–38
"associated production," 62–63
asymptotic freedom, 226, 304–5, 323; for strong force, 236–40, 242–44
Atomic Energy Commission, U.S., 36–37, 62, 125, 127, 247–48
atomism, 20–22, 323–24
atoms, atomic theory, 20–22, 323–24; Bohr's, 27–29; Dalton's contribution to, 20–22; etymology of, 18–19; kinetic theory of gases and, 21–22; Maxwell's Equations and, 27; Rutherford's, 24–27; Thomson's 23–24
Aubert, Jean-Jacques, 288–89, 299
Augustin, Jean-Eaudes, 299

Bacon, Francis, 20, 39
"bag model," 243–44
Barish, Barry, 143
baryon number, 69–72
baryons, 69–72, 229; charmed, 295–96; composite vs. elementary, 82; as elementary, 83; with fractional charges, 101–4, 114–16; hexagonal patterns of, 89; multiplicity of, 88–89
Baltay, Charles, 212
Becker, Ulrich, 265–66, 268–71, 273–74, 276–78, 314
Becquerel, Henri, 23
beta decay, 44–45, 67–68
Bethe, Hans, 54
Bevatron, 66–67, 75–76, 118
Big Bang theory, 114, 342
binding energy, 110–11
Bjorken, James, 136–55, 160–61, 164, 171, 179, 182, 184, 210, 227, 236, 239, 250, 254, 282, 294, 300, 337, 355, 368
"Bjorken limit," 239
Bloom, Eliott, 147–48, 173, 177–178
Bodek, Arie, 172–74, 177–78, 180–82, 238–39, 294
Böhm, Albrecht, 347
Bohr, Niels, 27–29, 33–38, 51, 67, 293–94
Bohr's Complementarity Principle, 34–38, 184
bootstrap model, 10, 82–86, 98–99, 324–25; equations of, 83–84; in-

elastic e-p scattering and, 143–44, 162; triplets and, 101; vector dominance and, 147
Born, Max, 32–34, 36
bosons, 66, 70–72, 85
see also kaons; photons; pions
Boyle, Robert, 20–21
Braque, Georges, 23
Breidenbach, Martin, 173–75, 255, 267, 278–80, 282, 285, 301–5, 317
Brodsky, Stanley, 286
broken symmetries, 92, 201–4
Brookhaven National Laboratory, 10, 62, 296–301, 316
Brown, Robert, 22
Brownian motion, 22, 343–44
bubble chambers, 74–78
Buddha, 90–91
Butler, Clifford, 59

Callan, Curtis, 158, 235
Callan-Symanzik equation, 235
Caltech, 48–52
Capra, Fritjof, 324–25
Carroll, Lewis, 36, 100, 121, 194, 293
"cascade decay," 309–10
"cascade particles," 64, 71
Cavendish Labs, 42, 126–27
CDHS (CERN, Dortmund, Heidelberg, and Saclay) collaboration, 337–39
CEA (Cambridge Electron Accelerator), 128–29, 132, 140–41, 174, 247
CEA Bypass, 251–59
CERN (European Center for Nuclear Research), 73–74, 93–94, 104–5, 107–14, 169–73, 185–86, 209, 224–26, 329
CERN Courier, 108, 176
Chadwick, James, 42–44, 67
Chamberlain, Owen, 66–67
charge, 64–65; conservation of, 63, 197–98; fractional, *see* fractional charges; quantum numbers for, 69–70; strong force and, 88
charmed quarks, 10, 210–11, 294–321
"charmonium," 300, 304, 307, 309
chemistry, 20–22
Chen, Min, 268–71, 274, 277–78, 286, 297
Chew, Geoffrey, 83–86, 98–99, 118, 233
Chinowski, William, 281
chi (χ) particles, 89, 310
Cline, David, 213, 219–23
Close, Frank, 169, 288–89
cloud chambers, 30, 48–52, 59–60; bubble chambers vs., 75
Cockcroft, John, 42
Coleman, Sidney, 203, 205, 235, 304
collisions, 12; billiard-ball analogy of, 129–30, 190–91, 226; "diffractive," 162–64, 172; elastic, 128–35, 137–38; hadron, "S-matrix" theory and, 84–85, 98–99; inelastic, 130–35, 137–39, 142–43, 145–48, 170, 186; partons and, *see* partons; proton-nucleon, 111; *see also* electron-proton scattering; scattering
color, 229–32, 237, 323, 336, 340–42
complementarity, 34–38, 184
Compton, Arthur H., 30, 36
"concrete block model," 104–10
conservation laws: electric charge, 63, 197–98; energy, 57–58, 67–68, 197; in inferring neutral, uncharged particles, 76; reactions suppressed by, 103–4; of strangeness, 63–64
cosmic rays, 48, 50–52, 59, 62, 112–14
cosmic-ray telescopes, 113
Cosmotron, 62–63, 66, 75–76, 78
Cowan, Clyde, 67–68, 137, 170
Coward, David, 132, 173
Cronin, James, 195
current algebra, 105–6, 136–40, 148; for hadrons, 105–6, 136–40, 227–28; parton theory and, 151–55; scaling in, 227
cyclotrons, 60–61, 123

Dalitz, Richard, 119, 169–70, 332
Dalton, John, 20–22

Darwin, Charles, 65
DASP (Double Arm Spectrometer), 308–11
Davisson, Clinton, 31, 36
de Broglie, Louis, 30–32, 36–37
de Broglie wavelengths, 129, 154
decay: beta, 44–45, 67–68; "cascade," 309–10; kaon, 195, 211; by most difficult path, 103; pion, 53, 170; weak force in, 63–64
decimets, 92–95, 101
deep inelastic scattering, 132–35, 137, 147–48, 154–55, 162–66
delta (Δ) resonances, 94
Democritus, 18–19, 191, 362
De Rujula, Alvaro, 304–7, 325
Descartes, René, 20
DeStaebler, Herbert, 142
DESY (Deutsches Electronen Synchrotron), 129, 132, 140, 164, 235, 247, 262–66, 279, 308–12, 327–28, 331–33, 335–36, 347–53, 357–58
detectors, 10; bubble chamber, 74–78; human eye vs., 18, 128, 263; modern, 12–13; neutrinos and, 68; signals of, 17; SLAC, 17–18, 128, 156–58
deuterium, 172, 175
Deutsch, Martin, 273, 286, 307, 314–17
Diebold, Robert, 286–87
diffraction, 29
Dirac, Paul, 32, 49
Dirac Equation, 49, 53–55; negative energy solution to, 66
D mesons, 312, 317–20, 327–28
DORIS (Double Ring Storage), 279, 308–12, 327, 331
Drell, Sidney, 160, 179, 227, 236, 241, 250, 257

Eddington, Arthur, 136
Eightfold Way, 82, 86–95, 98–99, 228; quark model vs., 118; symmetry in, 89–93, 106; "unitary spin" and, 92
Einstein, Albert, 10–11, 22–23, 28, 30–36, 39, 62, 149, 154, 159, 323, 335, 343, 359
elasticity, 128–35, 137–38
electricity, electric charge, 20, 45
electromagnetic force, 10, 45, 63, 148–49
electron microscopes, 16
see also SLAC
electron-neutron scattering, 161, 171–72
Electron-Photon Symposiums, 164–66, 180, 221
electron-positron annihilation, 231–32, 252–55, 299, 326–28, 345–53
electron-positron pairs, 57, 263, 265
electron-positron "storage rings," 231
electron-proton (e-p) scattering, 128–35, 137, 143–44, 159; inelastic, 130–35, 137–39, 142–43; "jets" in, 343–53
electrons: charge of, 40–41, 47–48, 114 (see also oil drop experiment); discovery of, 23–24; low-energy, 130–31; mass of, 23; orbits of, 27; polarized, 356; position vs. velocity of, 33–38; positive, *see* positrons; "self-energy" of, 53; size of, 58, 82; spin of, 41, 65–66; virtual particles and, 58, 137
electroweak force, *see* weak force
elementarity, size and, 82–83
Ellis, John, 322, 336, 348, 353, 358
energy conservation laws, 57–58, 67–68, 197
e-p scattering, *see* electron-proton scattering
European Center for Nuclear Research, *see* CERN
European Physical Society Conferences, 205, 309
Everhart, Glen, 268, 274
eye, human, 17–18, 128, 263

Fairbank, William, 343
Faissner, Helmut, 212, 216
Faraday, Michael, 197
F curve, *see* scaling
Feldman, Gary, 310, 326–28
Fermi, Enrico, 62, 66, 69, 78, 81, 93, 115

Fermi motion, 111, 150, 178; "smearing" effect of, 178
fermions, 66; baryons vs. leptons, 69–72; in pairs, 109
Feynman, Richard, 54–57, 87, 103–4, 118, 137, 148–56, 158, 161, 166–68, 177, 179, 181–82, 189–92, 199, 206, 229, 300, 338, 368
Feynman diagrams, 56–57, 137–38; for electron-positron annihilation, 253; perturbation theory and, 233–34
Fitch, Val, 195
flavor, 294–95, 299
Foelsche, Horst, 298
fractional charges, 101–4, 114–16, 160, 181–82, 187
Friedman, Jerome, 128, 134, 140–44, 147–48, 151, 154–58, 164, 166–67, 172–73, 176, 179–82, 236, 240, 294
Fritszch, Harald, 228–30, 238, 250, 252–54

Gaillard, Mary, 297
Gargamelle, 185–87, 212–24, 298
Garwin, Edward, 356–57
gases, kinetic theory of, 21–22
Gasiorowicz, Stephen, 119–20
gauge symmetry, 197–98, 202–3
gauge theory, 197–201, 204–7, 243
Geiger, Hans, 25–26, 157
Geiger counters, 265
Gell-Mann, Murray, 9, 63, 86–92, 94–95, 98–108, 117–21, 160, 179, 182, 186, 189–92, 206, 228–30, 232, 237–38, 244, 250, 252–54, 301, 343, 356, 368–69
Georgi, Howard, 304
Germer, Lester, 31
Gilman, Fred, 177–78
Gintzon, Edward L., 123–25
Glaser, Donald, 74–75
Glashow, Sheldon, 10, 198–207, 210–11, 230, 257, 294–322, 325, 328, 355–63, 365
gluon jets, 343–53
gluons, 184–85, 228, 301, 336, 339–42; colored, 237–38, 340–42
gold, 24–27
Goldhaber, Gerson, 272, 274, 279–85, 299, 303, 317–19, 327
Goldhaber, Maurice, 95
Goldman, Terence, 302, 305
"gold-plated event," 97, 296–97, 349
Goldstone bosons, 298
grand-unified theories, 359–62; baryon and lepton numbers in, 70*n*; discovery of quarks vs., 13; gauge theories in, 197–201, 204–7; gravity in, 11
gravity, 11, 20, 45, 63
Greenberg, O. W., 109–10, 230
Grodzins, Lee, 316
Gross, David, 158, 235–37
group theory, 88–89

hadrons, 71–72, 149; current algebra for, 105–6, 136–40, 227–28; general symmetry principle of, 88; as "mutual composites," 83; "S-matrix" theory and, 84–85, 98–99, 233; virtual photons and, 146–48; *see also* bootstrap model; partons
Han, M. Y., 230
Hand, Louis, 140, 143–44, 146–47, 240
Hansen, William W., 123–24
Hasert, Franz, 216
Heisenberg, Werner, 32–39, 73, 84, 198, 227
Heisenberg's Uncertainty Principle, 33–38, 57, 199, 207, 241, 318; quark theory and, 110; resonance lifetime and, 79, 268–70, 290–92; virtual photons and, 45–48
helium, 24; nuclei, *see* alpha particles; spectrum of, 29
HEPL (High Energy Physics Lab), 245–46
"HEPL alumni," 127–28, 132, 142
Higgs mechanisms, 202–4, 206, 242–43
high-y anomaly, 338
Hoffmann, Banesh, 27
Hofstadter, Robert, 82, 124–25, 128, 137, 141–42, 147
Holton, Gerald, 191

hydrogen: Bohr-Rutherford model of, 29; in bubble chambers, 74–76; heavy, 172, 175; liquid, 74–76, 96, 240; mass of, 23; spectrum of, 29; two states of, 54
hypercharge, 89, 91–92
hyperons, 97–98

Iliopoulos, John, 210–11, 260–61, 297–99, 319, 323
inelasticity, 130–35, 137–39, 142–43, 145–48, 170, 186
infrared radiation, 30
"interaction regions, " 231–32
intermediate vector bosons, 201
Intersecting Storage Rings (ISR), 247–50
intrinsic angular momentum, 65–67
isospin, 89, 91–92

Jaffe, Robert, 225–26, 228, 236, 244
Jean-Marie, Bernard, 298–99
Jentschke, Willibald, 222, 270–71, 316
Johnson, Kenneth, 243–44
Joule, James, 21
J/Ψ particles, 10, 284–92, 305–17, 322, 328–31, 337

Kadyk, John, 267
kaons, 59, 62–65, 71; decay of, 195, 211; in pairs, 62–65; spin of, 66; as "strange" mesons, 88
Kapitza, Peter, 42
Kendall, Henry, 128, 140–49, 151, 156–58, 172–73, 176, 179–82, 190, 206
kinetic theory of gases, 21–22
Kirkby, Jasper, 316
Kline, Franz, 76
klystrons, 16, 122–23
K-mesons, *see* kaons
Kuhn, Thomas, 11, 245
Kuti, Julius, 184–85, 228
Kycia, Ted, 285, 289

Lagarrigue, Andre, 212, 298
Lamb, Willis, 54
lambda (Λ) particles, 59, 62–64, 66, 296–97; as fundamental particle, 81–82
Lamb shift, 54–55
Larsen, Rudolf, 280
Lauritsen Laboratory, 191
Lavoisier, Antoine, 20–21
Lawrence, Ernest O., 60–61, 75, 214
Lawrence Berkeley Laboratory (LBL), 256
Lawrence Radiation Laboratory, 60–61, 118
Lederman, Leon, 111–12, 118, 249, 264–65, 272, 285, 328–32, 337, 351
"Lederman shoulder," 264–66, 270–71, 297, 328–29
Lee, Benjamin, 205, 297
Lee, Tsung Dao, 101, 109, 118, 140, 195, 270–71, 316
Lee, Y. Y., 273, 278
lepton-nucleon scattering, 171
lepton numbers, 69–72
Lepton-Photon Symposium, 310–12, 330
leptons, 69–72, 170–71
Leucippus, 18–19, 191
Leutwyler, Hans, 238
light: derivation of frequencies of, 32–33; frame of reference and, 23, 31; particulate properties of, 23, 29–31
Livingston, M. Stanley, 61, 74
Low, Francis, 211, 234
luminosity, 251, 275
Lynch, Harvey, 262, 279, 301, 317

McGreary, A. J., 286
Mach, Ernst, 343
McMillan, Edwin, 61
Maiani, Luciano, 210–11, 297
Manhattan Project, 54–55, 126
Mann, Alan, 213, 219–23
Mark I accelerator, 123, 256, 266–67, 271, 299, 310, 326–27, 348
Mark II accelerator, 123
Mark III accelerator, 123–24, 245–46
MARK-J detector, 350–52
Marsden, Ernest, 24–25, 157
mass, rest, 47

Massachusetts Institute of Technology (MIT), 128–35, 140–41
mass energy equivalence, 30–31, 47, 62
matrix mechanics, 32
Maxwell's Equations, 27, 30
measurement, 12; defined, 33–34; in 17th and 18th centuries, 20–21; subject vs. object and, 12; in 20th century, 12–13; *see also* Heisenberg's Uncertainty Principle
Mendeleev, Dmitri, 22, 89
"Mercedes events," 350
meson resonances, 81
mesons, 229; chi (χ), 89; D, 312, 317–20, 327–28; as elementary, 83; hexagonal patterns of, 89; K-, *see* kaons; multiplicity of, 88–89; μ-, *see* muons; partons as, 152, 158; phi (φ), 103–4; pi (π-), 53, 71; "singlet," 103; tau (τ-), 59; vector, *see* vector mesons
mesotrons, 51–52
Michelson-Morley experiment, 31
Millikan, Robert A., 40–41, 48–49, 51, 101, 114
Mills, Robert, 197–98, 200–201, 204–5, 226
MIT-SLAC collaboration, 130–35, 139, 141–49; rivalry in, 147–48
MIT-SLAC experiment, 134–35, 154
momentum distribution, 150–54
momentum transfers, 129, 131
Morpurgo, Giacomo, 115–16, 118–19, 332
Morrison, Douglas, 108
multiplets, 88–89, 91
mu (μ-) mesons, *see* muons
Munch, Edvard, 23
muon pairs, 57
muons, 53, 59–60, 71; deep inelastic scattering of, 240; spin of, 66
Musset, Paul, 212, 216–18, 220–21
Myatt, George, 221

"naked color," 300–301
Nambu, Yoichiro, 230, 237
National Accelerator Laboratory, 186, 213–16, 222–24, 240–41
Neddermeyer, Seth, 50–51
Ne'eman, Yuval, 86–87, 94–95, 106
neutral currents, 208–16, 294–95
neutrino beam, 170–71
neutrino-nucleon scattering, 208–10, 212–24
neutrinos, 67–68, 137, 170, 186–87, 208–10; as "left-handed," 195; muonless collisions of, 208–10, 212–24; "muon-" vs. "electron-type," 170
neutrino scattering, 170, 186–87
neutrons, 44, 66, 82, 175–76
Newman, Harvey, 351
Newton, Isaac, 15, 20, 29, 42
New York Times, 112, 178–80, 289–90, 304, 306, 318, 352
Nishijima, Kazuhiko, 63
Nolde, Emil, 23
nuclear democracy, 86, 101, 172, 175
nucleon resonances (Δ), 78–79
nucleons, 45, 53; pointlike, 170–71; structure of, 183, 186–87

"Occam's razor," 72, 84
Oersted, H. C., 214
oil drop experiment, 40, 101, 114
omega minus (Ω^-) particles, 94–97, 295–96, 304
O'Neill, Gerard, 246
Oppenheimer, Robert, 100
"orthocharmonium," 304, 307
Osborne, Louis, 255

Pagels, Heinz, 238
Palmer, Robert, 295–97, 306
Pancini, Ettore, 52
Panofsky, Wolfgang K. H., 125–28, 130, 134–35, 137, 140, 155–56, 169, 173, 179, 209, 246, 248, 281, 290, 301–2, 333
"paracharmonium," 304, 307
"paraquarks," 109–10, 230
parity violation, 194–96, 199–201
partons, 149–55; in experimentation, 158–68; as mesons, 152, 158; relativity and, 159; spin of, 185; *see also* quark-parton model
Paschos, Emmanuel, 149–50, 152, 160, 171, 184

Paterson, Ewan, 275, 278–79, 282
Pati, Jogesh, 359
Pauli, Wolfgang, 32, 35–36, 49, 67, 109, 198, 229
Pauli Exclusion Principle, 109, 229
Perkins, Donald, 169–71, 186–87, 208–9, 212, 215–16, 349
Perl, Martin, 325–28
perturbation expansions, 233
perturbation theory, 233–35, 242–44
PETRA (Positron-Electron Tandem Ring Accelerator), 333–34, 336, 347–53, 357–58
"phase stability" principle, 61
phi (ϕ) mesons, 103–4
phonons, 117–18
photoelectric effect, 23
photons, 30; spin of, 66; in transmission of strong force, 45–48; virtual, 45–48, 56–58, 146–48; *see also* light
Piccione, Oreste, 52
Pickering, Andrew, 183, 366–67
Pierre, François, 318
pi (π-) mesons, 53, 71
pion-kaon resonance, 81, 297
pion-pion resonance (ρ), 81
pions, 53, 60, 62; decay of, 53, 170; neutral, 126; in scattering, 78–81; spin of, 66; virtual, 59–60
Pipkin, Francis, 251–52, 262–63
Planck, Max, 23, 28
Planck's constant, 28–29, 31, 37, 47, 65
Plato, 19–20, 99, 120–22, 191, 362
Platonic solids, 19, 91–92, 362
Politzer, David, 235–37, 300, 304
Pollock, Jackson, 76
"positronium," 307
positrons, 49–51, 170; spin of, 66
"potentia," 19, 38–39
Powell, Cecil F., 42, 52–53, 59
Prentki, Jacques, 288
preons, 361–62, 365
Prescott, Charles, 356
probability functions, 32–33, 38, 57 *see also* Schrödinger Wave Equation
Project M, 124
propane, 75
protons: in collisions, 77; naming of, 44; partons and, 152–54; pointlike, 143–48; in QED, 58–59; quarks in, 110–12; size of, 82; spin of, 66; *see also* electron-proton scattering
psi (ψ), *see* Schrödinger Wave Equation
psi (ψ) particles, *see* J/Ψ particles
psi-prime (ψ′) particles, 303, 305, 309–17, 330–31

quantum chromodynamics (QCD), 11, 238, 244, 323; gluons in, 336; QED and, 339; quark confinement and, 336; scaling violations and, 336–39; *see also* asymptotic freedom
quantum electrodynamics (QED), 55, 58–60; in post–WW II period, 196–97; QCD and, 339; quantum field theory vs., 196–98; "radiative corrections" in, 142–43; strong force in, 58–60; strong interactions in, 84–85
quantum field theories: in post–WW II period, 196–97; QED vs., 196–98; of strong force, 59–60, 82; strong interactions and, 84–85
quantum mechanics, 57–58, 323; "Copenhagen interpretation" of, 34–36, 39; particle identity changes in, 146
quantum numbers, 41, 65–67, 177; for charge, 69–70; for particle classification, 69–72; for strangeness, 69–70
quantum of action, *see* Planck's constant
quark-antiquark pairs, 184
quark confinement, 336
quark-parton model, 152, 158, 171, 176–78, 182–85
quarks, quark theory, 10; "bag model" of, 243–44; binding energy of, 110–11; charmed, 10, 210–11, 294–321; color of, 229–32, 237, 323, 336; confinement of, 243–44; in cosmic rays, 112–14; discovery of, 9; "dressed

up," 343–44; in Earth's crust, 114–16; Eightfold Way vs., 118; experimental vs. theoretical science in, 12; flavor of, 294–95, 299; Heisenberg's Uncertainty Principle and, 110; "nonrelativistic," 119, 170; Pauli Exclusion Principle and, 109; "point-like," 139–40; in protons, 110–12; sea, 184–85; strong force and, 225–44; triplets in, 101–3; valence, 184–85, 187
Quinn, Helen, 304
"quorks," 101–3

Rabi, Isidor, 60
radar, 54
radiative corrections, 142–43, 150
radioactive fallout, 52
radio waves, 30
Ratcliffe, J. A., 43
Rau, Ronald, 267–68, 272–73, 278, 286, 298, 316
R-crisis, 258–61, 299, 328
Rees, John, 254–55
Regge, Tullio, 85
Regge trajectories, 85–86, 98
regular solids, 19, 91–92, 362
Reines, Frederick, 67–68, 137, 170
relativity, relativity theory, 57–58; antiparticles in, 66; hadron collisions in, 149; partons and, 159; Special Theory of, 23, 31
renormalization, 199–201, 204–5, 234
resonances, 78–81, 131–33; Uncertainty Principle and lifetime of, 79, 268–70, 290–92; *see also specific particles*
rest mass, 47
Resvanais, Leo, 284–85
Retherford, Robert, 54
Rhoades, Terence, 266, 268
Richter, Burton, 245–48, 250–51, 255, 257, 259, 267, 274–75, 278–85, 287, 290, 299, 300–304, 314–18, 320–21, 326, 345
Ritson, David, 247–48, 308
Rochester, George, 59
Rochester Conferences, 93, 118, 139, 164, 172, 176, 215–16, 220, 224, 232, 240, 244, 252, 319
Roentgen, Konrad, 23
Rosenfeld, Arthur, 98
Rosner, Jonathan, 297
Rubbia, Carlo, 213–16, 218–24, 338, 364–65
Rutherford, Ernest, 24–27, 42–45, 124, 127–28, 133, 139, 157, 241

Sakata, Shoichi, 81–82, 104
Sakurai, J. J., 147, 162–66, 232, 306
Salam, Abdus, 10, 203–5, 220–21, 321, 356–59, 361
Samios, Nicholas, 95–98, 102, 105, 118, 271, 295–97, 306, 337
scaling, 146–47, 150–54, 160–62, 165, 172, 227, 238–39; in current algebra, 227; violations of, 239–41, 336–39
scattering, 128–35, 137; pointlike, 143–48; *see also* deep inelastic scattering; *specific collisions*
Schrödinger, Erwin, 32–33, 36
Schrödinger Wave Equation, 32–33, 37, 49, 57, 227, 332
Schultz, Ingrid, 286
Schwartz, Melvin, 170, 207–10, 266–68, 272–73, 277, 285, 314–16, 337
Schwinger, Julian, 54–55, 57, 199–201, 306
Schwitters, Roy, 255–56, 271–72, 274–75, 278–81, 287, 299, 310–12, 317
scientific revolutions, 11–12, 245
sea quarks, 184–85
Segrè, Emilio, 66–67
Serber, Robert, 100–103, 105, 118
Shaw, Ronald, 203–4
"shopping-bag model," 244
Shutt, Ralph, 96–97
sigma (Σ) particles, 64, 66, 71, 82
sigma* (Σ*) resonances, 94
Sinclair, Charles, 356–57
"singlet" mesons, 103
SLAC (Stanford Linear Accelerator Center), 9, 13–18, 74, 122–34, 140–41; detectors of, 17–18, 128, 156–58; structure of, 15–18
"SLAC bag," 244

SLAC-LBL collaboration, 256–57, 307
SLAC-MIT-Caltech collaboration, 128–30
"S-matrix" theory, 84–85, 98–99, 233
Snow, C. P., 56
Soddy, Frederick, 24
Söding, Paul, 350, 352
SPEAR (Stanford Positron-Electron Asymmetric Rings), 245, 248, 250, 254–57, 262, 271, 275, 301–4, 308–15, 327–28
Special Theory of Relativity, 23, 31
Special Unitary group SU(3), 89–93, 365
spectrometers, 17, 128, 131
spectroscopy, 293–94, 313
spin, 65–68, 89; electron, 41; "unitary," 92
"SP" particles, 284–85
"Standard Model," 323–25, 328, 332–34, 355–58, 365–67
Stanford University, 15, 122–35
see also SLAC
Steinberger, Jack, 260, 337–39
"storage rings," electron-positron, 231–32
strangeness, 63–65, 229; paraquarks vs., 110; of pions and nucleons, 78–79; quantum numbers for, 69–70
strong force, 45–48, 63–64, 69–72 asymptotic freedom for, 236–40, 242–44; color as source of, 238; comparison of models of, 98–99; electric charge and, 88; in particle production, 64; perturbation theory and, 233–35, 242–44; in QED, 58–60; quantum numbers and, 69–72; quarks and, 225–44
structure functions, 144, 151–54, 157–58
subatomic "reflection," 194–96
Sulak, Larry, 218–20
Sullivan, Walter, 178–80, 289–90, 306, 318, 352
"sum rules," 137–40, 148
"supermultiplets," 91
Super Proton Synchrotron (SPS), 337–38
SU (Special Unitary group)3, 89–93, 365
Symanzik, Kurt, 235
symmetry: broken, 92, 201–4; gauge, 197–98, 202–3; hidden, 202–3; SU(3), 89–93

TASSO (Two-Armed Solenoidal Spectrometer), 348–53
tau (τ) particles, 59, 327–28, 332–34
Taylor, Richard, 128, 130–31, 134, 140–41, 144, 148, 155–58, 164–68, 172–80, 184, 248
Thomson, J. J., 23–24, 40, 42, 139
't Hooft, Gerard, 205, 209–11, 230, 235
Thorn, Charles, 243–44
three-pion resonance (ω), 81
Ting, Jeanne, 277
Ting, Samuel Chao Chung, 262–78, 285–92, 294, 297–301, 311–17, 320–21, 325, 329, 348, 351–53
Tomonoga, Sin'itiro, 55, 199
Touschek, Bruno, 247
Treiman, Samuel, 203
Trigg, George, 278, 289–90, 353
triplets, 101–3, 105–8
Tsai, Yung-Su, 150

"ultraviolet catastrophe," 28
ultraviolet radiation, 30
Uncertainty Principle, Heisenberg's, *see* Heisenberg's Uncertainty Principle
unified field theory, 10
Union of Concerned Scientists, 141
"unitary spin," 92
U particles, 327
upsilon (Υ) particles, 329–34

valence, 27, 184–85, 187
Van Hove, Leon, 109, 140
Varian, Russell and Sigurd, 123
vector bosons, intermediate, 201
vector (meson) dominance, 147–48, 162–66, 172, 232; generalized, 165–66
vector mesons, 81; neutral, 146–48, 263
Veltmann, Martin, 205, 209–10

vibrations, natural frequencies of, 79–80
virtual bosons, 85
virtual photons, 45–48, 56–58, 146–48; high-energy, 58
virtual pions, 59–60
vision, human vs. mechanical, 17–18, 128, 263
V-particles, 59–60, 62–65, 96–97

Walton, Ernest, 42
Ward, John, 204
water, 19, 21
wave-particle duality, 23, 29–36
weak force, 10, 63, 148–49, 194; parity violation and, 194–96, 199–201; renormalization of, 199–201, 204–5; in strange particle decay, 63–64
weak neutral currents, 208–16
Weinberg, Steven, 10, 198–207, 221, 230, 304, 321, 356–59, 361
"Weinberg angle," 357–58
Weisskopf, Victor, 102, 179, 184–85, 228, 244, 263, 270–71, 278, 289, 361
Wells, Eugene, 353
Wheeler, John Archibald, 40
Whitaker, Scott, 279
Whitehead, Alfred North, 39
Wiik, Bjorn, 308–10, 348–49
Wilczek, Frank, 235–37
Wilson, C. T. R., 48
Wilson, Kenneth, 232–36
Wilson, Robert R., 61, 214–16, 248, 329
Winter, Klaus, 353
Wölf, Gunther, 308–10, 351–52
Wu, Sau Lan, 285–86, 298, 314, 348–49

X bosons, 360
xi (Ξ) particles, 64, 66, 71, 82
xi* (Ξ*) resonances, 94, 95
X-particles, 50–51
X rays, 23, 30

Yang, Chen Ning, 69–72, 79, 81, 195–201, 204–5, 226, 306
Yang-Mills theory, 197–98, 200–201, 204–5, 211, 235, 238, 242
Yoh, John, 330
Yukawa, Hideki, 45–48, 51–53, 198, 242

Z° bosons, 203, 206–7, 301, 304, 307, 356
Zichichi, Antonio, 112–14, 118
Zweig, George, 103–8, 118, 160, 182, 186

ABOUT THE AUTHOR

Michael Riordan is the coauthor of *The Solar Home Book.* He was a research associate in physics at M.I.T. and is now affiliated with Stanford's Linear Accelerator Center.

PHOTO CREDITS

AIP Niels Bohr Library: 35, 43, 50, 54, 87
Courtesy of Ulrich Becker: 276
Lawrence Berkeley Laboratory: 77
Brookhaven National Laboratory: 97, 291, 296
Deutsches Electronen Synchrotron: 347
European Center for Particle Physics: 217
Courtesy of Joseph Faust: 319
Fermi National Accelerator Laboratory: 215, 351
Courtesy of Henry W. Kendall: 153
Courtesy of Vera Luth: 282, 283
MIT Museum: 269
Princeton University Press: 70
Stanford Linear Accelerator Center: 16, 126, 131, 256, 320
Courtesy of Sau Lan Wu: 350